# Causal Analysis

# Causal Analysis

Impact Evaluation and Causal Machine
Learning with Applications in R

Martin Huber

The MIT Press
Cambridge, Massachusetts
London, England

The MIT Press would like to thank the anonymous peer reviewers who provided comments on drafts of this book. The generous work of academic experts is essential for establishing the authority and quality of our publications. We acknowledge with gratitude the contributions of these otherwise uncredited readers.

This book was set in Sabon by Westchester Publishing Services. Printed and bound in the United States of America.

Library of Congress Cataloging-in-Publication Data

Names: Huber, Martin, 1980– author.
Title: Causal analysis : impact evaluation and causal machine learning
    with applications in R / Martin Huber.
Description: Cambridge, Massachusetts : The MIT Press, [2023] |
    Includes bibliographical references and index.
Identifiers: LCCN 2022033285 (print) | LCCN 2022033286 (ebook) |
    ISBN 9780262545914 (paperback) | ISBN 9780262374927 (epub) |
    ISBN 9780262374910 (pdf)
Subjects: LCSH: Social sciences—Research. | Causation. | Inference. |
    Econometrics.
Classification: LCC H62 .H713 2023 (print) | LCC H62 (ebook) |
    DDC 300.72—dc23/eng/20221223
LC record available at https://lccn.loc.gov/2022033285
LC ebook record available at https://lccn.loc.gov/2022033286

10 9 8 7 6 5 4 3 2

*To Margarita, Kalina, and Nina, to my parents and brothers, to my parents-in-law and brother-in-law, to all the colleagues, researchers, and analysts I have met along my quest for causal reasoning, including my former Ph.D. supervisor, Michael, and to the voices in my head that made me write this book.*

*The author thanks Mario Fiorini, Michael Knaus, Fabian Krüger, Jannis Kück, Michael Oberfichtner, Franco Peracchi, Regina Riphahn, Jonathan Roth, Aurélien Sallin, Anthony Strittmatter, and Hannes Wallimann for helpful comments and suggestions.*

# Contents

# 1

## Introduction

### 1.1 About Causality and This Book

From the beginning of our lives as human beings until the end, we are permanently confronted with questions about causes and effects: that is, the consequences of doing one thing versus another. Would I rather eat croissants or muesli for breakfast? What would I enjoy more? Would I rather go skiing or snowboarding to benefit the most from the current snow conditions (snowboarding on icy slopes can be a hassle, as the fans of winter sports among us might know)? Should I study for my statistics exam next week, or will I pass even if I don't? This also applies to broader (and possibly socially more relevant) questions concerning politics, business or work life, health, and society in general, such as: Will more education increase my salary? Does a discount on a product or service increase sales? Does smoking or drinking kill? Does harsher punishment reduce crime? Do mothers work more when childcare is available for free? Does trade and globalization increase or reduce wealth and/or income equality? Does free education foster a more egalitarian society in terms of opportunities?

In short, reasoning about the impact of specific options or actions is an integral part of our lives as human beings. To conjecture about such causal effects, we frequently take into account past observations, be they based on personal experiences, information (and also misinformation) in the media or social media, or the opinions of other people. Depending on where our information comes from and how we process it, our judgments on causality might be more or less biased in one or the other direction, but there is little doubt that learning (or aiming to learn) from empirical observation is one of the most remarkable human qualities. This statement also applies to social sciences and statistics, where a large and growing number of researchers and analysts investigate empirical data (i.e., systematically recorded observations) in order to causally evaluate the impacts of a multitude of human behaviors and policies based on quantitative methods.

The last several decades have seen important methodological advances in data-based causal analysis, including its combination with artificial intelligence (specifically so-called machine and deep learning) that appear very relevant in the light of the ongoing digitization and availability of more comprehensive data. In addition, such methods became more and more standard for evaluating the impacts of the actions or policies of public administrations (e.g., social policies), international institutions (e.g., development aid), companies (e.g., price policies or marketing strategies), or health-care providers (e.g., medical treatments), to name only a few examples. At the center of interest is the causal effect of a particular intervention or treatment, such as training for job seekers, a new type of surgery, or an advertising campaign, on an outcome of interest, such as employment, health, or sales, by inferring what the outcome would be in the presence and the absence of a specific intervention.

This textbook gives an introduction to the causal analysis of empirical data. It presents the most important quantitative methods for evaluating causal effects, along with the statistical assumptions they rely on, which are ultimately assumptions about human behavior. For this reason, the discussion puts a great deal of emphasis on conveying the ideas and intuitions behind the various methods, as well as their similarities and differences, using examples and graphical illustrations. At the same time, it also formally explores the key concepts using statistical notation, albeit not always with the greatest level of detail. Readers with a basic knowledge of statistics, including topics like probability theory, averages, covariances, hypothesis tests, and linear regression, should be able to follow all or most of the formal discussion without prohibitive difficulty. Depending on each reader's prior knowledge and focus, some of the material might even be skipped, such as the rehearsal of linear regression and its properties in the context of causal analysis offered in sections 3.2 and 3.3. This book, therefore, is suitable for PhD, master's, and advanced bachelor's degree courses, as well as for autodidacts with basic knowledge about and an interest in statistics and causal analysis.

It is worth mentioning that this work covers several methodological developments not yet considered in other textbooks, like approaches combining causal analysis with machine learning. Therefore, it is comparably comprehensive regarding methods of impact evaluation. At the same time, this textbook aims to be "clean and lean": while offering fair treatment of the topics, it avoids going into the smallest details, so the discussion may sometimes appear less extensive than those in other textbooks. Furthermore, it sometimes emphasizes conceptual analogies or overlaps between various methods to avoid redundancy and burdensome statistical notation. All of this serves the purpose of appropriately balancing breadth and depth while making the reader familiar with the most important concepts of causal analysis in a compact way, inspired by Albert Einstein's quote: "Everything must be made as simple as possible. But not simpler."

Last but not least, this book also provides a range of applications of the various methods to empirical data using the open-source software R, which is currently one of the most powerful options for causal analysis due to the swift implementation of even the most recent methods. This is useful for students, researchers, and analysts who want to be able to immediately apply such methods of impact evaluation and causal machine learning. Introductory knowledge in R is sufficient to follow the applications, which are based on user-friendly commands and consist only of a few lines of code.

## 1.2  Overview of Topics

This textbook starts with an introduction to the concept of causality in chapter 2, particularly the difference between the causal effect of an intervention or treatment on an outcome of interest and a mere noncausal association between the two. For instance, we might observe that individuals with a higher education earn on average more than individuals with less education. However, it is unclear whether this positive association between earnings and education is indeed solely caused by education, or also by the fact that higher- and lower-educated individuals also differ in other background characteristics likely relevant for earnings (e.g., intelligence, motivation, or other personality traits).

Chapter 3 introduces the probably most intuitive method for causal analysis: namely, the social experiment. The latter assigns the treatment (e.g., a vaccine versus a placebo treatment) randomly, such as by flipping a coin. Under successful randomization, implying, for instance, that no one can manipulate the result of the coin flip, experiments can (at least under a sufficiently large number of study participants) generate treated and nontreated groups that are similar in their background characteristics. In this case, differences in average outcomes (e.g., health) across treated and nontreated groups can be credibly attributed to the treatment, as the groups are otherwise comparable. The discussion of the experiment-based estimation of differences in average outcomes is based on linear regression, one of the most popular statistical methods out there, and also covers desirable properties that any data-based method should have.

For instance, we usually want a method to get the causal effect right on average when applied to many data sets (e.g., customer surveys) coming from the same population (e.g., the total number of customers), a property known as *unbiasedness*. We also want a method more likely to come close to the true effect, as we increase the size (i.e., number of observations) in the data, a property known as *consistency*. Furthermore, we will also look into the concepts of hypothesis tests (and so-called statistical inference), which we typically use to judge the likelihood that an effect found in the data is coincidental (or spurious) in the sense that it does not really exist (which may happen if a customer survey is by chance not quite representative for all customers). In the context of the social experiment, the discussion provides a fair amount of technical details (e.g., a definition and proof of unbiasedness) about these concepts and properties. The latter also carry over to other, nonexperimental methods considered in this book, for which we spare such details (e.g., we refrain from proving unbiasedness for further methods) and rather focus on the key assumptions and underlying intuition of each approach.

Even though experiments are frequently regarded as the gold standard for evaluating causal effects, nonexperimental methods are a very important cornerstone of causal inference and frequently applied in practice. One reason is that many interesting research questions cannot be investigated by means of experiments, such as due to financial constraints or ethical reasons. As an example, consider the effect of education on earnings, where it would be hard to imagine to randomly admit students to different levels of schooling. One of the most popular nonexperimental strategies is based on the assumption that the researcher or analyst can measure (and therefore observe) all characteristics that influence both the decision to receive a treatment (e.g., a discount) and the outcome of interest (e.g., buying behavior). Such characteristics, commonly referred to as covariates, might include age, income, or gender and might be asked in a questionnaire or come from administrative data sources.

Chapter 4 presents a range of different methods for causal analysis under such a "selection-on-observables-assumption," including so-called regression, matching, weighting, and doubly robust estimators. The idea underlying any of these approaches is to only compare the outcomes of subjects exposed and not exposed to the treatment that are similar in terms of covariates. This shall guarantee that one compares "apples with apples" when assessing the causal effect, in order to avoid that the treatment effect is mixed up with any impact of differences in the characteristics. The aim is thus to mimic the experimental context with the help of observed information: After finding groups with and without treatment that are similar in observed characteristics, differences in the outcomes are assumed to be exclusively caused by differences in the treatment. The "selection-on-observables assumption," therefore, implies that among subjects with the same characteristics, the policy is as good as randomly assigned. We will also discuss a range of extensions of the standard evaluation framework with a binary treatment (just taking the values 1 and 0), such as treatments that take several values, sequential treatments where several treatments take place in different points of time (like a job application training, followed by information technology training), or the evaluation of causal mechanisms by analyzing through which intermediate variables the treatment affects the outcome.

In the light of ongoing digitization and the ever-increasing availability of data and observed covariates (e.g., shopping behavior on online platforms), the question arises of how to optimally exploit this wealth of information for the purpose of causal analysis. To this end, chapter 5 combines the concepts of the previous chapters with a subfield of artificial intelligence (namely, machine learning). When invoking a selection-on-observables assumption, such causal machine learning algorithms can learn in a data-driven way in which covariates affect the treatment and the outcome in a significant way to make sure that we compare "apples with apples." This appears particularly useful in big data contexts where the number of observed covariates is very large—maybe even larger than the number of observations. The latter scenario can make it very hard, if not impossible, for an analyst or researcher to manually select the covariates in which treated and nontreated groups should be comparable. In addition, such a manual approach may jeopardize desirable statistical properties like unbiasedness.

Another important domain of causal machine learning is the detection of important effect heterogeneities across subgroups in the data, be it in the experimental or nonexperimental context. Let us, for instance, consider the case in which a marketing intervention is more successful in promoting a product among younger customers than among older ones, implying that the causal effect of marketing is heterogeneous in (i.e., different across) age groups. Causal machine learning algorithms can learn in a data-driven way which covariates (like age) importantly drive the size and heterogeneity of the effect, which can be useful for customer segmentation. Very closely related to this argument is a further variant of causal machine learning called *optimal policy learning*, which learns in a data-driven way the optimal assignment of a treatment (e.g., a marketing intervention) to specific subgroups defined in terms of observed characteristics (e.g., younger or older customers), in order to target those groups in which the effect is largest. This aims at optimizing the overall effect of a specific treatment that may come with a cost or budget constraint (e.g., a marketing intervention with a fixed budget).

It is worth noting that any of these causal approaches are distinct from conventional predictive machine learning algorithms, which are not suitable for causal analysis per se, but take such algorithms as ingredients. Examples include so-called decision trees, random forests, lasso/ridge regression, boosting, support vector machines, and neural networks. Chapter 5 very briefly describes the basic intuition underlying some of these predictive machine learners (e.g., based on contrasting them with linear regression), but for a more comprehensive discussion, the interested reader is encouraged to address one of the many textbooks or (often free) online courses on predictive machine learning. Finally, we will also look at algorithms for optimally designing repeated experiments (e.g., repeated advertisements on an online platform), depending on the relative effectiveness of alternative treatments in past experiments, which is called *reinforcement learning*. On the one hand, this approach aims at learning over time which treatment is most effective (e.g., which advertisement generates the most revenue). On the other hand, it tries to drop treatments that turn out to be ineffective at an early stage of the experiment for the sake of overall effectiveness.

Chapter 6 introduces a further approach to causal inference based on so-called instrumental variables. These variables possess the property that they affect the treatment whose causal effect is of interest, but not directly the outcome. This is easiest described in the context of a broken (i.e., failed) experiment, in which some subjects deviate from their assigned treatment. Let us assume that employees are randomly assigned to be eligible for a training program, but some of those assigned to it decide not to take it. While the assignment is random, and therefore satisfies the experimental context, the decision to actually participate is not. Those subjects who are not participating despite being eligible might be less motivated than others. In this case, the motivation level of actual participants and nonparticipants (rather than eligible and ineligible individuals) differs, such that comparing the wages of both groups would mix up the effect of the training with that caused by differences in motivation.

In this context, the assignment may be used as an instrumental variable if it induces at least some subjects to participate in the training but does not directly affect wages through mechanisms other than training participation. Such mechanisms to be ruled out include, for instance, an effect of mere assignment (rather than actual participation) on the motivation at work (e.g., when feeling discouraged or disappointed due to not being assigned to training). If assignment meets these conditions, the following strategy can be applied: First, one measures the effect of an assignment on the outcomes by comparing the outcomes of the groups assigned and not assigned to the training. In the absence of a direct effect, this corresponds to the impact of the assignment on training participation times the effect of training participation on the outcome. Second, one measures the effect of the assignment on training participation by comparing the participation decisions of the groups assigned and not assigned to the training. Finally, dividing (or scaling) the first effect by the second produces the impact of training participation on the outcome. Starting from this setup, we will also consider various extensions, possible modifications of the assumptions in the presence of observed covariates, and how previously mentioned estimation methods, including regression, matching, weighting, doubly robust (DR) estimation, and causal machine learning (CML), can be adapted to instrumental variable approaches.

Chapter 7 considers difference-in-differences (DiD) and related approaches, which require that the outcome of interest is observed over time—that is, prior to and after the introduction of a treatment—and that the treatment is introduced for one group but not another. The DiD method relies on the assumption that in the absence of the treatment, the outcomes of the groups with and without treatment would have experienced the same change over time: that is, they would have followed a common trend. Let us consider a labor market reform as a treatment, which increases the unemployment benefits for job seekers who are at least 60 years old but not for younger groups. Simply comparing the employment outcomes of older and younger groups after the reform does not show the causal effect of the reform because differences in employment might be caused by both the treatment or age-related factors. Likewise, simply comparing the outcomes of job seekers aged 60 or older before and after the reform may not give the causal effect of the reform either, because differences in employment might be caused by both the treatment effect and a general time trend in employment: for example, due to a change in the economic conditions (or the business cycle) over time.

However, if such a time trend in employment can be assumed to be the same across age groups, we may measure it by the before-and-after treatment comparison in the outcomes among younger individuals who are not affected by the reform. In this case, subtracting from the before-and-after treatment difference in employment among the 60+ group (which consists of the policy effect plus the time trend) the before-and-after difference among those below 60 (which consists of the time trend alone) yields the treatment effect. That is, taking the difference in (before-and-after) differences across groups allows for evaluating the reform. Chapter 7 also discusses several extensions, such as when covariates are controlled for or the treatment is introduced in different periods across different groups. Finally, we will look at a method that replaces the common trend assumption by an alternative restriction on the stability of someone's outcome rank over time (e.g., someone's rank in the wage distribution before and after introduction of the treatment) when not receiving the treatment, which is known as the *changes-in-changes approach*.

Chapter 8 presents the *synthetic control method*, another approach that relies on the observability of outcomes prior and after the treatment introduction that was originally developed for case study setups with only one treated and many nontreated units. It is based on taking the difference between the treated unit's outcome and a weighted average of the nontreated units' outcomes, which serves as a synthetic imputation of what the treated unit's outcome would have been without the treatment. The importance that a nontreated unit receives in the computation of the average depends on how similar it is to the treated unit before the treatment introduction in terms of observed outcomes (and possibly covariates as well). Therefore, the assumption underlying the synthetic control method is that generating a mix of nontreated units with comparable outcomes (and covariates) like the treated unit in the periods prior to treatment permits assessing the effect on the treated unit after the treatment introduction.

As an example, let us consider the effect of German reunification after the fall of the Berlin Wall on economic growth in West Germany, which was the only European country experiencing such a reunification at that time. The synthetic control method aims at generating a weighted average of other European countries (such as Austria and the Netherlands) that closely matches the economic conditions in West Germany prior to reunification. This is for the purpose of evaluating whether the reunification induced a differential development of economic growth between West Germany and the weighted average of nontreated countries. We will also discuss several modifications, including machine learning approaches and extensions to multiple treated units.

Chapter 9 discusses so-called regression discontinuity designs, which aim at mimicking the experimental context of a randomly assigned treatment locally at a specific threshold of an index or running variable that determines access to the treatment. To fix ideas, let us assume that a university admits only applicants who earn a minimum score on an admission test. If neither applicants nor examiners manipulate the test scores, then students just passing (by obtaining the minimum score) are arguably quite similar in terms of intelligence and other characteristics to those just failing due to a slightly lower score (i.e., receiving 1 point less than the required minimum). Locally, at the threshold of the test score, we can therefore compare the earnings outcomes of admitted and nonadmitted students to assess the causal effect of admittance, as the context resembles an experiment with a comparable treated group (just attaining the threshold) and nontreated group (just failing to attain the threshold).

The described setup corresponds to the so-called sharp regression discontinuity design, which assumes that everyone above and no one below the threshold receives the treatment. Furthermore, and very much in the spirit of the instrumental variable approach mentioned previously, the framework can also be adapted to the context of a broken experiment, implying that not everyone admitted to a university might decide to attend it. This entails a so-called fuzzy regression discontinuity design, which assumes that the threshold changes treatment participation for some subjects, but not necessarily for all of them.

A related and yet different approach is the so-called regression kink design for assessing treatments that are continuous (i.e., can take many values, not just 1 or 0) and change their association with the running variable at a specific threshold of the running variable. For instance, we might be interested in assessing the causal effect of unemployment benefits (a treatment) on unemployment duration (an outcome). The benefits may amount to a specific share (such as 60 percent) of the previous earnings, which is the running variable, but they may be capped at (and therefore not go beyond) a specific threshold of previous earnings. This entails a kinked relation between unemployment benefits and the running variable and may be used to assess causal effects around the threshold. Finally, we will take a look at so-called bunching designs, which also rely on thresholds in a running variable. But in contrast to regression continuity and kink designs, they consider the case that subjects can manipulate or decide whether they will be above or below the threshold: for example, a specific income tax bracket as a function of employment. Under specific assumptions, we can exploit this setup to estimate the size of the selection (or bunching) effect just above or below the threshold, or to correct for nonrandomness in treatment assignment.

Chapter 10 is concerned with the question of how stable or robust causal effects are when specific assumptions that underlie the previously mentioned selection-on-observables or instrumental variable framework are not satisfied. It shows that if we drop some of or all these assumptions, we obtain a range or set of possible causal effects rather than a single number. Such a so-called partial identification approach might nevertheless be interesting and sensible if the satisfaction of stronger assumptions appears implausible in the empirical problem at hand. Rather than dropping specific assumptions altogether, a second approach consists of assessing the robustness of causal effects to minor or even stronger deviations from the questionable assumptions, which serve as the default, by means of a so-called sensitivity analysis. We will consider several causal problems for which partial identification and sensitivity analyses can be fruitfully applied, but the text will not exhaustively cover all possible approaches.

Chapter 11 addresses the important issue that, in a world of social interactions and interference, an outcome of interest might be affected not only by someone's own treatment, but also by the treatments received by others, such as family members, friends, or even society as a whole. For instance, providing a student with a textbook on causal analysis might not exclusively affect the student's own learning process, but also spill over to other students if the book is used by a group. Likewise, a welfare payment to a poor household might not exclusively affect the household's own income, but also that of other people or the general society due to increased spending on goods or services by the welfare-receiving household.

We will therefore consider several distinct approaches for separating the so-called direct effects of someone's own treatment from so-called interference effects that come from the treatment of others: for example, due to spillover effects or social interactions. One class of methods, for instance, assumes that interference effects take place within, but not across clusters like geographic regions, which is known as *partial interference*. As an alternative, another class of methods based on so-called exposure mappings assumes that anyone's social network relevant to the transmission of interference effects is observable and can be used for defining a tractable number of types of interference (e.g., having no versus one or more treated friends in a social network). We will discuss such approaches in a number of causal contexts when invoking selection-on-observables or instrumental variables assumptions, but without being exhaustive in the coverage of the methods available.

Chapter 12, the final chapter, gives a brief outlook on potential future trends in the field of causal analysis, in particular causal discovery, which aims at learning causal relations between possibly many variables in a data-driven way. So the best in causal analysis might yet be to come!

# 2

# Causality and No Causality

## 2.1 Potential Outcomes, Causal Effects, and the Stable Unit Treatment Value Assumption

The fundamental problem of causality is rooted in the fact that given a specific point in time, we cannot observe the world with and without a particular intervention whose causal effect we intend to evaluate. To illustrate this issue with an example, let us assume that we are interested in the effect on wages of training (such as a course in causal analysis). To measure the wage effect for any individual who is of interest to us, such as a training participant, we would like to compare the individual's wage with and without training participation. However, at any point in time, an individual has either participated or not participated in the training, but never both. Due to the impossibility of observing individuals at the same time in two mutually exclusive participation states, causal effects generally cannot be evaluated for a specific individual, an issue that Holland (1986) calls the "fundamental problem of causal inference."

To discuss the concept of causality and the difficulty of evaluating causal effects more formally, we now introduce some statistical notation. Throughout this textbook, we will use capital letters for denominating random variables and lowercase letters to express specific values of these variables whenever appropriate. Let us, for instance, denote by $D$ our intervention, which we henceforth refer to as "treatment" (inspired by medical interventions). In our example, the treatment is an indicator for training participation: that is, $D = 1$ if someone participates in the training and $D = 0$ if not. Furthermore, we denote by $Y$ the observed outcome variable (the wage, in our example), for which we would like to know how it is affected by $D$. To give a graphical illustration, figure 2.1 provides a causal graph, which displays the causal effect we are interested in by means of an arrow going from treatment $D$ to outcome $Y$. Such graphs, and more specifically, directed acyclic graphs (DAGs), which rule out cyclic or simultaneous relations (like arrows going both from $D$ to $Y$ and $Y$ to $D$), are very popular for displaying causal structures; see, for instance, the examples provided in Pearl (2000), Cunningham (2021), and Huntington-Klein (2022).

**Figure 2.1**
The causal effect of the treatment on the outcome.

To see that the effect of $D$ on $Y$ cannot be observed on the individual level, we introduce some further notation known as the *potential outcome framework*, as proposed by Neyman (1923) and advocated by Rubin (1974), Imbens and Rubin (2015), and many other contributions to causal analysis. Let $Y(1)$ and $Y(0)$ denote the potential outcomes hypothetically realized if treatment $D$ were set to 1 and 0, respectively. In our example, $Y(1)$ is the potential wage under training participation and $Y(0)$ the potential wage under nonparticipation. Importantly, denoting the potential outcomes $Y(1)$ and $Y(0)$ as a function of someone's own treatment status $D$ alone implicitly imposes the assumption that someone's potential outcomes are not affected by the treatment status of others. This is known as the *stable unit treatment value assumption* (SUTVA); see, for instance, the discussion in Rubin (1980) and Cox (1958).

To better understand the implications of the SUTVA, let us for the moment express the potential outcomes of an individual, indexed by $i$, as a function of that individual's own treatment state $D_i$ (where the subscript now makes the reference to individual $i$ more explicit), which can be either 1 or 0, and the treatments assigned to all other subjects but individual $i$ in the population of interest, denoted by $\mathcal{D}_{-i}$. The potential outcome under individual $i$'s treatment assignment $D_i = d_i$, with $d_i$ being 1 or 0, and all others' assignments $\mathcal{D}_{-i} = \mathbf{d}_{-i}$, with $\mathbf{d}_{-i}$ being a vector containing the treatments of everyone else in the population, is thus given by $Y_i(d_i, \mathbf{d}_{-i})$. The SUTVA implies, for any individual $i$ in the population, that

$$Y_i(d_i, \mathbf{d}_{-i}) = Y_i(d_i) \text{ for } d_i \in \{0, 1\} \text{ and any assignment } \mathbf{d}_{-i}. \tag{2.1}$$

By equation (2.1), someone's potential outcomes are not influenced by the treatment assignments among others, $\mathcal{D}_{-i}$, but only a function of his or her own treatment, $d_i$. We can thus refrain from using the subscripts $i, -i$ when defining the potential outcomes, which simplifies the notation. However, it is important to acknowledge that ruling out any spillover or interference effects from the treatment of others to a subject's own outcome might not appear plausible in all contexts. For instance, if more individuals obtain training, then the supply or availability of a certain skill in the labor market increases. This may negatively affect the wage of an individual independent of her or his own training participation because the companies can now choose among a larger pool of trained individuals. Therefore, whether other labor market participants are trained may have an impact on one's own labor market outcomes. However, if the number of trained individuals is small relative to the total demand for and supply of individuals with that skill, then the SUTVA might at least come close to being satisfied. The plausibility of this assumption, therefore, needs to be assessed in the empirical context at hand. We will assume the SUTVA throughout this book unless stated otherwise in chapter 11, which explicitly considers causal analysis under violations of this assumption.

After having understood the implications of the SUTVA for the definition of potential outcomes, let us now verify how the latter are associated with the observed outcome $Y$: that is, the wage. Quite naturally, $Y$ corresponds to the potential outcome under treatment, $Y(1)$, for individuals actually participating in the training $(D = 1)$, and to the potential outcome under nontreatment, $Y(0)$, for individuals actually not participating $(D = 0)$. It is impossible to observe both potential outcomes at the same time, as individuals either participate or not, but never both at once. We can formally express this relation between $Y$ on the one hand and $Y(1)$, $Y(0)$, $D$ on the other hand by means of the following equation:

$$Y = Y(1) \cdot D + Y(0) \cdot (1 - D) \text{ for } D \text{ being either 1 or 0.} \tag{2.2}$$

Equation (2.2) is equivalent to equation (2.3), in which the observed outcome is expressed as the potential outcome without treatment, to which $Y(1) - Y(0)$ (i.e., the difference in the potential outcome with and without treatment) is added in the case of actual treatment:

$$Y = Y(0) + \underbrace{(Y(1) - Y(0))}_{\text{causal effect}} \cdot D. \tag{2.3}$$

The difference in potential outcomes $Y(1) - Y(0)$ is the causal effect of $D$ on $Y$ of interest as graphically displayed in figure 2.1, corresponding to the individual change in the wage due to participating versus not participating in the training. However, as either $Y(1)$ or $Y(0)$ is not observed depending on the value of $D$, the effect cannot be measured for any individual, which creates the previously mentioned fundamental problem of causal inference.

Even though causal effects are fundamentally unidentifiable at an individual level, we may under specific statistical assumptions evaluate them on more aggregate levels: that is, based on groups of treated and nontreated individuals. One interesting parameter in this context is the average causal effect, also known as the average treatment effect (ATE): that is, the average effect of $D$ on $Y$ in a predefined population, consisting, for instance, of all employees in a region or country where the training is offered. The ATE, which we henceforth denote by $\Delta$, corresponds to the difference in the average potential outcomes $Y(1)$ and $Y(0)$ for in the population of interest. It is formally defined as

$$\Delta = E[Y(1)] - E[Y(0)], \tag{2.4}$$

where $E[\ldots]$ stands for "expectation," which is simply the average in the population. Furthermore, we could also be interested in the ATE in a specific subpopulation, in particular, the subpopulation of employees who actually participated in the training($D = 1$), rather than all employees. This average treatment effect on the treated (ATET), which we henceforth denote by $\Delta_{D=1}$, is formally defined as

$$\Delta_{D=1} = E[Y(1)|D = 1] - E[Y(0)|D = 1], \tag{2.5}$$

where $|D = 1$ is to be read as "conditional on $D = 1$," which means "only for those who actually participated." That is, $|$ is to be interpreted as an if condition such that the average effect refers only to the subpopulation satisfying this if condition. Analogously, we can also define the average treatment effect on the nontreated (ATENT), denoted by $\Delta_{D=0}$:

$$\Delta_{D=0} = E[Y(1)|D = 0] - E[Y(0)|D = 0]. \tag{2.6}$$

## 2.2   Treatment Selection Bias

After having defined aggregate effects like the ATE and the ATET, a natural question is how to properly assess them. Can we simply compare the outcomes of individuals receiving the treatment, commonly referred to as the "treatment group," with those not receiving the treatment, the "control group?" In general, unfortunately, the answer is no. The reason is that the treatment and control groups may differ in background characteristics that affect the outcome. In this case, the differences in $Y$ across the treatment and control groups may not only reflect the treatment effect, but they might also be driven by differences in the background characteristics. Let us assume, for instance, that individuals participating in training are on average more motivated than those who do not. If motivation affects wages, which appears likely, then comparing the average wages of training participants and nonparticipants will not give the causal effect of training, but rather a mixture of the effects of training and motivation.

Indeed, we are comparing apples with oranges when basing our analysis on treatment and control groups that are not comparable in terms of motivation. Even though we might observe a statistical association like a correlation between $Y$ and $D$, implying that the level of $Y$ systematically varies with different levels of $D$, this does not correspond to the causal effect of $D$. Figure 2.2 shows this issue by means of a causal graph, in which $U$ denotes motivation or other characteristics affecting both training participation and wages, as indicated by the dotted arrows going from $U$ to $D$ and $Y$. Using dotted (rather than solid) arrows emphasizes that such characteristics $U$ are frequently not observed in the data, such that their causal effect on $D$ and $Y$ cannot

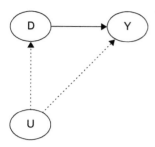

**Figure 2.2**
Treatment selection bias.

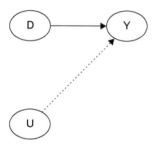

**Figure 2.3**
No treatment selection bias.

be assessed. For this reason, we generally cannot infer causal relations from patterns observed in data, as argued by Haavelmo (1943), for instance.

For a proper evaluation of the causal effect of $D$ on $Y$, the treated and control groups must be comparable in terms of any background characteristics $U$ that affect $Y$, which is known as the *ceteris paribus condition* ("everything else equal" apart from $D$) in econometrics and statistics. Graphically, this implies that if $U$ affects $Y$, it must not affect $D$, as displayed in figure 2.3. For instance, if motivation affects wages, it must not influence training participation, in order to satisfy the ceteris paribus condition that motivation does not systematically differ across treatment and control groups. Then, differences in $Y$ across groups with $D = 1$ and $D = 0$ can be attributed to differences in the treatment alone: that is, correspond to the causal effect of $D$.

For a more formal discussion, let us denote by $E[Y|D = 1]$ the average outcome in the treatment group (recalling that $|D = 1$ reads as "conditional on $D = 1$") and by $E[Y|D = 0]$ the average outcome in the control group. Under which condition does the mean difference in the observed outcomes across groups, $E[Y|D = 1] - E[Y|D = 0]$, correspond to the ATE, $\Delta = E[Y(1)] - E[Y(0)]$, the average effect of $D$?

To answer this question, we first note that

$$E[Y|D=1] = E[Y(1)|D=1], \quad E[Y|D=0] = E[Y(0)|D=0]$$

because for the treatment group, $Y = Y(1)$, while for the control group, $Y = Y(0)$. This follows from our discussion of the association of observed and potential outcomes in section 2.1, and in particular from equation (2.2). Therefore,

$$E[Y|D=1] - E[Y|D=0] = E[Y(1)|D=1] - E[Y(0)|D=0].$$

We can now see that $E[Y(1)|D=1] - E[Y(0)|D=0]$ equals $\Delta = E[Y(1)] - E[Y(0)]$ only if the following conditions hold:

$$E[Y(1)|D=1] = E[Y(1)] \text{ or, equivalently, } E[Y(1)|D=1] - E[Y(1)] = 0, \qquad (2.7)$$

$$E[Y(0)|D=0] = E[Y(0)] \text{ or, equivalently, } E[Y(0)|D=0] - E[Y(0)] = 0.$$

In other words, $E[Y|D=1] - E[Y|D=0]$ generally corresponds to the ATE only if (i) the average of $Y(1)$ in the treatment group and in the total population are the same and (ii) the average of $Y(0)$ in the control group and in the total population are the same. This rules out that background characteristics $U$ that affect the potential outcome $Y(1)$ are on average different across the treatment group and the total population, which also contains the control group. It also rules out that any $U$ affecting $Y(0)$ is on average different across the control group and the total population, which also contains the treatment group. In short, the treatment and control groups must be comparable with regard to $U$, at least on average, for satisfying equation (2.7).

Conversely, if the conditions in expression (2.7) do not hold, then $E[Y|D=1] - E[Y(1)]$ corresponds to a so-called treatment selection bias. This is the error that arises when measuring $E[Y(1)]$ based on $E[Y|D=1]$, and it is due to the noncomparability in $U$ across the treatment and control groups. Likewise, $E[Y|D=0] - E[Y(0)]$ corresponds to the selection bias when measuring $E[Y(0)]$. Finally, a combination of both gives the treatment selection bias due to differences in $U$ when aiming at assessing the ATE based on the mean difference $E[Y|D=1] - E[Y|D=0]$. That is, from $E[Y|D=1] \neq E[Y(1)]$ and/or $E[Y|D=0] \neq E[Y(0)]$, it generally follows that

$$E[Y|D=1] - E[Y|D=0] \neq E[Y(1)] - E[Y(0)],$$

such that $\underbrace{E[Y|D=1] - E[Y(1)] - E[Y|D=0] + E[Y(0)]}_{\text{selection bias when assessing } E[Y(1)] - E[Y(0)]} \neq 0.$ \qquad (2.8)

It is worth noting that the size (or even the existence) of the treatment selection bias cannot be verified in the data, as $E[Y(1)]$ and $E[Y(0)]$ are not observed because either $Y(1)$ or $Y(0)$ is not known for any subject in the population. Whether a statistical method properly measures a causal effect or gives a noncausal parameter flawed by treatment selection bias is typically unknown, as it hinges on the satisfaction of particular, generally nonverifiable conditions, like the satisfaction of expression (2.7). In the chapters to follow, we will consider a range of alternative methods that are distinct in terms of the conditions they impose for identifying causal effects, frequently referred to as *identifying assumptions*. As the latter are often not verifiable (or testable) in the data, a key task in any causal analysis is to scrutinize whether such assumptions are likely satisfied or violated in an empirical problem at hand: for example, based on theoretical arguments or previous empirical evidence.

# 3
## Social Experiments and Linear Regression

### 3.1 Social Experiments

Experiments such as those considered in Fisher (1935) are probably the most intuitive, and from a statistical perspective, often the most convincing approach to causal analysis. They consist of randomly (i.e., coincidentally) granting or denying subjects access to a treatment like a training program, such as one based on flipping a coin. Because the members of the treatment and control groups are exclusively chosen by luck (the result of a coin flip) rather than based on their background characteristics, the latter are comparable across the treatment and control groups, at least when the number of participants in the experiment is sufficiently large. Therefore, the causal effect of the training can be assessed by simply comparing the outcomes (e.g., the wages) of both groups, as there are no systematic differences in other important characteristics (such as motivation, age, etc.) that could also have an influence on the outcome, previously denoted as $U$. Graphically, random assignment of intervention $D$ corresponds to the framework in figure 2.3 where $U$ does not affect $D$ because $D$ is exclusively determined by the coin flip.

More formally, random assignment of $D$ rules out that subjects with specific, particularly high or low potential outcomes $Y(1)$ or $Y(0)$ (and thus $U$, which is the only element affecting the outcome apart from $D$) are more likely than others to either receive the treatment or not. In statistical parlance, we say that $D$ is statistically independent of $Y(1)$, $Y(0)$, as expressed by the following independence assumption:

$$\{Y(1), Y(0)\} \perp D, \tag{3.1}$$

where $\perp$ denotes statistical independence. A consequence of the potential outcomes being independent of the treatment (i.e., comparable across treatment and control groups) is that their means (as well as variances and distributions in general) are comparable. Therefore, expression (2.7) in section 2.2 is satisfied under statistical independence. This implies that by their randomized design, experiments are (if properly conducted) not prone to treatment selection bias, such that the average treatment effect (ATE) corresponds to the mean difference in the outcomes of treated and nontreated observations:

$$\Delta = E[Y(1)] - E[Y(0)] = E[Y|D=1] - E[Y|D=0].$$    (3.2)

We note that the mean difference in equation (3.2) and any other parameter discussed so far, like the ATE $\Delta$, refers to the total population. In practice, however, experiments (or other causal approaches) are typically conducted in a sample that is to be randomly drawn from the population of interest to be representative. For instance, we might randomly select 10,000 individuals in a country to participate in an experiment, with the goal that the chosen individuals represent the country's population well in terms of any characteristics like education or age. As we cannot observe the average outcomes $E[Y|D=1]$ or $E[Y|D=0]$ in the population, but rather aim at approximating or estimating them in the sample, our statistical notation must distinguish between the population and the sample perspective. Let us for this reason denote by $n$ the number of observations in the sample and by $i$ the index of any such observation, implying that $i \in \{1, 2, \ldots, n\}$. For instance, $i=1$ for the first observation and $i=n$ for the last observation in the sample. Furthermore, we denote by $Y_i$ and $D_i$ the outcome and treatment of observation $i$ in our sample.

Equation (3.2) suggests that we may estimate the ATE by the mean differences of the treated and nontreated observations in our sample:

$$\frac{\sum_{i=1}^{n} Y_i \cdot D_i}{\sum_{i=1}^{n} D_i} - \frac{\sum_{i=1}^{n} Y_i \cdot (1-D_i)}{\sum_{i=1}^{n} (1-D_i)}.$$    (3.3)

The summation $\sum_{i=1}^{n} Y_i \cdot D_i$ corresponds to adding up $Y_1 \cdot D_1 + Y_2 \cdot D_2 + \cdots + Y_n \cdot D_n$ and thus yields the sum of treated outcomes in the sample, because $Y_i \cdot D_i$ is $Y_i$ if $D_i = 1$ and zero if $D_i = 0$. An alternative way of writing this summation is $\sum_{i:D_i=1} Y_i$, such that we only choose treated outcomes to be added up. Furthermore, the summation $\sum_{i}^{n} D_i = D_1 + \cdots D_n$ corresponds to the number of treated observations in the sample. $\frac{\sum_{i=1}^{n} Y_i \cdot D_i}{\sum_{i=1}^{n} D_i}$ therefore yields the mean outcome among the treated in the sample, and by analogous arguments, $\frac{\sum_{i=1}^{n} Y_i \cdot (1-D_i)}{\sum_{i=1}^{n} (1-D_i)}$ gives the mean outcome among the nontreated.

Let us consider an empirical example for the estimation based on mean differences using the statistical open-source software R, provided by R Core Team (2015). We will analyze an experiment conducted between November 1994 and February 1996, which randomized access to Job Corps, a large US education program financed by the US Department of Labor that targets disadvantaged individuals aged sixteen to twenty-four. Schochet, Burghardt, and Glazerman (2001) and Schochet, Burghardt, and McConnell (2008) provide detailed discussions of the experimental design and the effects of random program assignment. These studies find that the Job Corps program increases educational attainment, reduces criminal activity, and increases employment and earnings, at least over some years after the program.

In our empirical example, we aim at assessing the average treatment effect of random program assignment on the weekly earnings in the fourth year after the assignment. To this end, we install and load the *causalweight* package for R provided by Bodory and Huber (2018) using the *install.packages* and *library* commands. The package includes the data set *JC* with 9,240 observations and 46 variables from the experimental Job Corps study, which we load into the R workspace using the *data* command. Applying the command *?JC* opens the corresponding help file, with more detailed information about the variables in the data set. We can see that *assignment* is a binary indicator for random assignment to Job Corps, which is our treatment *D*. *earny4* measures weekly earnings in US dollars (USD) in the fourth year, which we consider as outcome *Y*.

We can access these variables in the *JC* data by using $, which permits calling subobjects in R objects, to define *D* and *Y*: *D=JC$assignment* and *Y=JC$earny4*. Finally, we take the mean difference in the outcome across treatment groups. To this end, the expression *Y[D==1]* selects the treated outcomes because square brackets permit selecting observations that satisfy a specific condition *D==1*; that is, that the treatment is equal to 1. Note that we need to be sure to use the double equals sign in *D==1*, as we use it for checking a condition, while single equals signs are for defining R objects like variables. Wrapping the *Y[D==1]* expression with the *mean* command yields the mean outcome among the treated, and proceeding analogously with the nontreated and taking differences yield the ATE estimate corresponding to equation (3.3): *mean(Y[D==1])-mean(Y[D==0])*. The box here provides the R code for each steps.

```
install.packages("causalweight")   # install causalweight package
library(causalweight)               # load causalweight package
data(JC)                            # load JC data
?JC                                 # call documentation for JC data
D=JC$assignment                     # define treatment (assignment to JC)
Y=JC$earny4                         # define outcome (earnings in fourth year)
mean(Y[D==1])-mean(Y[D==0])         # compute the ATE
```

Running this R code yields the following output:

```
[1] 16.05513
```

Our ATE estimate suggests that on average, access to Job Corps increases weekly earnings in the fourth year after assignment by roughly 16 USD, thus pointing to a positive earnings effect of the educational program. Congratulations—we just ran our first causal analysis in R!

## 3.2   Effect Identification by Linear Regression

The mean difference in the outcomes of treated and nontreated individuals provided in equation (3.2) for the population and in equation (3.3) for the sample can also be expressed by a so-called linear regression. The latter was first suggested by Gauss (1809) and is one of the most popular techniques for analyzing statistical associations, be they causal or noncausal. So, importantly, linear regression can be applied to data even if the conditions for causal analysis like the independence assumption in expression (3.1) are not satisfied, but in this case, the output of linear regression will in general not yield a causal effect. However, if these conditions are met, then linear regression provides a convenient framework for assessing desirable statistical properties of causal analysis in empirical data, as discussed in section 3.3, like unbiasedness, consistency, and asymptotic normality. For this reason, we will subsequently express ATE evaluation under a randomized treatment as a regression problem, which requires a somewhat more formal discussion (which might be skipped by readers already familiar with the properties of linear regression). To this end, let us reconsider the characterization of the observed outcome in equation (2.3) in section 2.1 and take the conditional expectation of this expression given $D$ in the population; that is, the average of the observed outcome in a particular treatment group:

$$E[Y|D] = E[Y(0) + (Y(1) - Y(0)) \cdot D|D] \tag{3.4}$$
$$= E[Y(0)|D] + \{E[Y(1)|D] - E[Y(0)|D]\} \cdot D.$$

The second equality in equation (3.4) follows from the fact that $E[D|D] = D$: that is, the average of any variable conditional on the very same variable is the variable itself. Furthermore, equation (2.2) implies that $E[Y|D=0] = E[Y(0)|D=0]$ and $E[Y|D=1] = E[Y(1)|D=1]$.

We notice that by the independence assumption in expression (3.1), $E[Y(0)] = E[Y(0)|D] = E[Y(0)|D=0] = E[Y|D=0]$   and   $E[Y(1)] = E[Y(1)|D] = E[Y(1)|D=1] = E[Y|D=1]$, such that equation (3.4) becomes

$$E[Y|D] = \underbrace{E[Y|D=0]}_{\alpha} + \underbrace{(E[Y|D=1] - E[Y|D=0])}_{\beta} \cdot D.$$

Therefore, the conditional mean of outcome $Y$ given treatment $D$, such as average sales in the presence or absence of a marketing campaign, can be expressed as a linear function of the so-called coefficients $\alpha$ and $\beta$. $\alpha$ corresponds to the mean outcome among the nontreated $E[Y|D=0]$, such as the average sales without marketing, which under the independence assumption in expression (3.1) equals the mean potential outcome under nontreatment $E[Y(0)]$. $\beta$ corresponds to the mean difference of treated and nontreated outcomes $E[Y|D=1] - E[Y|D=0]$, which under equation (3.1) equals the ATE, $\Delta = E[Y(1) - Y(0)]$: that is, the causal effect of the marketing campaign. Finally, $\alpha + \beta$ corresponds to the mean outcome among the treated, $E[Y|D=1]$, such as the average sales when conducting the marketing campaign, which under equation (3.1) equals $E[Y(1)]$.

We now have a representation of the potential outcomes and the causal effect by means of a linear regression, but it still must be demonstrated how linear regression actually determines the values of coefficients $\alpha$ and $\beta$. Readers not interested in such technical details might skip the subsequent discussion, but it seems interesting for better understanding why and how linear regression works in the experimental context with a binary treatment. Let us introduce yet a further variable $\varepsilon$, which represents the difference between the observed outcome $Y$ (e.g., sales) and its respective conditional mean in a specific treatment group (e.g., average sales among the treated), $E[Y|D]$. $\varepsilon$ is commonly referred to as an *error term* or *residual* and is formally defined as follows:

$$\varepsilon = Y - \underbrace{(\alpha + \beta D)}_{E[Y|D]}. \tag{3.5}$$

Solving equation (3.5) for $Y$ by rearranging terms shows that the observed outcome can be expressed as the sum of the average outcome in a specific treatment state and its deviation thereof, represented by the error term

$$Y = \underbrace{\alpha + \beta D}_{E[Y|D]} + \varepsilon. \tag{3.6}$$

Figure 3.1 illustrates this graphically by depicting parameters $\alpha$ and $\beta$, as well as the error term $\varepsilon$ for one of the many subjects in the population, whose values of $Y$ (on the $y$-axis) and $D$ (on the $x$-axis) are presented by a scatterplot of dots.

To compute coefficients $\alpha$ and $\beta$, linear regression is based on exploiting two specific properties of error term $\varepsilon$ that are known as *moment conditions*, as they refer to means, which are the first moments of the distribution of some variable. The first moment condition is that the average of $\varepsilon$ equals zero, which is shown by taking expectations (i.e., population averages) in equation (3.5):

$$E[\varepsilon] = E[Y] - E[E[Y|D]] = E[Y] - E[Y] = 0. \tag{3.7}$$

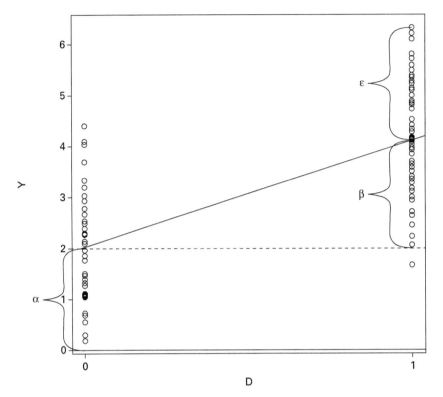

**Figure 3.1**
Linear regression.

We note that the second equality follows from the law of iterated expectations, which says that the mean of a variable corresponds to the mean of the conditional means of that variable. For instance, the average sales in the population equal the appropriately averaged average sales under treatment (e.g., a marketing campaign) and nontreatment. Quite intuitively, equation (3.7) states that deviations from a variable's mean must average to zero.

The second moment condition is a consequence of the independence assumption in expression (3.1), which permits replacing $E[Y|D]$ in equation (3.5) with functions of mean potential outcomes as provided in equation (3.4):

$$\varepsilon = Y - E[Y(0)] - \{E[Y(1)|D] - E[Y(0)|D]\} \cdot D$$
$$= Y - E[Y(1)] \cdot D - E[Y(0)] \cdot (1 - D)$$
$$= (Y(1) - E[Y(1)]) \cdot D + (Y(0) - E[Y(0)]) \cdot (1 - D), \tag{3.8}$$

where the last equality follows from the definition of the observed outcome $Y$ in equation (2.2) in section 2.1. As $E[Y(1) - E[Y(1)]] = 0$ and $E[Y(0) - E[Y(0)]] = 0$, it is easy to verify that $E[\varepsilon|D] = 0$. This in turn implies the second moment condition underlying linear regression:

$$E[D \cdot \varepsilon] = 0. \tag{3.9}$$

The reason is that by the law of iterated expectations, $E[D \cdot \varepsilon] = E[D \cdot E[\varepsilon|D]]$. We also note that it can be shown that $E[D \cdot \varepsilon] = E[(D - E[D]) \cdot \varepsilon] = Cov(\varepsilon, D)$, where $Cov$ denotes the covariance. Therefore, the independence assumption in expression (3.1) implies that the covariance of the treatment and the error term in the population is zero: $Cov(\varepsilon, D) = 0$.

In the next step, we can solve the first and second moment conditions (equations (3.7) and (3.9)) for $\alpha$ and $\beta$ to demonstrate how linear regression identifies the ATE and the mean potential outcomes. By using the definition of $\varepsilon$ in equation (3.5), the first moment condition implies

$$E[\varepsilon] = E[Y - \alpha - \beta D] = 0$$
$$\Leftrightarrow \alpha = E[Y] - \beta E[D]. \tag{3.10}$$

Furthermore, the second moment condition implies

$$\begin{aligned} E[D \cdot \varepsilon] &= E[D \cdot (Y - \alpha - \beta D)] = 0 \\ &= E[D \cdot (Y - E[Y] - \beta(D - E[D]))] = 0 \\ \Leftrightarrow \beta &= \frac{E[D \cdot (Y - E[Y])]}{E[D \cdot (D - E[D])]} = \frac{E[(D - E[D]) \cdot (Y - E[Y])]}{E[(D - E[D]) \cdot (D - E[D])]} = \frac{Cov(D, Y)}{Var(D)}, \end{aligned} \tag{3.11}$$

where the second equality follows from the definition of $\alpha$ in equation (3.10). With a little bit of effort, we have thus proved that linear regression identifies the ATE of a binary treatment in experiments by $\beta$, which is computed as the ratio of the covariance ($Cov$) of $Y$ and $D$ and the variance ($Var$) of $D$. This covariance-variance-ratio thus corresponds to $E[Y|D=1] - E[Y|D=0]$. Finally, plugging $\beta$ into equation (3.10) identifies $\alpha = E[Y|D=0] = E[Y(0)]$: that is, the mean potential outcome under nontreatment.

An alternative but related way of expressing linear regression for identifying the coefficients $\alpha$ and $\beta$ is to consider it as the solution to the following optimization problem:

$$\alpha, \beta = \arg\min_{\alpha^*, \beta^*} E[\underbrace{Y - \alpha^* - \beta^* D}_{\varepsilon}]^2. \tag{3.12}$$

$\alpha^*$ and $\beta^*$ represent a range of candidate values for $\alpha$ and $\beta$, and ultimately those values are selected that minimize the expectation of the squared error terms. Here, the moment conditions $E[\varepsilon] = 0$ and $E[D \cdot \varepsilon] = 0$ arise as first-order conditions (i.e., when taking the first derivatives of the function to be minimized in equation (3.12) with regard to $\alpha^*$ and $\beta^*$) and picking $\alpha^*$ and $\beta^*$ such that those derivatives are zero (as implied by the minimum of the function in equation (3.12)).

While this ends this discussion on why and how linear regression identifies the ATE in the population under random treatment assignment, let us make a final remark concerning the method's name. Even though we have considered linear regression, no linear relationship whatsoever is actually imposed when the treatment is binary. Linearity of the outcome in the treatment would imply the restriction that increasing the treatment by one unit (e.g., 1 hour of training) had always the same effect on the outcome, no matter what the initial treatment (e.g., 0 or 10 hours of training) was. This restriction, however, is irrelevant for binary treatments that can only take the value 1 or 0, as we currently consider. We will, however, also consider multivalued treatments further in section 3.5.

## 3.3   Estimation by Linear Regression and Its Properties

So far, our discussion of the linear regression has focused on the identification of the ATE in the population. As mentioned before, however, a social experiment is typically based on a sample rather than the entire population, implying that we need to estimate $\alpha$ and $\beta$ in the data. For this reason, we will denote the respective parameter estimates in a sample by a "hat" symbol (ˆ) in order to distinguish them from the true parameters in the population. As already discussed in section 3.1, let us assume that we have a randomly drawn sample of $n$ observations, where each observation is indexed by $i \in \{1, 2, \ldots, n\}$ and denote by $Y_i$ and $D_i$ the outcome and treatment for observation $i$ in the sample, respectively. In analogy to the minimization problem in the population provided in equation (3.12), we obtain an estimate of the ATE based on linear regression as the solution of the minimization problem in the sample:

$$\hat{\alpha}, \hat{\beta} = \arg \min_{\alpha^*, \beta^*} \sum_{i=1}^{n} (Y_i - \alpha^* - \beta^* D_i)^2. \tag{3.13}$$

$\alpha^*$ and $\beta^*$ are candidate values for the coefficient estimates $\hat{\alpha}$ and $\hat{\beta}$ and ultimately chosen such that sum of squared residuals (i.e., the sum of squared deviations between observed outcomes and the estimated regression line) is minimized. For this reason, linear regression is also known as *ordinary least squares* (OLS). We could also minimize the mean rather than the sum of the squared residuals in the sample—namely, $\hat{\alpha}, \hat{\beta} = \arg \min_{\alpha^*, \beta^*} \frac{1}{n} \sum_{i=1}^{n} (Y_i - \alpha^* - \beta^* D_i)^2$—in order to more closely match equation (3.12), where the mean squared error (MSE) in the population is minimized. For the optimization problem, however, this does not make any difference. The values $\alpha^*, \beta^*$ minimizing the sum in equation (3.13) will also minimize the mean, as the mean is just the sum divided by the sample size $n$.

Analogous to equation (3.10) in section 3.2 providing the population parameter $\beta = E[Y|D=1] - E[Y|D=0]$, the estimate $\hat{\beta}$ in equation (3.13) is the sample covariance ($\widehat{Cov}$) of $Y_i$ and $D_i$ divided by the sample variance ($\widehat{Var}$) of $D_i$:

$$\hat{\beta} = \frac{\widehat{Cov}(Y_i, D_i)}{\widehat{Var}(D_i)}, \text{ where} \qquad (3.14)$$

$$\widehat{Cov}(Y_i, D_i) = \frac{1}{n-1} \sum_{i=1}^{n} \left( Y_i - \frac{1}{n} \sum_{i=1}^{n} Y_i \right) \left( D_i - \frac{1}{n} \sum_{i=1}^{n} D_i \right) \text{ and}$$

$$\widehat{Var}(D_i) = \frac{1}{n-1} \sum_{i=1}^{n} \left( D_i - \frac{1}{n} \sum_{i=1}^{n} D_i \right)^2 .$$

$\hat{\beta}$ corresponds to the mean difference in the outcomes of the treated and nontreated groups in the sample, and is therefore numerically equivalent to equation (3.3). And in analogy to equation (3.10), which yields the average outcome under nontreatment, $\alpha = E[Y|D=0]$, the estimate $\hat{\alpha}$ corresponds to

$$\hat{\alpha} = \frac{1}{n} \sum_{i=1}^{n} Y_i - \hat{\beta} \frac{1}{n} \sum_{i=1}^{n} D_i. \qquad (3.15)$$

To ease notation, let us henceforth denote the sample average of the treatment as $\bar{D} = \frac{1}{n} \sum_{i=1}^{n} D_i$.

It is important to note that the sample-based estimates $\hat{\beta}$ and $\hat{\alpha}$ may generally differ from the true $\beta$ and $\alpha$ in the population. The reason is that even a randomly drawn sample may not be fully representative of the population and the independence assumption in expression (3.1) might not hold exactly in the sample. Even if estimates $\hat{\beta}$ and $\hat{\alpha}$ might somewhat (and hopefully not too strongly) differ from their respective true values in a given sample, one desirable property for any statistical method is that it hits the true values on average when applied to infinitely many randomly drawn samples. This would, for instance, imply that if we could draw a very large number of samples from a country's population and run a marketing intervention in each of these samples, the mean of the causal effects estimated in each of the samples corresponds to the true effect in the population. This property of "getting it right on average" is known as *unbiasedness* and more formally means that the expectations of estimates $\hat{\beta}$ and $\hat{\alpha}$ correspond to the true parameters $\beta$ and $\alpha$, respectively:

$$E[\hat{\beta}] = \beta, \quad E[\hat{\alpha}] = \alpha. \tag{3.16}$$

Unbiasedness can be best described by the following (admittedly old) joke about statisticians: A chemist, a biologist, and a statistician are out shooting with bow and arrow on a target. The biologist shoots and misses the target 2 meters to the left. The chemist shoots and misses 2 meters to the right. The statistician, delighted by the average performance, yells "It's a hit!"

A second desirable property is consistency, implying that if one draws a larger rather than a smaller sample, then the probability of obtaining an estimate $\hat{\beta}$ that is substantially different to the true effect $\beta$, say by more than an absolute value $\epsilon$, goes down. This appears attractive because it implies that in large enough samples, the probability of obtaining an estimate that is far from the truth is low. Formally, consistency is satisfied if

$$\Pr(|\hat{\beta} - \beta| > \epsilon) \to 0 \text{ for any } \epsilon > 0 \text{ when } n \to \infty, \tag{3.17}$$

where Pr denotes probability, $\to$ reads as "converges to," and $|\ |$ stands for the absolute value. Consistency thus states that estimate $\hat{\beta}$ is more and more likely to be close to the true $\beta$ as the sample size becomes very large, and ultimately collapses to $\beta$ when the sample size goes to infinity. A different but equivalent way of stating consistency is that $\beta$ is the so-called probability limit (plim) of $\hat{\beta}$:

$$plim(\hat{\beta}) = \beta. \tag{3.18}$$

Consistency holds, for instance, if $\hat{\beta}$ is unbiased and its variance decreases as the sample size increases. Analogous arguments apply concerning the consistency of $\hat{\alpha}$.

A third desirable property is *asymptotic normality*. Reconsidering the case of drawing many random samples and estimating $\beta$ and $\alpha$ in each sample, asymptotic normality means that the pooled estimates $\hat{\beta}$ and $\hat{\alpha}$ obtained from those samples follow a normal distribution, given that the size of each sample is very large. The beauty of this property is that with a sufficiently large number of observations, it enables us to well approximate the distribution of an estimate across many samples, even if we only have a single sample (rather than many) at hand as is usually the case in empirical applications. Knowing the distribution of an estimate is very useful for so-called statistical inference, considered in section 3.4. This includes judging the error probability with which a causal effect of zero can be rejected in the population (type I error) based on the results obtained in the sample, or defining an interval of values likely containing the true effect.

Under the satisfaction of the independence assumption in expression (3.1), a randomly drawn sample in which both treated and nontreated observations are available, and a binary treatment, linear regression (or OLS) is an unbiased, consistent, and asymptotically normal estimator. Therefore, the OLS-based estimates $\hat{\beta}$ and $\hat{\alpha}$ of the ATE and the mean potential outcomes satisfy all three desired properties. This will be formally shown in the subsequent discussion, which might be skipped by readers not interested in these technical details and proofs, which may be a bit more tedious. We also note that there are many more estimators of causal effects that, under particular identifying assumptions, satisfy all three of the desirable properties, or at least a subset (e.g., only consistency and asymptotic normality) of them. We will get to know a range of estimation approaches in the chapters to follow and sometimes make comments about their properties, such as unbiasedness, but without going into as many technical details as we do for linear regression.

To formally show the unbiasedness of ATE estimate $\hat{\beta}$, let us reconsider the definition of $\hat{\beta}$ in equation (3.14) and replace $Y_i$ by the definition of the outcome provided in equation (3.6), which yields

$$
\begin{aligned}
\hat{\beta} &= \frac{\widehat{Cov}(\alpha + \beta D_i + \varepsilon_i, D_i)}{\widehat{Var}(D_i)} \\
&= \beta \frac{\widehat{Var}(D_i)}{\widehat{Var}(D_i)} + \frac{\widehat{Cov}(\varepsilon_i, D_i)}{\widehat{Var}(D_i)} \\
&= \beta + \frac{\widehat{Cov}(\varepsilon_i, D_i)}{\widehat{Var}(D_i)} \\
&= \beta + \frac{\frac{1}{n-1}\sum_{i=1}^{n}(\varepsilon_i - \frac{1}{n}\sum_{i=1}^{n}\varepsilon_i)(D_i - \bar{D})}{\frac{1}{n-1}\sum_{i=1}^{n}(D_i - \bar{D})^2} \\
&= \beta + \frac{\sum_{i=1}^{n}\varepsilon_i \cdot (D_i - \bar{D})}{\sum_{i=1}^{n}(D_i - \bar{D})^2}.
\end{aligned}
\tag{3.19}
$$

To see this result, we note that the second line in equation (3.19) follows from the fact that $\alpha$ has a covariance of zero because it is a constant, and the covariance of $D_i$ with itself is equal to the variance. The third line follows from $\frac{\widehat{Var}(D_i)}{\widehat{Var}(D_i)} = 1$ and shows that ATE estimate $\hat{\beta}$ corresponds to the true ATE $\beta$ plus the sample covariance of the treatment and the residual divided by the sample variance of the treatment. The fourth line provides the formulas of these covariance and variance terms. The fifth line is obtained by the fact that $\frac{1}{n-1}\sum_{i=1}^{n}(\varepsilon_i - \frac{1}{n}\sum_{i=1}^{n}\varepsilon_i)(D_i - \bar{D}) = \frac{1}{n-1}\sum_{i=1}^{n}\varepsilon_i \cdot (D_i - \bar{D})$ and by $\frac{1}{n-1}$ canceling out in the numerator and denominator of the second expression on the right side.

Taking expectations of the terms in equation (3.19) (i.e., averaging over many samples) yields

$$E[\hat{\beta}] = \beta + E\left[\frac{\sum_{i=1}^{n} \varepsilon_i \cdot (D_i - \bar{D})}{(D_i - \bar{D})^2}\right]$$

$$= \beta + E\left[\frac{\sum_{i=1}^{n} \overbrace{E[\varepsilon_i|D_i]}^{=0} \cdot (D_i - \bar{D})}{(D_i - \bar{D})^2}\right]$$

$$= \beta, \tag{3.20}$$

where the second equality follows from the law of iterated expectations. Unbiasedness holds because the errors have an expectation (or mean) of zero in either treatment group: $E[\varepsilon_i|D_i] = 0$.

To show the unbiasedness of $\hat{\alpha}$, we take expectations in equation (3.15) to obtain

$$E[\hat{\alpha}] = E\left[\frac{1}{n}\sum_{i=1}^{n} Y_i - \hat{\beta}\frac{1}{n}\sum_{i=1}^{n} D_i\right]$$

$$= \frac{1}{n}\sum_{i=1}^{n} E[Y_i] - \frac{1}{n}\sum_{i=1}^{n} E[\hat{\beta} D_i]$$

$$= \frac{n}{n}E[Y] - \beta\frac{n}{n}E[D] = E[Y] - \beta E[D] = \alpha; \tag{3.21}$$

see the definition of $\alpha$ in equation (3.10). We also note that the third line follows from the fact that $E[\hat{\beta}] = \beta$, which has been shown in equation (3.20), and $\sum_{i=1}^{n} E[Y_i] = n \cdot E[Y]$ as well as $\sum_{i=1}^{n} E[D_i] = n \cdot E[D]$. That is, the sum of $n$ identical averages is simply $n$ times the average.

To formally show the consistency of $\hat{\beta}$, we reconsider the third line of equation (3.19) and verify to which expressions the parameters converge as the sample size goes to infinity, by using an application of the *plim* operator previously seen in equation (3.18):

$$plim(\hat{\beta}) = plim(\beta) + plim\left(\frac{\widehat{Cov}(\varepsilon_i, D_i)}{\widehat{Var}(D_i)}\right)$$

$$= \beta + \frac{Cov(\varepsilon, D)}{Var(D)} = \beta, \tag{3.22}$$

where the second equality follows from the fact that $\beta$ is a constant (namely, the ATE in the population), and covariances and variances in samples converge to the respective covariances and variances in the population by the so-called weak law of large numbers. The third equality follows from $Cov(\varepsilon, D) = 0$; see the discussion just after equation (3.8) in section 3.2, such that consistency holds. To show the consistency of $\hat{\alpha}$, we consider the probability limits of equation (3.15) to obtain

$$plim(\hat{\alpha}) = plim\left(\frac{1}{n}\sum_{i=1}^{n}Y_i - \hat{\beta}\sum_{i=1}^{n}D_i\right) = E[Y] - \beta E[D] = \alpha, \tag{3.23}$$

where the last equality follows from the definition of $\alpha$ in equation (3.10).

To formally show the asymptotic normality of $\hat{\beta}$, we reconsider the fifth line of equation (3.19) and bring $\beta$ to the left side:

$$\hat{\beta} - \beta = \frac{\sum_{i=1}^{n}\varepsilon_i \cdot (D_i - \bar{D})}{\sum_{i=1}^{n}(D_i - \bar{D})^2}$$

$$= \frac{\frac{1}{n}\sum_{i=1}^{n}\varepsilon_i \cdot (D_i - \bar{D})}{\frac{1}{n}\sum_{i=1}^{n}(D_i - \bar{D})^2}$$

$$\Leftrightarrow \sqrt{n}(\hat{\beta} - \beta) = \frac{\frac{1}{\sqrt{n}}\sum_{i=1}^{n}\varepsilon_i \cdot (D_i - \bar{D})}{\frac{1}{n}\sum_{i=1}^{n}(D_i - \bar{D})^2}. \tag{3.24}$$

The second line follows from multiplying the numerator and denominator of the right expression by $\frac{1}{n}$, the third from multiplying by $\sqrt{n}$ and the fact that $\frac{\sqrt{n}}{n} = \frac{\sqrt{n}}{\sqrt{n}\sqrt{n}} = \frac{1}{\sqrt{n}}$. Based on this expression, asymptotic normality can be shown by the central limit theorem, a fundamental and very useful law in statistics going back to De Moivre (1738), Lyapunov (1901), Lindeberg (1922), Lévy (1937), and others.

The central limit theorem implies that as the sample size $n$ increases, the sum of a randomly sampled, zero mean variable converges to a normal distribution with a zero mean and a variance that corresponds to the variance of that variable times the sample size. More formally, it holds for any randomly sampled variable, henceforth denoted by $W$, that has a mean of zero ($E[W] = 0$) and a bounded variance, that

$$\sum_{i=1}^{n}W_i \rightarrow^d N(0, n \cdot Var(W_i)),$$

$$\Leftrightarrow \frac{1}{\sqrt{n}}\sum_{i=1}^{n}W_i \rightarrow^d N(0, Var(W_i)), \tag{3.25}$$

where $\rightarrow^d$ reads as "converges in distribution to." Furthermore, $N$ stands for the normal distribution, with the first argument being the mean ($E[W_i] = 0$) and the second the variance, with $Var(W_i) = E[W_i^2]$ for a variable with zero mean. The second line in equation (3.25) follows by multiplying by $\frac{1}{\sqrt{n}}$ and noting that this fraction enters the variance formula in squared form, $\frac{1}{n}$, such that $\frac{n}{n} \cdot Var(W_i)$ becomes $Var(W_i)$.

We now apply the central limit theorem by defining $W_i$ in expression (3.25) to be the numerator of equation (3.24), $\varepsilon_i \cdot (D_i - \bar{D})$. The latter expression is zero in expectation (i.e., $E[\varepsilon_i \cdot (D_i - \bar{D})] = 0$) because $E[\varepsilon_i|D_i] = 0$ and the law of iterated expectations, as already discussed in the context of equation (3.20). By the central limit theorem, this expression converges in distribution to a normal distribution:

$$\frac{1}{\sqrt{n}} \sum_{i=1}^{n} \varepsilon_i \cdot (D_i - \bar{D}) \to^d N(0, E[\varepsilon^2 \cdot (D - E[D])^2]). \tag{3.26}$$

To see this result, we note that

$$Var(W_i) = E[W_i^2] = E\left[\varepsilon_i^2 \cdot (D_i - \bar{D})^2\right] = E[\varepsilon^2 \cdot (D - E[D])^2]. \tag{3.27}$$

However, to obtain the asymptotic distribution of equation (3.24), we need to also consider the probability limit of the denominator $\frac{1}{n}\sum_{i=1}^{n}(D_i - \bar{D})^2$ in equation (3.24), which is $Var(D)$. By a statistical rule about the convergence of random variables called *Slutsky's theorem* (see Slutsky (1925)), it holds that $E\left[\frac{W_i}{Var(D_i)}\right] = 0$ because $E[W_i] = 0$ and $Var\left(\frac{W_i}{Var(D_i)}\right) = E\left[\frac{W_i^2}{(Var(D_i))^2}\right]$. Therefore,

$$\sqrt{n}(\hat{\beta} - \beta) \to^d N(0, \frac{E[\varepsilon^2 \cdot (D - E[D])^2]}{(Var(D))^2}). \tag{3.28}$$

We have demonstrated that $\sqrt{n}$ times the difference between the estimated and the true ATE converges to a normal distribution with a zero mean and a specific variance. This in turn implies that the difference between the estimate $\hat{\beta}$ and the true effect $\beta$ converges to zero, with a speed or convergence rate of $\frac{1}{\sqrt{n}}$ as the sample size $n$ increases. Put differently, the estimate $\hat{\beta}$ converges to the true ATE $\beta$ with a convergence rate of $\frac{1}{\sqrt{n}}$. This provides an idea of how fast the discrepancy between the estimated and true effects decays as the sample size gets larger. For instance, if the sample size is quadrupled, then the discrepancy decreases by $\frac{1}{\sqrt{4}} = \frac{1}{2}$: that is, by half. This behavior is known as $\sqrt{n}$-consistency and corresponds to the fastest convergence rate that any estimator of causal effects can possibly attain. We will see many estimation approaches that are $\sqrt{n}$-consistent in the chapters to follow, however, without going into the technical details again.

In the next step, we divide expression (3.28) by $\sqrt{n}$ and add the true effect $\beta$ to obtain

$$\hat{\beta} \to^d N\left(\beta, \frac{E[\varepsilon^2 \cdot (D - E[D])^2]}{n \cdot (Var(D))^2}\right). \tag{3.29}$$

This ultimately shows that the estimate $\hat{\beta}$ converges to a normal distribution whose mean is the true effect $\beta$ and whose variance, denoted by $Var(\hat{\beta})$, is

$$Var(\hat{\beta}) = \frac{E[\varepsilon^2 \cdot (D - E[D])^2]}{n \cdot (Var(D))^2}. \tag{3.30}$$

By showing asymptotic normality, we also obtained an expression of the estimator's variance as a by-product, which will turn out to be useful for characterizing the uncertainty with which we estimate a causal effect, as discussed further next.

Equation (3.30) corresponds to the so-called heteroscedasticity-robust variance formula, going back to contributions by Eicker (1967), Huber (1967), and White (1980). Robustness to heteroscedasticity allows the variance of the error term $\varepsilon$ (i.e., the expected squared deviation of an outcome from its conditional mean given the treatment) to vary across treatment states. This appears to be a plausible scenario in many empirical contexts. Considering, for instance, the effect of training on wages, it may well be the case that the wages among the nontrained have a different variance than the wages among the trained, implying that the training changes the dispersion of wages (rather than their average alone).

We note that if we do not allow for heteroscedasticity (i.e., varying variances of $\varepsilon$ across $D$), then the variance of $\varepsilon$ is restricted to be the same under either treatment state, a scenario called *homoscedasticity*. In this case, equation (3.30) for the variance of $\hat{\beta}$ simplifies to

$$\frac{E[\varepsilon^2 \cdot (D - E[D])^2]}{n \cdot (Var(D))^2} = \frac{E[\varepsilon^2] \cdot \overbrace{E[(D - E[D])^2]}^{Var(D)}}{n \cdot (Var(D))^2} = \frac{E[\varepsilon^2]}{n \cdot Var(D)}, \tag{3.31}$$

because $E[\varepsilon^2]$, the variance of $\varepsilon$, does not depend on $D$. In general, there appears to be no good argument for assuming homoscedasticity in empirical applications, as it imposes the restriction that the treatment does not affect the dispersion of the outcome around its treatment-specific mean. Even though the homoscedasticity-based variance formula in equation (3.31) is frequently introduced first in statistics classes and might therefore be perceived as the default option, it is actually the heteroscedasticity-robust variance formula in equation (3.30) that is more universal because it does a priori not restrict the treatment effect to not concern the variance of the outcome. For this reason, relying on equation (3.30) rather than equation (3.31) when assessing the variance of ATE estimation appears more appropriate from a practical perspective.

To show the asymptotic normality of $\hat{\alpha}$, we could follow an analogous strategy as for $\hat{\beta}$. We will, however, consider a somewhat less tedious approach, which is based on the fact that $\alpha = E[Y|D = 0]$. To this end, we apply the insights of the central theorem in equation (3.25) to the subsample of nontreated observations only and define $W_i = Y_i - \alpha$. The latter satisfies $E[Y_i - \alpha|D_i = 0] = 0$ and thus has an expectation of zero under nontreatment. Furthermore, let us denote by $n_0$ the sample size of nontreated observations. Applying the central limit theorem gives

$$\frac{1}{\sqrt{n_0}} \sum_{i:D_i=0} (Y_i - \alpha) \to^d N(0, Var(Y|D=0)). \tag{3.32}$$

It is easy to see that $\frac{1}{\sqrt{n_0}} \sum_{i:D_i=0}(Y_i - \alpha) = \frac{\sqrt{n_0}}{n_0} \sum_{i:D_i=0}(Y_i - \alpha) = \sqrt{n_0}(\hat{\alpha} - \alpha)$ because $\hat{\alpha} = \frac{1}{n_0} \sum_{i:D_i=0} Y_i$ is the average outcome among the nontreated in the sample. By rearranging terms in an analogous way as for $\hat{\beta}$, we can therefore show that $\hat{\alpha}$ converges to a normal distribution with mean $\alpha$ and a variance of $\frac{Var(Y|D=0)}{n_0}$:

$$\hat{\alpha} \to^d N(\alpha, \frac{Var(Y|D=0)}{n_0}). \tag{3.33}$$

After showing specific statistical properties and deriving the variance of our estimates, we conclude this section by introducing a measure for the overall accuracy or error committed by a method, which depends on both its variance (i.e., the uncertainty of estimation across different samples) and the bias (i.e., the average error across samples). We consider the mean squared error (MSE)—that is, the average of the squared difference between the estimate $\hat{\beta}$ and the true effect $\beta$ when applying linear regression to infinitely many randomly drawn samples, which is formally defined as follows:

$$
\begin{aligned}
MSE(\hat{\beta}) &= E[(\hat{\beta} - \beta)^2] \\
&= E[(\hat{\beta} - E[\hat{\beta}] + E[\hat{\beta}] - \beta)^2] \\
&= E[(\hat{\beta} - E[\hat{\beta}])^2] + 2 \cdot \underbrace{(E[\hat{\beta}] - E[\hat{\beta}])}_{=0} \cdot (E[\hat{\beta}] - \beta) + E[(E[\hat{\beta}] - \beta)^2] \\
&= \underbrace{E[(\hat{\beta} - E[\hat{\beta}])^2]}_{\text{variance}} + \underbrace{(E[\hat{\beta}] - \beta)^2}_{\text{squared bias}},
\end{aligned}
\tag{3.34}
$$

where the second equality follows from subtracting and adding $E[\hat{\beta}]$.

Equation (3.34) demonstrates that the MSE can be decomposed into an estimate's variance, which corresponds to $\frac{E[\varepsilon^2 \cdot (D-E[D])^2]}{n \cdot (Var(D))^2}$ for the OLS estimate $\hat{\beta}$, and its squared bias, which in our context of social experiments equals zero, because $\hat{\beta}$ is an unbiased estimate, as formally shown in equation (3.20). However, there also are estimators for which unbiasedness does not hold, at least not in small samples, but only when the sample size becomes very (infinitely) large. For such cases, the MSE is a very useful concept for considering trade-offs between the influence of the variance and the bias on the expected overall error (characterized by the discrepancy between the estimate and the true effect). As we will discuss later in this chapter, some methods can be tweaked in a way that the bias goes up while the variance goes down or vice versa, and the question then is how this affects the overall MSE, which we would like to be as small as possible to minimize the estimation error.

## 3.4   Variance Estimation, Inference, and Goodness of Fit

In section 3.3 we got familiar with the normal distributions and variances of the coefficient estimates $\hat{\alpha}$ and $\hat{\beta}$ in large samples. Assessing the variance is important because even if the previously discussed properties of unbiasedness and consistency hold, the estimate of the ATE in our sample might differ from the true ATE in the population due to its variance across samples. As briefly mentioned before, knowing the variance and the distribution is useful for statistical inference that aims at quantifying the precision or (when phrasing it in a negative way) the uncertainty with which we can estimate the true ATE in our sample. This permits, for instance, answering the following two interesting questions: With which error probability can we rule out that the ATE is equal to zero (or some other value we are interested in) in the population, given the ATE estimate in our sample? What is the range or interval of values that likely includes the ATE in the population, given the findings in our sample?

Focusing on the $\hat{\beta}$ estimate in the subsequent discussion, we would ideally directly exploit the variance formula provided in equation (3.30) from section 3.3 to evaluate the uncertainty of the estimate based on its distribution provided in expression (3.29). Unfortunately, this is infeasible, as equation (3.30) contains parameters that refer to the population and are therefore not directly observed in our sample. This concerns, for instance, $E[D]$, the mean (or share) of treatment in the population, and $\varepsilon$, the true error term in the population. Even if we do not know such population parameters, however, we may estimate them in the sample (which is similar in spirit to estimating the true $\beta$ by $\hat{\beta}$ in the data) to ultimately obtain an estimate of the variance in equation (3.30).

To estimate the variance, we denote by $\hat{\varepsilon}_i$ the estimate of the true error term $\varepsilon_i$ for observation $i$, also known as residual. In analogy to the definition of $\varepsilon_i$ in equation (3.5) from section 3.2 based on linear regression in the population, $\hat{\varepsilon}_i$ is obtained from linear regression in the sample and corresponds to the difference between observation $i$'s outcome and the conditional sample average of the outcome given the treatment. The conditional sample average is also known as *prediction* and denoted by $\hat{E}[Y|D_i]$, as it is an estimate of the conditional mean of $E[Y|D=D_i]$ in the population. Formally, the residual thus corresponds to

$$\hat{\varepsilon}_i = Y_i - \underbrace{(\hat{\alpha} + \hat{\beta}D_i)}_{\hat{E}[Y_i|D_i]}. \tag{3.35}$$

In the next step, we may use $\hat{\varepsilon}_i$ as an estimate of the unknown $\varepsilon_i$ to obtain a consistent estimator of $Var(\hat{\beta})$, the true asymptotic variance of $\hat{\beta}$ given in equation (3.30).

Formally, the variance estimator, henceforth denoted by $\widehat{Var}(\hat{\beta})$, corresponds to

$$\widehat{Var}(\hat{\beta}) = \frac{\frac{1}{n}\sum_{i=1}^{n}\hat{\varepsilon}_i^2 \cdot (D_i - \bar{D})^2}{n \cdot (\widehat{Var}(D_i))^2}, \tag{3.36}$$

where the population mean $E[D]$ has been replaced by the sample mean $\bar{D}$.

We are now ready for diving into statistical inference. The first concept that we will consider is hypothesis testing, which aims at assessing whether the true ATE in the population is likely different, smaller, or larger than a specific value (e.g., zero), given the estimate and its variance obtained in the sample. To formalize the idea of hypothesis testing, let us reconsider the asymptotically normal distribution of $\hat{\beta}$ in expression (3.29) of the previous section: $\hat{\beta} \to^d N(\beta, Var(\hat{\beta}))$, with $Var(\hat{\beta}) = \frac{E[\varepsilon^2 \cdot (D-E[D])^2]}{n \cdot (Var(D))^2}$. We normalize this distribution such that it turns into a standard normal distribution that has a zero mean and a variance of 1. To this end, we subtract $\beta$ from $N(\beta, Var(\hat{\beta}))$ and also divide by $Var(\hat{\beta})$, which yields $N(0, 1)$. Importantly, and similar to equations containing equals signs, the subtraction and division also needs to be conducted on any expression to the left of the convergence in distribution sign $\to^d$.

Furthermore, we need to pay attention to the fact that division by any number outside a variance formula correspond to divisions by the square of that number within a variance formula. Put differently, dividing by a positive number in a variance formula corresponds to dividing by its square root outside the variance. For this reason, any expression to the left of $\to^d$ is to be divided by $\sqrt{Var(\hat{\beta})}$ rather than $Var(\hat{\beta})$ when normalizing the distribution to have a variance of 1. Finally, we note that the square root of a variance is called a *standard deviation*, which we henceforth denote by $sd(\hat{\beta}) = \sqrt{Var(\hat{\beta})}$. Putting all these arguments together, we obtain the following result:

$$\hat{\beta} \to^d N(\beta, Var(\hat{\beta})),$$

$$\Leftrightarrow \frac{\hat{\beta} - \beta}{sd(\hat{\beta})} \to^d N(0, 1), \tag{3.37}$$

with

$$sd(\hat{\beta}) = \sqrt{\frac{E[\varepsilon^2 \cdot (D - E[D])^2]}{n \cdot (Var(D))^2}}. \tag{3.38}$$

This implies that in large enough samples, the z-statistic provided in the second line of expression (3.37), which consists of the true ATE $\beta$ and the standard deviation $sd(\hat{\beta})$, closely follows a standard normal distribution. (As a side remark, if the error term $\varepsilon$ is normally distributed, which is a strong assumption, this even holds in small samples.) Of course, we do not know what the true ATE is (otherwise, we would not bother to estimate it), but this result can be used for hypothesis testing—that is, checking the plausibility of hypothesized (i.e., hypothetically assumed) values of $\beta$. For instance, our default or null hypothesis (denoted by $H_0$) might be that on average, the treatment has no effect at all, which implies that our alternative hypothesis (denoted by $H_1$) states the opposite—namely, that the ATE is different to zero. Formally,

$$H_0 : \beta = 0, \quad H_1 : \beta \neq 0. \tag{3.39}$$

If the null hypothesis $H_0$ stating that $\beta = 0$ were true, then the z-statistic $\frac{\hat{\beta} - \beta}{sd(\hat{\beta})}$ would follow a standard normal distribution (across many different, randomly drawn samples), and in addition simplify to $\frac{\hat{\beta}}{sd(\hat{\beta})}$ because $\beta = 0$. This permits assessing how likely it is that the ATE in the population is different from zero, given the value of $\frac{\hat{\beta}}{sd(\hat{\beta})}$ in the sample, which measures the size of the estimated ATE $\hat{\beta}$ in terms of (or normalized by) the standard deviation $sd(\hat{\beta})$ as unit of estimation uncertainty.

To highlight this with a numerical example, let us assume that the ATE is truly zero in the population, such that the null hypothesis $H_0 : \beta = 0$ is satisfied. Under a standard normal distribution, the probability of obtaining a value of $\frac{\hat{\beta}}{sd(\hat{\beta})}$ that is larger than 1.96 in a sample is just 2.5 percent. Likewise, the probability of obtaining a value smaller than $-1.96$ is also just 2.5 percent. Put differently, if the true ATE in the population is zero, then the probability of observing a value of $\frac{\hat{\beta}}{sd(\hat{\beta})}$ in a sample that is rather extreme (i.e., more positive than 1.96 or more negative than $-1.96$), is just 5 percent, as graphically displayed in figure 3.2. That is, if we could draw many samples, we would observe such extreme estimates only in 5 percent of the samples given the satisfaction of $H_0$.

Put in yet another way, the error probability of claiming that a (rather extreme) sample statistic $\left| \frac{\hat{\beta}}{sd(\hat{\beta})} \right| > 1.96$ (where $| \, |$ stands for the absolute value) points to a nonzero effect is smaller than 5 percent. We thus reject $H_0$ and accept $H_1$ with an error probability below 5 percent. This is the case simply because the probability of such an extreme statistic is less than 5 percent given the satisfaction of $H_0$. We can therefore decide to maintain or reject $H_0$ based on this error probability of incorrectly rejecting $H_0 : \beta = 0$ or equivalently, incorrectly accepting $H_1 : \beta \neq 0$, the type I error. The lower this error probability is, the more confident we are in rejecting $H_0$. One conventionally even predefines a maximum admissible error probability for rejecting $H_0$, such as 5 percent.

However, one issue we face with this approach to hypothesis testing is that the asymptotic standard deviation $sd(\hat{\beta})$ provided in equation (3.38), which enters the

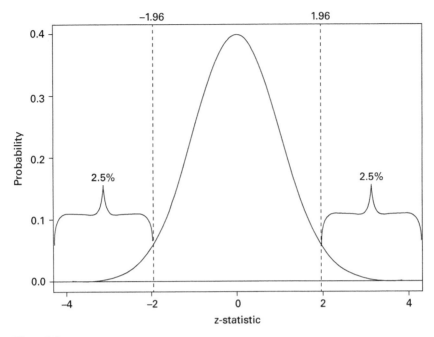

**Figure 3.2**
Standard normal distribution.

denominator in expression (3.37), is typically unknown, as it, just like the asymptotic variance in equation (3.30), relies on population parameters, such as $E[D]$ and $\varepsilon$. For this reason, we replace $sd(\hat{\beta})$ with an estimate obtained in the sample—namely, the square root of the estimated variance provided in equation (3.36). This estimate of the standard deviation is commonly referred to as *standard error*, henceforth denoted as *se*, and formally defined as

$$se(\hat{\beta}) = \sqrt{\frac{\frac{1}{n}\sum_{i=1}^{n}\hat{\varepsilon}_i^2 \cdot (D_i - \bar{D})^2}{n \cdot (\widehat{Var}(D_i))^2}}. \tag{3.40}$$

Replacing $sd(\hat{\beta})$ in expression (3.37) by $se(\hat{\beta})$ yields the t-statistic, $\frac{\hat{\beta}-\beta}{se(\hat{\beta})}$. It is worth mentioning that the latter does not follow a standard normal distribution in smaller samples, but rather a t-distribution. However, a t-distribution converges to a normal distribution as the sample size increases. For a sample of roughly 120 observations, a t-distribution is practically indistinguishable from a normal one, at least if only a few parameters are to be estimated as our two, $\hat{\alpha}$ and $\hat{\beta}$, which imply in statistical parlance a large number of degrees of freedom. Furthermore, we also note that the standard error converges to the true standard deviation as the sample size increases.

All of this implies that our t-statistic also converges to a standard normal distribution as the sample size increases:

$$\frac{\hat{\beta} - \beta}{se(\hat{\beta})} \rightarrow^d N(0, 1), \tag{3.41}$$

similar to the z-statistic in expression (3.37) with a known standard deviation. In large enough samples (with, say, 120 observations or more), we can thus compute the type I error probability associated with the ATE estimate $\hat{\beta}$ in our sample as the probability of values that are as large or even more extreme (i.e., further away from zero) than $\frac{\hat{\beta} - \beta}{se(\hat{\beta})}$, which simplifies to $\frac{\hat{\beta}}{se(\hat{\beta})}$ under the null hypothesis that $\beta = 0$. This error probability of incorrectly rejecting the null hypothesis based on the estimate in our sample is known as the *p-value* and formally defined as follows:

$$\text{p-value} = \Pr\left(|A| \geq \left|\frac{\hat{\beta} - \beta}{se(\hat{\beta})}\right|\right), \tag{3.42}$$

where $A$ denotes a random variable following a t-distribution (which, as we know by now, converges to a standard normal distribution).

Our considerations about hypothesis testing suggest the following procedure for maintaining or rejecting the null hypothesis in two-sided hypothesis tests, which aim to see whether $\beta$ likely differs from some specific value (e.g., zero) given the estimate $\hat{\beta}$ in the sample:

1. Define the null and alternative hypotheses $H_0$ and $H_1$. In causal analysis, we are typically interested in testing the presence versus absence of an ATE, implying that hypotheses are defined as provided in expression (3.39). But we could also test whether the ATE is likely different from some other value, such as 1, such that $H_0 : \beta = 1$, $H_1 : \beta \neq 1$, and the corresponding t-statistic is $\frac{\hat{\beta}-1}{sd(\hat{\beta})}$.
2. Define the significance level denoted by $\alpha$—that is, the maximally accepted type I error probability of incorrectly rejecting $H_0$ and accepting $H_1$. $\alpha = 0.05$ implies that the error probability must not exceed 5 percent, but other conventional levels of significance are 0.01 (1 percent) or 0.1 (10 percent).
3. Compute the critical value, denoted as $c$, which is the value or quantile in the standard normal or t-distribution that corresponds to $\alpha$. If, for instance, $\alpha = 0.05$, then $c = 1.96$ because the probabilities of values larger than 1.96 and smaller than $-1.96$ add up to 5 percent in the standard normal distribution—that is, under the satisfaction of $H_0$.

4. Verify if $\left|\frac{\hat{\beta}-\beta}{se(\hat{\beta})}\right| \geq c$, or equivalently, if the p-value is less than or equal to $\alpha$. If the absolute value of the t-statistic is greater than or equal to the critical value, then one rejects $H_0$ and accepts $H_1$ because the error probability of rejection is less than or exactly $\alpha$. In this case, $\hat{\beta}$ is said to be statistically significantly different from the $\beta$ hypothesized under $H_0$ at the $\alpha$ level of significance. If the absolute value of the t-statistic is less than the critical value, one keeps $H_0$ and does not accept $H_1$ because the error probability of rejection is greater than $\alpha$. In this case, $\hat{\beta}$ is said to be statistically insignificantly different from the $\beta$ hypothesized under $H_0$. Alternatively, the p-value directly gives the type I error probability related to the effect estimate in the sample at hand, and one rejects $H_0$ if the p-value is less than or equal to the maximally admitted type I error probability $\alpha$, but keeps $H_0$ if the p-value exceeds $\alpha$.

It is important to bear in mind that such hypothesis testing can only reject the validity of a null hypothesis, but never confirm it. This implies that the nonrejection of a null hypothesis does not automatically imply that it is correct, but simply that we cannot rule out that it is correct given our data.

Besides two-sided hypothesis tests, we may also be interested in one-sided hypothesis tests, such as whether the ATE found in the sample is statistically significantly larger (rather than just different) than zero or any other value of interest. In this case, the null hypothesis to be tested is that the effect is either zero or less than zero, while the alternative hypothesis is that the effect is greater than zero:

$$H_0 : \beta \leq 0, \quad H_1 : \beta > 0. \tag{3.43}$$

Accordingly, we modify the p-value to suit one-sided hypothesis testing. It now corresponds to the probability that a random variable $A$ following a t-distribution has a value that is equal to or larger than the t-statistic under the satisfaction of $H_0$:

$$\text{p-value} = \Pr\left(A \geq \frac{\hat{\beta}-\beta}{se(\hat{\beta})}\right). \tag{3.44}$$

The condition for a rejection of the null is $\frac{\hat{\beta}-\beta}{se(\hat{\beta})} \geq c$—that is, that the t-statistic is greater than or equal to a specific critical value $c$ suitable for one-sided tests, such as $c = 1.64$, for a significance level of $\alpha = 0.05$. An equivalent condition is that the p-value in equation (3.44) is at most as large as $\alpha$.

Likewise, one could be interested in whether the effect estimated in the sample is statistically significantly smaller than zero or any other value of interest. In this case, the null hypothesis to be tested is that the effect is either zero or greater than zero, while the alternative hypothesis is that the effect is less than zero:

$$H_0 : \beta \geq 0, \quad H_1 : \beta < 0. \tag{3.45}$$

The p-value then corresponds to the probability that the random variable $A$ has a value that is equal to or smaller than the t-statistic under the satisfaction of $H_0$:

$$\text{p-value} = \Pr\left(A \le \frac{\hat{\beta} - \beta}{se(\hat{\beta})}\right). \tag{3.46}$$

The condition for a rejection of the null is now $\frac{\hat{\beta} - \beta}{se(\hat{\beta})} \le c$—for instance, $c = -1.64$ for $\alpha = 0.05$—or equivalently, that the p-value in equation (3.46) is at most as large as $\alpha$. In the remainder of this book, however, we will stick to two-sided tests of zero versus nonzero causal effects.

A further concept of inference that is related to hypothesis testing and the p-value is the *confidence interval*. It provides a range or interval of ATE values such that the true ATE $\beta$ is included with probability $1 - \alpha$ based on the estimated ATE $\hat{\beta}$ and the standard error $se(\hat{\beta})$ obtained in the sample. To be concise, the confidence interval is constructed in such a way that in the hypothetical case that we could draw many samples and construct confidence intervals in all those samples, a share of $1 - \alpha$ confidence intervals would include the true $\beta$. For $\alpha = 0.05$, for instance, this would be the case in 95 percent of samples, such as $1 - 0.05 = 0.95$. To formally discuss the construction of a confidence interval, let us denote by $\underline{\beta}$ and $\bar{\beta}$ its lower and upper bound, respectively—that is, the minimum and maximum values in the interval (e.g., 10 and 20). We can compute these lower and upper bounds by either subtracting from or adding to the estimated ATE $\hat{\beta}$ the product of the standard error $se(\hat{\beta})$ and the critical value $c$ of a two-sided hypothesis test. More formally, the confidence interval, which we denote by $CI$, corresponds to

$$CI = [\underline{\beta}, \bar{\beta}], \text{ with}$$

$$\underline{\beta} = \hat{\beta} - c \cdot se(\hat{\beta}), \quad \bar{\beta} = \hat{\beta} + c \cdot se(\hat{\beta}). \tag{3.47}$$

For instance, when setting $\alpha = 0.05$, it follows that $c = 1.96$, as this choice results in a confidence interval that has a coverage rate of 95 percent—that is, it includes the true $\beta$ with 95 percent probability across many samples. Confidence intervals therefore indeed provide us with some confidence about the range of values that the true effect in the population could likely take. It probably does not seem too surprising that whenever an estimate $\hat{\beta}$ is judged to be not statistically significantly different from zero by a two-sided hypothesis test, then the corresponding confidence interval (based on the same $\alpha$ as the test) includes the zero. And analogously, whenever the estimate is statistically significantly different from zero, the corresponding confidence interval does not include the zero, such that the upper and lower bounds are either both positive or both negative.

Before concluding our discussion on statistical inference, we will look at an alternative and increasingly popular method in statistics for computing confidence

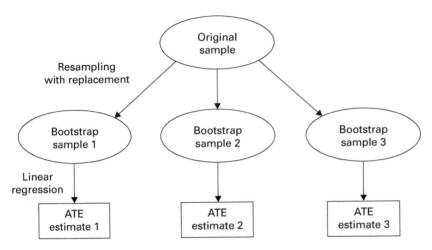

**Figure 3.3**
Bootstrapping.

intervals and p-values, which does not rely on the formula for computing the standard error provided in equation (3.40): bootstrapping, as suggested by Efron (1979). It is based on repeatedly generating so-called bootstrap samples of the original sample size $n$, where each sample is generated by randomly drawing $n$ observations from the original data with replacement. The latter implies that the same subject, in principle, can be drawn several times.

On average, the bootstrap samples match the original data, but they differ from the data and one another, because in any bootstrap sample, some subjects may randomly appear several times or not at all. This mimics the situation that the original data are a random sample from the population and a newly drawn sample might differ from the first one. Therefore, the idea of bootstrapping is to reestimate the ATE in each bootstrap sample, as illustrated in figure 3.3 (for just three bootstrap samples), to obtain the standard error as the standard deviation of the ATE estimates across all bootstrap samples. This quite cleverly approximates the approach of randomly drawing many samples from the population and considering the standard deviation of the ATE estimates across all samples, which is practically infeasible because we have usually only one sample at hand.

To discuss the bootstrap more formally, let us denote by $B$ the number of bootstrap samples that we randomly draw from the original data, which should ideally be large—say, not smaller than 999. Furthermore, $b$ is the index of a specific bootstrap sample such that $b \in \{1, 2, \ldots, B\}$, while $\hat{\beta}^b$ denotes the ATE estimate in the bootstrap sample $b$. Then, the standard error can be computed by

$$se(\hat{\beta}) = \sqrt{\frac{1}{B-1} \sum_{b=1}^{B} \left( \hat{\beta}^b - \frac{1}{B} \sum_{b=1}^{B} \hat{\beta}^b \right)^2}, \tag{3.48}$$

which is simply the standard deviation of all ATE estimates obtained in the various bootstrap samples. We may now use this bootstrap-based standard error for computing the t-statistic in expression (3.41) and proceeding with statistical inference as previously discussed.

However, there is a further bootstrap approach for directly computing the p-value without even using the t-statistic. Taking the two-sided hypothesis test as an example, it consists of verifying how extreme the ATE estimate $\hat{\beta}$ in the original data is relative to the distribution of the bootstrap-based ATEs $\hat{\beta}^b$ given the satisfaction of the null hypothesis. This can be assessed by counting how often the absolute value of the difference $\hat{\beta}^b - \hat{\beta}$ is larger than the absolute value of $\hat{\beta}$:

$$\text{p-value} = \frac{1}{B} \sum_{b=1}^{B} I\{|\hat{\beta}^b - \hat{\beta}| > |\hat{\beta}|\}, \tag{3.49}$$

where $I\{|\hat{\beta}^b - \hat{\beta}| > |\hat{\beta}|\}$ denotes an indicator function that is equal to 1 whenever $|\hat{\beta}^b - \hat{\beta}| > |\hat{\beta}|$ holds and zero whenever $|\hat{\beta}^b - \hat{\beta}|$ is not larger than $|\hat{\beta}|$.

To understand the intuition of this approach, it is important to see that the so-called recentered bootstrap-based ATE $\hat{\beta}^b - \hat{\beta}$ has a mean of zero due to the subtraction of $\hat{\beta}$, as the bootstrap ATEs must on average correspond to the ATE in the original data from which the bootstrap samples were drawn. For this reason, the distribution of $\hat{\beta}^b - \hat{\beta}$ mimics the distribution of the true ATE under the null hypothesis when there is no effect. Therefore, we can compute the p-value by verifying how often the absolute value of $\hat{\beta}^b - \hat{\beta}$ exceeds the absolute value of $\hat{\beta}$. If this is rarely the case, such that $\hat{\beta}$ appears rather extreme when comparing it with the distribution of the recentered bootstrap-based effects, which mimics the distribution under the null hypothesis, then the latter is rejected. As a final remark on the bootstrap, using the absolute value for computing p-values for two-sided tests is only appropriate for parameters with symmetric distributions (such as the normal or t-distribution) like $\hat{\beta}$, but the method can be easily adjusted to also yield the p-value for parameters following a nonsymmetric distribution like an F-statistic.

It is worth pointing out that there are further variants of bootstrapping than the one considered here, such as the wild bootstrap suggested by Wu (1986). Rather than repeatedly drawing observations in terms of their outcome and treatment values, the wild bootstrap is based on first estimating the treatment effect in the total sample and then repeatedly drawing functions of the residuals $\hat{\varepsilon}_i$ to generate the bootstrap samples. More thorough discussions of alternative bootstrap methods are provided in MacKinnon (2006) and Horowitz (2019).

In this section and the previous section, we have discussed the estimation of the magnitude and variance of the ATE and how to use these statistics for statistical inference like the construction of hypothesis tests and confidence intervals. However, a further interesting parameter, on top of the effect size and its statistical significance, is the relative importance of the treatment in explaining the outcome in the sample when compared to the residuals, which reflect the effects of any other characteristics on the outcome. This so-called goodness of fit can be judged by the proportion of the sample variation in the outcome $Y$ (e.g., the variation in wages), which is explained by the sample variation in $D$, a statistic known as R squared ($R^2$). To obtain $R^2$, let us rearrange equation (3.35) by solving it for the outcome to obtain

$$Y_i = \hat{E}[Y_i|D_i] + \hat{\varepsilon}_i, \tag{3.50}$$

which demonstrates that outcome $Y_i$ for some observation $i$ in the sample can be decomposed into two components. The first component $\hat{E}[Y_i|D_i]$ is the conditional sample average of the outcome, given the treatment of observation $i$—that is, the estimated part of the outcome that is explained by the treatment, which is also known as *prediction*. The second component $\hat{\varepsilon}_i$ corresponds to the residual—that is, the estimated part of the outcome explained by other, possibly unobserved characteristics.

It can be shown that the variance of $Y_i$ simply corresponds to the sum of the variances of these two components (because the covariance between $\varepsilon_i$ and $D_i$—and thus, $\hat{E}[Y_i|D_i]$—is zero). Formally,

$$Var(Y_i) = Var(\hat{E}[Y_i|D_i]) + Var(\hat{\varepsilon}_i). \tag{3.51}$$

Dividing by $Var(Y_i)$ in equation (3.51) yields

$$1 = \underbrace{\frac{Var(\hat{E}[Y_i|D_i])}{Var(Y_i)}}_{R^2} + \frac{Var(\hat{\varepsilon}_i)}{Var(Y_i)}. \tag{3.52}$$

That is, the variances of the parts of the outcome that are explained by the treatment and the residuals, respectively, sum to 1 (i.e., 100 percent of the variance of $Y_i$). Therefore, $\frac{Var(\hat{E}[Y_i|D_i])}{Var(Y_i)}$ corresponds to the share in the variation of $Y_i$ that is caused by the treatment (i.e., $R^2$). $R^2$ close to 1 means that almost 100 percent of the outcome variation is caused by the treatment, such that other characteristics play a minor role. $R^2$ close to zero (or 0 percent) implies that the treatment is responsible for little variation in the outcome relative to other characteristics captured by the residuals. It needs to be emphasized that $R^2$ is conceptually different from the magnitude of the ATE. For instance, a treatment like training might have a sizeable ATE on an outcome like wages, but still explain only little of the variation in wages relative to other characteristics like education, labor market experience, and personality traits.

After discussing effect and variance estimation, statistical inference, and $R^2$ in the context of linear regression, let us now consider an application in R. To this end, we reconsider our empirical example based on the Job Corps experimental study introduced at the end of section 3.1 and assess the ATE of random assignment on weekly earnings in the fourth year based on linear regression. Assuming that all packages have been previously installed, we load the *causalweight*, *lmtest* (provided by Zeileis and Hothorn (2002)), and *sandwich* (provided by Zeileis, Köll, and Graham (2020)) packages using the *library* command. The latter two packages permit computing heteroscedasticity-robust standard errors in regressions based on the variance formula in equation (3.30).

Next, we load the Job Corps data using *data(JC)* into the R workspace and define the treatment $D$ and outcome $Y$ in the same way as in section 3.1: *D=JC$assignment* and *Y=JC$earny4*. Then, we run a regression of $Y$ on $D$ using the *lm* command (where *lm* stands for linear model), which has a fairly simple syntax that only requires typing in the outcome and treatment variables, separated by $\sim$ (tilde): *ols=lm(Y$\sim$D)*. We store the output of the regression in a newly created R object named *ols*. Finally, we wrap the latter with the *coeftest* command to display the regression output and also include *vcov=vcovHC* as the second argument in the command to obtain heteroscedasticity-robust standard errors, t-statistics, and p-values. The box here provides the R code for each of the steps.

```
library(causalweight)        # load causalweight package
library(lmtest)              # load lmtest package
library(sandwich)            # load sandwich package
data(JC)                     # load JC data
D=JC$assignment              # define treatment (assignment to JC)
Y=JC$earny4                  # define outcome (earnings in fourth year)
ols=lm(Y~D)                  # run OLS regression
coeftest(ols, vcov=vcovHC)   # output with heteroscedasticity-robust se
```

This yields the following output:

```
              Estimate Std. Error t value  Pr(>|t|)
(Intercept) 197.9258     3.0726  64.4164  < 2.2e-16 ***
D            16.0551     4.0740   3.9408  8.18e-05 ***
---
Signif. codes:  0 '***' 0.001 '**' 0.01 '*' 0.05 '.' 0.1 ' ' 1
```

The first column (*Estimate*) yields the coefficient estimates. The first row *(Intercept)*, which is an alternative denomination for the constant, corresponds to $\hat{\alpha}$, the mean outcome among the nontreated, which amounts to almost 198 USD per week in the fourth year. The second row (*D*) provides the coefficient on the treatment, $\hat{\beta}$, which corresponds to the estimated ATE of the program assignment. It amounts to roughly 16 USD and is numerically equivalent to the mean difference previously computed at the end of section 3.1. However, the regression output provides more information than the mean differences. The second column (*Std. Error*) yields the heteroscedasticity-robust standard errors of the coefficient estimates—see equation (3.40) for that of $\hat{\beta}$.

The third column (*t value*) provides the t-statistics (see expression (3.41)), under the null hypothesis that the coefficients equal zero. Therefore, the respective t-statistic corresponds to the ratio of the respective coefficient to its standard error. The fourth column (*Pr(> |t|)*) contains the corresponding p-values for two-sided hypothesis tests, again under the null hypothesis that the coefficients equal zero. The p-value of $\hat{\beta}$ is $8.18e - 05 = 0.0000818$, and thus very close to zero. We can safely reject the null hypothesis that Job Corps assignment has an average effect of zero at any conventional level of statistical significance. This is also indicated by the three stars \*\*\*, implying statistical significance at the 0.001 (or 0.1 percent) level according to the significance codes (*Signif. codes*). The estimate of $\hat{\alpha}$ is highly statistically significantly different from zero, too, with its p-value being very close to zero.

The standard errors, t-statistics, and p-values in the previous R exercise are based on the asymptotic variance formula provided in equation (3.30). Let us now alternatively estimate the standard errors by bootstrapping, as considered in equation (3.48), even though this is admittedly somewhat more involved in terms of coding. To this end, we use the *function* command to specify a bootstrap function named *bs* for drawing bootstrap samples and estimating the coefficients in those samples. The *bs* function contains two input arguments provided in round brackets after the *function* command—namely, the *data* used in the procedure and the *indices* of the observations randomly sampled (with replacement) from the *data* to be part of a specific bootstrap sample. In curly brackets follow the commands to be executed by the *bs* command. First, we define the data set to be considered as observations (or rows) that have been randomly drawn from the data with replacement according to the R object *indices*, by using the square brackets: *dat=data[indices, ]*. Setting the first argument in the brackets to *indices* only selects the rows provided in *indices*, which contains randomly chosen numbers between 1 and the number of observations, such that observations from the original *data* are picked at random. We note that this random number selection is conducted by the *boot* command described next, which therefore determines the observations in the object *indices*. The second argument in *data[indices, ]*, after the comma refers to the columns of the data and is left blank, implying that all columns (or variables) should be sampled for the chosen observations.

In the next step, we apply *lm(dat)* to run a linear regression based on the bootstrap data *dat*, where the first column in the data is automatically considered as the outcome variable and the remaining ones as regressors. Furthermore, appending *$coef* to *lm(dat)* allows us to exclusively select the coefficients of that regression, which are the parameters we would like to collect from each bootstrap sample. (More generally, the *$* operator permits retrieving subobjects in R objects.) Finally, we store the coefficient estimates in an R object named *coefficients* and use the *return* command to provide this object as the output of our *bs* function. The box here provides the R code for each step in the *bs* function.

```
bs=function(data, indices) {       # defines function bs for bootstrapping
  dat=data[indices, ]              # creates bootstrap sample according to indices
  coefficients=lm(dat)$coef        # estimates coefficients in bootstrap sample
  return(coefficients)             # returns coefficients
}                                  # closes the function bs
```

We now apply our bootstrap function for linear regression to the Job Corps sample, and to this end also load the *boot* package developed by Canty and Ripley (2021) using the *library* command. To prepare the data, we use the *data.frame* command to append *Y* and *D* columnwise to form a data matrix, which we name *bootdata*. For running the bootstrap estimations, we apply the *boot* command, which consists of three arguments: the *data* to be analyzed (in our case *bootdata*), the *statistic* to be computed in each bootstrap sample as defined in our *bs* function, and *R* for the number of bootstrap replications, which we set to 1999. We store the output in an R object named *results* and call the latter to investigate the estimates; see the box here for the R code of the various steps.

```
library(boot)                    # load boot package
bootdata=data.frame(Y,D)         # data frame with Y,D for bootstrap procedure
results = boot(data=bootdata, statistic=bs, R=1999) # 1999 bootstrap estimations
results                          # displays the results
```

This yields the following output:

```
Bootstrap Statistics :
        original        bias      std. error
t1*  197.92584     0.02480312      3.013465
t2*   16.05513    -0.02075945      3.954810
```

While the first column of the output (*original*) contains the coefficient estimates $\hat{\alpha}$ and $\hat{\beta}$ in the original data, the third column (*std. error*) contains the respective standard errors based on the coefficients' distribution across the 1999 bootstrap samples. They are in fact very similar to the standard errors obtained from the asymptotic variance formula in the previous R example. We also note that the standard errors can differ somewhat each time we run the *boot* command, as bootstrap samples randomly differ in their included observations. Setting a so-called seed prior to bootstrapping permits replicating the definition of the bootstrap samples, and thus of the results. For instance, running the command *set.seed(1)* prior to *boot(data=bootdata, statistic=bs, R=1999)* always results in standard errors of 3.013 and 3.955.

To compute the p-value for the ATE estimate, we first retrieve the second coefficient from the bootstrap output, which corresponds to $\hat{\beta}$, by using *results$t0[2]*, as well as the distribution of the coefficient across all bootstrap samples by using *results$t[,2]*. Wrapping the latter by the *sd* function for computing standard deviations yields the bootstrap standard error $se(\hat{\beta})$, as provided in equation (3.48). We then compute the t-statistic based on expression (3.41) and store it in an R object called *tstat*. Finally, we use the *abs* command for computing the absolute value of the t-statistic, and the *pnorm* command to compute the p-value based on equation (3.42) when assuming a standard normal distribution of the t-statistic, which is justified by our sufficiently large sample and the result in expression (3.41). The R code for these steps is provided in the box shown here.

```
tstat=results$t0[2]/sd(results$t[,2])    # compute t-statistic
2*pnorm(-abs(tstat))                      # compute the p-value
```

This yields the following output:

```
           D
4.914718e-05
```

The bootstrap approach replicates the finding of the previous R example—namely, that the ATE is highly statistically significant, as the bootstrap-based p-value is very close to zero, amounting to only $4.914718e - 05 = 0.00004914718$. It is thus considerably lower than any conventional level of statistical significance for the maximum error probability when rejecting the null hypothesis of a zero ATE, like 0.1, 0.05, or 0.01.

## 3.5   Extensions to Multiple or Continuous Treatments

The discussion in the previous sections has focused on a binary treatment that only takes the value 1 or 0. However, there are many empirical questions where the interest lies in the effects of several, potentially competing treatments. For this reason, we will subsequently adapt our causal analysis to such frameworks with multivalued treatments that are discrete in the sense that they can take only a limited number of different values. More formally, we consider a treatment that can take values $D \in \{0, 1, 2, \ldots, J\}$, where $J$ denotes the number of treatments, in addition to no treatment, which is coded as zero ($D = 0$). This may either cover the case of an ordered amount of a specific treatment, such as $1 = 1$ week of training and $2 = 2$ weeks of training, or of multiple unordered treatments, where a higher treatment value is not necessarily more treatments but just a different treatment, such as $1 = $IT course and $2 = $sales training. If nontreatment and all the various treatments $1, \ldots, J$ are successfully randomized in a social experiment, the independence assumption introduced in expression (3.1) from section 3.1 can be adapted to hold for any treatment value:

$$\{Y(0), Y(1), Y(2), \ldots, Y(J)\} \perp D. \tag{3.53}$$

We may analyze the ATEs of each nonzero treatment based on linear regression by including binary variables, also called "dummies" or dummy variables, for all nonzero treatment values. Formally, we denote by $D_1 = I\{D = 1\}, D_2 = I\{D = 2\}, \ldots, D_J = I\{D = J\}$ the binary variables, where $I\{A\}$ is the indicator function, which is equal to 1 if event $A$ (in our case a particular treatment value) occurs and 0 otherwise. This implies the following regression model in the population, where the coefficients $\beta_1, \beta_2, \ldots, \beta_J$ correspond to the ATEs of the various treatments when compared to no treatment—that is, $E[Y(1) - Y(0)], E[Y(2) - Y(0)], \ldots, E[Y(J) - Y(0)]$:

$$E[Y|D] = \underbrace{\alpha}_{E[Y|D=0]} + \underbrace{\beta_1}_{E[Y|D=1]-E[Y|D=0]} D_1 + \underbrace{\beta_2}_{E[Y|D=2]-E[Y|D=0]} D_2$$

$$+ \cdots + \underbrace{\beta_J}_{E[Y|D=J]-E[Y|D=0]} D_J. \tag{3.54}$$

Similar to the discussion in section 3.2, it is important to note that the linear regression model in equation (3.54), which we may estimate in the data to obtain ATE estimates and conduct statistical inference, does not impose any linear relationship between $Y$ and $D$. The reason is that the treatment values are flexibly coded by means of multiple binary variables, which entails pairwise comparisons of the average outcomes between any group with a nonzero treatment and the nontreated control group.

Next, we consider a continuously (rather than discretely) distributed treatment $D$, which may take even infinitely many values that respect cardinality. The latter implies that a higher treatment value actually means more, like expenditures on employee training measured in a currency like USD or CHF. We therefore adapt the independence assumption in equation (3.53) to hold for any values the continuous treatment $D$ can possibly take:

$$Y(d) \perp D \text{ for any value } d \text{ that treatment } D \text{ might take.} \tag{3.55}$$

One way to analyze a continuous treatment is to discretize it by generating binary indicators for specific brackets of values, which then entails the same regression model as in equation (3.54). For instance, defining $D_1 = I\{D \leq 1000\}$, $D_2 = I\{1000 < D \leq 2000\}$,... permits analyzing the ATEs of various expenditure brackets in steps of 1,000 CHF, thus providing insights into the average effect of any expenditure bracket. However, this approach cannot capture the average effect of a marginal increase in the continuous treatment, such as by one single CHF, which may also be of interest to the analyst or researcher.

One potential approach that comes closer to the assessment of marginal effects is to decrease the range of values of each bracket and, in the extreme case, creating dummies for every possible value of the treatment. Depending on the empirical problem, however, this might be practically infeasible if the sample size is limited and the treatment can take (infinitely) many different values. An alternative approach consist of directly including $D$ rather than any discretized version thereof in the linear regression, which models $Y$ as a linear function of $D$ and an error term $\varepsilon$, in analogy to equation (3.5):

$$Y = \alpha + \beta D + \varepsilon. \tag{3.56}$$

Under the independence assumption in expression (3.55), a linear regression based on equation (3.56) permits evaluating the average effect of a marginal increase in $D$ on $Y$, which conveniently corresponds to coefficient $\beta$. This can be roughly interpreted as the average increase in the outcome (measured in units of $Y$) due to an increase in $D$ by 1 unit, such as 1 CHF.

To better see this result, let us denote by $\mu_d = E[Y|D = d]$ the conditional mean of $Y$, given a specific value $d$ of treatment $D$. Furthermore, we note that $\mu_d = E[Y(d)]$, which follows from expression (3.55), ruling out treatment selection bias across values of $d$. Therefore, the first derivative of $\mu_d$ with regard to the treatment, denoted by $\nabla \mu_d = \frac{\partial \mu_d}{\partial d}$, tells us how much the mean potential outcome $E[Y(d)]$ changes in reaction to a marginal change in the continuous treatment $D$ at treatment value $d$. To obtain the average of such marginal effects in the population, we take the average derivative $E[\nabla \mu_D]$ in equation (3.56), which yields

$$E[\nabla \mu_D] = \beta. \tag{3.57}$$

Interestingly, the result in equation (3.57) holds even if the outcome model postulated in equation (3.56) is incorrect, in the sense that $Y$ is not a linear function of $D$, meaning that the marginal effect of the treatment may not be constant across different treatment doses. The latter implies that $\nabla \mu_d$ generally differs across treatment values of $d$. Even in this case, $\beta$ nevertheless yields the average marginal effect.

It is important to see that this average effect is based on an increase in $D$ for everyone in the population, and thus it averages the effects over subjects with distinct treatment values as they occur in the population. This approach yields the average effect of increasing spending on employee training by, say, 1 CHF in the total population of employees, without distinguishing the effect by how much has previously been spent on employee training. However, this baseline level (or point of departure) of $D$ could play a role in the size of the marginal effect. For instance, marginally increasing spending might on average be particularly beneficial for employees who received particularly little or particularly much training before (such that their $D$ is comparably large or small). This suggests that not only does the average marginal effect across all values of $D$ in the population appear interesting, but the marginal treatment effect at a particular value of $D = d$ does as well. For instance, we might want to evaluate the average effect of one additional CHF conditional on $d = 1000$ or $d = 2000$—that is, if employees previously participated in training activities worth either 1,000 or 2,000 CHF.

Such a marginal effect corresponds to the derivative $\nabla \mu_d$ at a specific baseline value of interest, $d$. In general, the marginal effect $\nabla \mu_d$ differs from the average marginal effect $E[\nabla \mu_D]$, with the exception of the special case that the marginal effect is the same for everyone across all baseline values $d$—that is, independent of the previously received treatment. The marginal effect is then said to be constant or homogeneous, which can be formally stated as follows:

$$\frac{\partial E[Y(d')]}{\partial d'} = \frac{\partial E[Y(d)]}{\partial d} \text{ for any values } d' \neq d \text{ that } D \text{ might take.} \tag{3.58}$$

The equality in equation (3.58) implies that when picking any two arbitrary and different baseline treatment values $d$ and $d'$ (e.g., training expenditures), a small increase in the treatment in $D$ (e.g., by 1 USD) has the same effect, no matter how small or big $d$ and $d'$ are, which is thus homogeneous. Very much in contrast to our previous applications of linear regression, such a homogeneous effect implies that the conditional mean outcome $E[Y|D]$ is truly linear in $D$. This is illustrated in figure 3.4, in which the outcomes are plotted against a continuously distributed treatment and the solid line corresponds to the mean potential outcome for specific values of the treatment. Only under homogeneous effects, and thus a linear association between

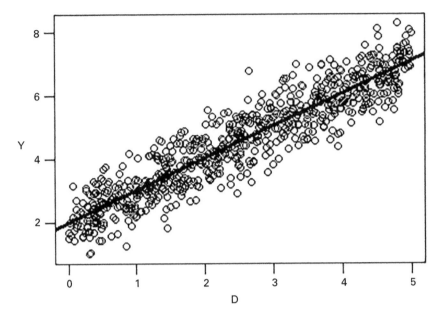

**Figure 3.4**
Linear association of the outcome and treatment.

$E[Y|D]$ and $D$, does the linear regression in equation (3.56) permit identifying the marginal effect at a specific value of $d$, simply because $\frac{\partial E[Y(d)]}{\partial d} = \frac{\partial E[Y(d')]}{\partial d'} = \beta$ for any $d$ and $d'$ under the condition in equation (3.58). In this case, a linearity assumption indeed underlies the linear regression approach, which was not the case when considering binary or multiple discrete treatments.

However, in many empirical settings, the causal relation of the outcome and a continuous treatment might be nonlinear, implying that marginal effects are not homogeneous, but rather differ depending on the baseline values of the treatment. Figure 3.5 illustrates such a nonlinear relationship, with the solid line characterizing the mean potential outcome as a function of the treatment. In this example, the outcome-treatment relation is even nonmonotonic because the mean potential outcome first increases in the treatment for comparably small values of $d$ but then decreases after $D$ surpasses a specific value. If we want to permit such nonlinear associations and, thus, heterogeneous effects rather than imposing linearity, we may still use linear regression, but we can make it more flexible by also allowing nonlinearities. This is obtained by also including higher-order terms of $D$ in the regression, such as its quadratic term $D^2$:

$$E[Y|D] = \underbrace{\alpha}_{E[Y|D=0]} + \beta_1 D + \beta_2 D^2. \tag{3.59}$$

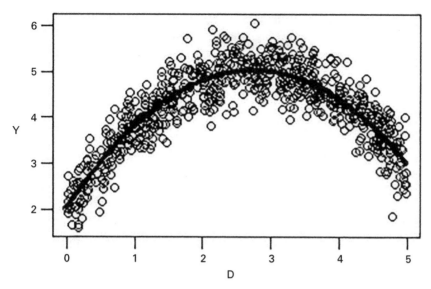

**Figure 3.5**
Nonlinear association of the outcome and treatment.

In equation (3.59), two coefficients (rather than one) are relevant for the computation of the marginal effect (namely, $\beta_1$ and $\beta_2$). We can easily see this by taking the first derivative of $E[Y(d)]$ with regard to $d$ based on the regression model in this equation:

$$\frac{\partial E[Y(d)]}{\partial d} = \beta_1 + 2\beta_2 d. \tag{3.60}$$

Equation (3.60) shows that the marginal effect coming from our nonlinear model now depends on the treatment value $d$, which heterogeneously affects the magnitude of the marginal effect through the factor $2\beta_2$. We might therefore estimate equation (3.59) in the sample and use the coefficient estimates to compute any marginal effect of interest based on equation (3.60). However, including even more higher-order terms like $D^3$ (i.e., cubic or even higher) in the regression further increases the model's flexibility to incorporate nonlinearities and heterogeneous effects. Yet, picking too many higher-order terms, in particular when they are actually not relevant for correctly modeling the association between $E[Y|D]$ and $D$, comes with the cost of increasing the variance of effect estimation, in particular if the sample size is limited. Ideally, we would like to cleverly choose the amount (or order) of higher-order terms in a way that minimizes the overall estimation error, in particular the MSE considered in equation (3.34) at the end of section 3.3.

As we have already discussed, the MSE consists of both a variance component, due to including too many irrelevant higher-order terms, and a bias component, due to including too few higher-order terms that are relevant for appropriately describing effect heterogeneity. In fact, we can choose the optimal order aiming at minimizing the MSE and thus optimally trade off bias and variance in a data-driven way called *cross-validation*; see, for instance, Stone (1974). We will discuss cross-validation in more detail in section 4.2 in chapter 4. Such an estimation approach based on first determining the optimal number of higher-order terms and then estimating the marginal effects of interest is also known as *series regression*. It is an example of nonparametric methods aiming at a flexible estimation of regression functions without imposing restrictions like the linearity condition in equation (3.58).

An alternative nonparametric approach is *kernel regression*, which is more thoroughly considered in section 4.2. It estimates the conditional mean $E[Y|D=d]$ or the marginal effect $\frac{\partial E[Y(d)]}{\partial d}$ based on a weighted average of outcomes in the sample, in which more importance (or weight) is given to observations with treatment values that are close to the value $d$ of interest. Also in this case, cross-validation can be used for finding the weighting scheme that is optimal for computing the weighted average, in the sense that it minimizes the MSE. As with most things in life, however, the use of nonparametric approaches does not come entirely for free. The price to pay for their attractiveness in terms of flexibility is that they tend to have a higher variance and, thus, a higher level of uncertainty when estimating the causal effects. In particular, the estimated marginal effects converge to the true effects at a slower pace than the optimal rate of $\frac{1}{\sqrt{n}}$ discussed in section 3.3, but estimation nevertheless satisfies asymptotic normality and consistency under specific conditions. In large enough samples where the costs in terms of variance are small relative to the gains in flexibility, nonparametric methods therefore appear to be an attractive alternative to linear regression in the case of a continuous treatment.

Let us conclude this section with two empirical examples that apply regression with multiple treatments and nonparametric regression, respectively, with a continuous treatment in R. Assuming that all packages have been properly installed, we load the *causalweight*, *lmtest*, and *sandwich* packages using the *library* command. In the next step, we use the *data* command to load the *wexpect* data set provided in the *causalweight* package and previously analyzed in Fernandes, Huber, and Vaccaro (2021). By using *?wexpect* to address the help file, we see that the data contain the wage expectations of 804 Swiss college or university students, as well as dummies for two mutually exclusive treatments that were randomly assigned before students answered a questionnaire on their wage expectations, among others. The first treatment, named *treatmentinformation*, included a graph with information on the monthly gross private-sector earnings by age and gender in the questionnaire, which could arguably affect a student's wage expectations. The second treatment, *treatmentorder*, reversed the order of questions about professional (e.g., workplace-related) and personal (e.g., family-related) preferences. This permits checking whether asking personal rather than professional questions first matters for wage expectations due to so-called framing effects, implying that the questions' order influences the perceived importance of private versus professional life.

We extract the information and order treatments from the *wexpect* data and store them in separate R objects, *D1* and *D2*, by running *D1=wexpect$treatmentinformation* and *D2=wexpect$treatmentorder*. Likewise, we define the expectations about the monthly gross wages three years after studying as outcome Y, which are measured in brackets of 500 CHF (such that an increase by 1 unit means having 500 CHF more): *Y=wexpect$wexpect2*. We then run a linear regression using the *lm* command with *D1* and *D2* as treatments, which need to be separated by a +, and store the output in an R object called *ols*: *ols=lm(Y~D1+D2)*. Finally, we inspect the regression output using the *coeftest* command. The R code for these steps is provided in the box shown here.

```
library(causalweight)             # load causalweight package
library(lmtest)                   # load lmtest package
library(sandwich)                 # load sandwich package
data(wexpect)                     # load wexpect data
?wexpect                          # call documentation for wexpect data
D1=wexpect$treatmentinformation   # define first treatment (wage information)
D2=wexpect$treatmentorder         # define second treatment (order of questions)
Y=wexpect$wexpect2                # define outcome (wage expectations)
ols=lm(Y~D1+D2)                   # run OLS regression
coeftest(ols, vcov=vcovHC)        # output with heteroscedasticity robust se
```

Running the code yields the following output:

```
             Estimate  Std. Error  t value  Pr(>|t|)
(Intercept)  9.40761   0.15873     59.2679  <2e-16 ***
D1           0.34529   0.24291     1.4214   0.1556
D2          -0.17341   0.23391    -0.7413   0.4587

___
Signif. codes:  0 '***' 0.001 '**' 0.01 '*' 0.05 '.' 0.1 ' ' 1
```

The ATE estimate of the information treatment *D1* amounts to roughly 0.345 (or roughly 173 CHF when calculating $0.345 \times 500$), which suggests an increase in expected monthly gross wages. However, the effect is not statistically significant at the 10 percent level, as the p-value is roughly 0.156 (or 15.6 percent). We therefore fail to reject the null hypothesis at any conventional level of significance. The ATE estimate for the order treatment is even less statistically significant, with the p-value amounting to roughly 45.9 percent and also closer to zero in absolute terms. Therefore, we do not find compelling statistical evidence for nonzero ATEs of our two interventions.

For our second empirical example considering a continuous treatment, we load the *datarium* package by Kassambara (2019), which contains the *marketing* data set, and the *np* package by Hayfield and Racine (2008) for nonparametric kernel regression, as discussed in more detail in section 4.2. After loading the *marketing* data using the *data* command, we address the help file *?marketing*. The sample consists of 200 observations with information on sales and advertising budgets, containing (among others) a variable called *newspaper*, which measures the budget of advertising in newspapers in thousands of USD. This variable is our continuous (and assumably randomly assigned) treatment variable, and we store it in an R object named $D$. Our outcome variable, which we define as $Y$, are the *sales*, whose exact unit of measurement, however, is not revealed in the help file.

We use the *npregbw* command for nonparametrically estimating the association of $Y$ and $D$, where the latter two variables need to be separated by $\sim$, just as in the *lm* command. The *npregbw* procedure uses cross-validation, as explained in section 4.2, to estimate the association in a way that minimizes the average of the MSE across all treatment values in the sample. After storing the regression output in the R object *results*, we use the *plot* command to plot the regression function: that is, the estimate of the conditional mean outcome $E[Y|D]$ as a function of the values of the treatment $D$. The first argument in the *plot* command is the regression output *results*. As the second argument (separated from the first one by a comma), we specify *plot.errors.method="asymptotic"*, which also plots confidence intervals based on the analytic formula (rather than the bootstrap) for the standard error. See the box here for the R code of the various steps.

```
library(datarium)                    # load datarium package
library(np)                          # load np package
data(marketing)                      # load marketing data
?marketing                           # call documentation for marketing data
D=marketing$newspaper                # define treatment (newspaper advertising)
Y=marketing$sales                    # define outcome (sales)
results=npregbw(Y~D)                 # kernel regression
plot(results, plot.errors.method="asymptotic") # plot regression function
```

Running the code yields the graph in figure 3.6. The y-axis in the latter gives the estimates of the conditional mean outcomes $E[Y|D = d]$, while the x-axis provides the value $d$ of the advertising treatment $D$. By and large, the solid regression line suggests

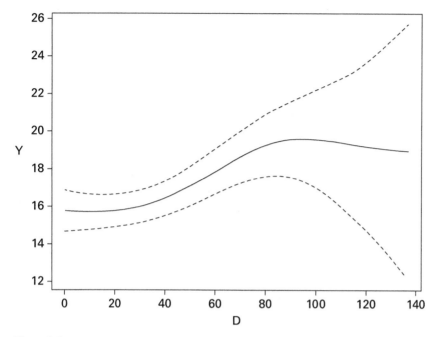

**Figure 3.6**
Estimation of the conditional mean outcome.

that newspaper advertising positively affects average sales up to a budget of roughly 90,000 USD, while the association is rather flat (and even slightly decreasing) for even higher budgets. This suggests a nonlinear relation between the outcome and the treatment. However, it needs to be pointed out that the association between average sales and spending for newspaper advertising is not very precisely estimated for larger values of the treatment. The 95 percent confidence intervals (shown by the dashed line) become very large beyond budgets of 90,000 USD due to the small number of observations with such high spending.

The *np* package also gives us the opportunity to plot the marginal effects $\frac{\partial E[Y(d)]}{\partial d}$ (rather than the conditional mean outcomes) across value $d$ of the advertising treatment $D$, which is the first derivative of the solid line in figure 3.6. To do so, we simply add the argument *gradients=TRUE* to the previous use of the *plot* command:

```
plot(results, gradients=TRUE, plot.errors.method="asymptotic") # plot effects
```

Running the code generates the graph in figure 3.7. We can see that the marginal effects on average sales of slightly increasing the advertising budget are statistically

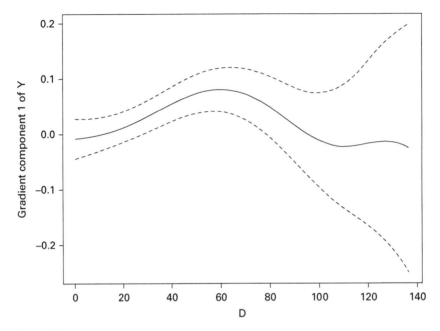

**Figure 3.7**
Estimation of the marginal effects.

significantly positive over a limited range of baseline budget values around 60,000 USD—namely, whenever the 95 percent confidence intervals (dashed line) do not include a zero effect. In contrast, the marginal effects are statistically insignificant when considering larger baseline values of the treatment, as the confidence intervals are rather wide and always include a zero effect.

## 3.6   Including Covariates

Under the satisfaction of the independence assumption in equation (3.1), as in experimental studies with a successfully randomized treatment, treated and nontreated groups are comparable in terms of any background characteristics that affect the outcome. For this reason, we need not include (i.e., control for) any characteristics that are observed in the data in our regression to unbiasedly and consistently estimate the ATE. Nevertheless, controlling for observed characteristics, commonly referred to as *covariates*, might be beneficial in terms of reducing the variance (and thus, uncertainty) of treatment effect evaluation. To consider this possibility, let us denote by $X = (X_1, X_2, \ldots, X_K)$ a vector of covariates measured at or before treatment assignment, with $X_1, X_2, \ldots, X_K$ denoting the first to the $K$th covariate and $K$ being the number of observed covariates.

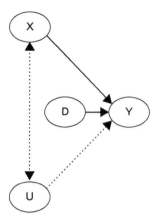

**Figure 3.8**
Pretreatment covariates.

As the covariates $X$ are measured prior to treatment, they cannot be affected by $D$ unless subjects anticipate their future treatment in a way that influences their covariates even before the treatment, a case that we rule out in the subsequent discussion. Assuming treatment $D$ to be binary and denoting by $X(1)$ and $X(0)$ the potential covariates as a function of the treatment (in analogy to the definition of potential outcomes), this implies that $X(1) = X(0) = X$. As we will see later in this chapter, this property is crucial because otherwise, $X$ would be a function of the treatment and thus an outcome itself (just as $Y$ is) and in general, controlling for variables affected by the treatment might jeopardize the randomization of the latter. Figure 3.8 provides such a causal framework, where pretreatment covariates $X$ may influence $Y$, but (due to random treatment assignment) neither influence nor are influenced by $D$. In contrast, $X$ may influence or may be influenced by unobserved characteristics $U$ that affect the outcome (but not $D$ due to treatment randomization), as indicated by the bidirectional dotted causal arrow.

Let us reconsider equation (3.6) in section 3.2, and now additively include the covariates on the right side, as regressors, to control for them:

$$Y = \alpha + \beta_D D + \beta_{X_1} X_1 + \cdots + \beta_{X_K} X_K + \varepsilon. \tag{3.61}$$

This models the conditional mean outcome, given the treatment and the covariates, by the following linear function:

$$E[Y|D, X] = \alpha + \beta_D D + \beta_{X_1} X_1 + \cdots + \beta_{X_K} X_K. \tag{3.62}$$

We note that some of the variation in $Y$ that was previously part of the error term in equation (3.6) is now captured by $X$ in equation (3.61), such as age or gender if observed in the data. Therefore, the number of unobservables included in $U$ goes down, and relatedly, the residuals $\hat{\varepsilon}_i$ estimated in the sample tend to decrease in absolute magnitude: that is, they generally get closer to zero. Put differently, some of the unexplained part in $Y$ that is due to unobserved characteristics is shifted to the explained part due to the inclusion of $X$.

To discuss this more formally, let us reconsider the definition of the outcome in the sample provided in equation (3.50) and add the covariates, along with their coefficient estimates in analogy to equation (3.61):

$$Y_i = \underbrace{\hat{\alpha} + \hat{\beta}_D D_i + \hat{\beta}_{X_1} X_{i1} + \cdots + \hat{\beta}_{X_K} X_{iK}}_{\hat{E}[Y_i|D_i, X_i]} + \hat{\varepsilon}_i, \tag{3.63}$$

where $X_{i1}, \ldots, X_{iK}$ denotes the covariate values of observation $i$ in the sample and $\hat{E}[Y_i|D_i, X_i]$ is an estimate (or prediction) of the conditional mean outcome $E[Y|D, X]$. Reconsidering the formula for the goodness-of-fit criterion $R^2$ provided in equation (3.52), and now including covariates, yields the following definition of $R^2$:

$$R^2 = \frac{Var(\hat{E}[Y_i|D_i, X_i])}{Var(Y_i)}. \tag{3.64}$$

It is easy to see that in equation (3.64), the share of the sample variation in $Y$ explained by $D$ and $X$, is larger than $\frac{Var(\hat{E}[Y_i|D_i])}{Var(Y_i)}$ in equation (3.52), the sample variation in $Y$ explained by $D$ alone, whenever $X$ partly explains $Y$, while $\frac{Var(\hat{\varepsilon}_i)}{Var(Y_i)}$ is smaller due to reduced variance of $\hat{\varepsilon}_i$. This in turn implies that the estimated variance of $\hat{\beta}_D$ given in equation (3.36) is reduced, entailing a decrease in the standard error $se(\hat{\beta})$, and thus an increase in the t-statistic (expression (3.41)) and a reduction in p-value (equation (3.42)) for any nonzero ATE estimate. In short, estimation uncertainty goes down, while the likelihood (or statistical power) to detect ATEs that are different from zero in the population goes up.

It is important to note that even if the influence of $X$ on $Y$ is not linear, such that equation (3.62) is in fact an incorrect (or misspecified) model for the outcome, we can nevertheless consistently estimate the causal effect of $D$. While this approach may introduce some estimation bias in small samples (of a few hundred observations or fewer) as demonstrated in Freedman (2008), the bias quickly goes to zero as the sample size grows. In light of the discussion on linear and nonlinear associations in section 3.5, this result may seem surprising, but it comes from the fact that $D$ is not associated with $X$ due to randomization: that is, $D \perp X$. For this reason, the error of incorrectly assuming a linear association between $Y$ and $X$ does not spill over to the evaluation of the ATE and therefore does not introduce any asymptotic (i.e., large sample) bias in $\hat{\beta}_D$. This would not be the case if the treatment were not fully randomized, but rather were associated with $X$, as discussed in chapter 4.

In contrast to our discussion so far, let us now consider the case where $X$ is affected by $D$, such that $X(1) \neq X(0)$. In general, controlling for $X$ does no longer allow for assessing the causal effect of the treatment for two reasons. First, part of the causal effect of $D$ on $Y$ might operate via $X$, such that controlling for the latter switches off or conditions away this effect. As an empirical example, let us consider the effect of mothers' smoking behavior during pregnancy ($D$) on children's health outcomes ($Y$), such as postnatal infant mortality, when using birth weight as the control variable ($X$). Having a higher or lower birth weight, however, can already reflect part of the negative effects of smoking on a child's health. In this case, controlling for birth weight (e.g., by only considering newborns with a low birth weight) conditions away part of the negative effect on postnatal health, such that only the direct impact of $D$ on $Y$ that does not operate via $X$ remains. Section 4.10 in chapter 4 provides a more detailed discussion of the distinction between direct and indirect effects operating via intermediate variables.

The second reason for the nonidentifiability of the treatment effect is that controlling for posttreatment covariates $X$ affected by $D$ destroys the randomization of the treatment in the likely case that $X$ is influenced by some unobservables $U$ that also influence $Y$. In fact, if both $D$ and $U$ have a causal effect on $X$, then subjects with different treatment values ($D = 1$ or $D = 0$) that have the same value in $X$ generally differ in terms of $U$; otherwise, they would not have the same value in $X$. If $U$ also affects $Y$ (e.g., postnatal health), this implies that the independence assumption in expression (3.1) no longer holds when considering treated and nontreated units with the same values in $X$. Therefore, controlling for $X$ introduces a statistical association between $D$ and $U$ that initially (i.e., without controlling for $X$) does not exist. This specific form of selection bias is discussed in Rosenbaum (1984), and it is also known as "collider bias" in statistics (see Pearl (2000)).

The graph in figure 3.9 illustrates such a causal framework, in which $D$ partly affects $Y$ via $X$, such that controlling for the latter conditions part of the effect away. Furthermore, both $D$ and $U$ have a causal effect on $X$, such that controlling for the latter introduces a statistical association between $D$ and $U$. For this reason, $X$ is referred to as a "collider" in statistics because the effects of $D$ and $U$ collide when conditioning on $X$. Due to these issues, covariates affected by $D$ are also called "bad controls"—for instance, see Angrist and Pischke (2008)—indicating that they should not be considered as control variables when assessing causal effects.

Reconsidering our empirical example, collider bias implies that those newborns having a low birth weight ($X$) because the mother was smoking during pregnancy ($D = 1$) are not comparable to newborns of nonsmoking mothers ($D = 0$), because the low birth weight of the latter group is necessarily caused by other characteristics ($U$) besides smoking. An example of such other characteristics is increased incidence

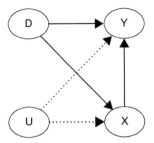

**Figure 3.9**
Posttreatment covariates that are bad controls.

of birth defects, which also affect postnatal mortality (Y). By comparing low-birth-weight children of smoking and nonsmoking mothers, one incorrectly mixes the causal effect of smoking with that of the birth defects. As discussed in such papers as Wilcox (2001) and Hernandez-Diaz, Schisterman, and Hernan (2006), one might then paradoxically find a negative association between smoking and the mortality conditional on having low birth weight, which apparently contradicts the generally negative effects of smoking found in health studies. But in reality, this result is driven by differences in birth defects or other characteristics across treatment groups due to ignoring the selection problem created by controlling for X.

Let us now consider an empirical example in R and run a linear regression that includes pretreatment covariates. To this end, we analyze an experimental informa-tion campaign conducted by the Ecosystem Europe nongovernmental organization in Bulgaria. It was based on randomly distributing leaflets to high school and university students with information about the environmental and social implications of cof-fee production and measuring their awareness of environmental issues after receiving or not receiving the information. We load the *causalweight* package, which contains the *coffeeleaflet* data with 522 observations on Bulgarian students, as well as the *lmtest* and *sandwich* packages (for heteroscedasticity-robust standard errors) using the *library* command. In the next step, we load the data set into the R workspace using *data(coffeeleaflet)* and apply the *attach* command to the *coffeeleaflet* data. The latter defines each variable in the data as an own R object, thus avoiding the use of the $ extension for addressing any variable in *coffeeleaflet*.

By *?coffeeleaflet*, we access the documentation for the 48 variables. The binary variable *treatment* contains the random assignment to receiving versus not receiving the leaflet, which we define as D: D=treatment. Furthermore, we consider *aware-waste*, a student's awareness of waste production due to coffee production on a 5-point scale (1=not aware, ..., 5=fully aware), as outcome Y: Y=awarewaste. In the next step, we use the *cbind* command to append two covariates, mother's education and student's gender (which are measured prior to the treatment) column by column, in order to generate a covariate matrix $X$: $X=cbind(mumedu,sex)$. We then run a linear regression of outcome $Y$ on both $D$ and $X$ using the *lm* command, with $\sim$ separating the outcome from the regressors $D, X$ and $+$ separating $D$ and $X$, and store the results in an R object called *ols*: $ols=lm(Y\sim D+X)$. Finally, we apply *coeftest(ols, vcov = vcovHC)* to display the regression output in the R object *ols*. The box shown here provides the R code for each of the steps.

```
library(causalweight)      # load causalweight package
library(lmtest)            # load lmtest package
library(sandwich)          # load sandwich package
data(coffeeleaflet)        # load coffeeleaflet data
attach(coffeeleaflet)      # store all variables in own objects
?coffeeleaflet             # call documentation for coffeeleaflet data
D=treatment                # define treatment (leaflet)
Y=awarewaste               # define outcome (aware of waste production)
X=cbind(mumedu, sex)       # define covariates (grade, gender, age)
ols=lm(Y~D+X)              # run OLS regression
coeftest(ols, vcov=vcovHC) # output with heteroscedasticity robust se
```

This yields the following output:

```
              Estimate  Std. Error  t value  Pr(>|t|)
(Intercept)   1.186993    0.248823   4.7704  2.469e-06  ***
D             0.331801    0.096195   3.4493  0.0006136  ***
Xmumedu       0.272022    0.090448   3.0075  0.0027778  **
Xsex          0.136845    0.099917   1.3696  0.1714819

Signif. codes:  0 '***' 0.001 '**' 0.01 '*' 0.05 '.' 0.1 ' ' 1
```

Our results suggest that the information leaflet increases the awareness of coffee-induced waste production on average by 0.332 points (on a 5-point scale). The ATE is highly statistically significant, even at the 0.1 percent level, because the p-value is $0.0006 < 0.001$. We also find that mother's education is statistically significantly associated with the outcome at the 1 percent level conditional on the other regressors, while student's gender is not statistically significantly associated at any conventional level because its p-value of 0.171 (or 17.1 percent) exceeds 0.1 (or 10 percent). However, we bear in mind that the coefficients on the covariates generally cannot be interpreted as causal effects, first because $X$ is not randomized and may be correlated with unobservables (in contrast to $D$) and second, because the association of $Y$ and $X$ need not be linear.

# 4

# Selection on Observables

## 4.1 Identification under Selection on Observables

For many causal questions in the social sciences and life more generally, experiments are the exception, or even fully absent. In fact, a large share of empirical analyses are based on observational rather than experimental data, which may come from surveys (e.g., an online survey among customers), company data (e.g., product features and sales in stores), or administrative data (e.g., information on labor market performance and public transfer payments). Observational data typically contains one outcome (or even several outcomes), a range of observed covariates, and the treatment of interest, which is, however, not randomly assigned. For this reason, the previously imposed independence assumption in expression (3.1) from section 3.1 in chapter 3 appears generally implausible and a simple comparison of the mean outcomes of treated and nontreated groups seems inappropriate for assessing the average treatment effect (ATE). Are there scenarios where causal analysis works with observational data despite the absence of a proper experiment? The answer is yes, if the information on pretreatment covariates is rich enough to facilitate an alternative strategy for the evaluation of causal effects based on a selection-on-observables assumption.

The selection-on-observables assumption, also called *conditional independence*, unconfoundedness, or exogeneity, postulates that the covariates in the data are comprehensive enough to control for the influence of any confounders: that is, characteristics jointly affecting the treatment and the outcome. This is satisfied if we directly observe covariates with an effect on both the treatment and the outcome, or if controlling for the covariates blocks the effects on either the treatment or the outcome (or both) of any unobserved confounders (that would otherwise jointly affect the outcome and the treatment). Put differently, the treatment is assumed to be as good as if it were randomly assigned (as in an experiment) among those treated and nontreated subjects that are comparable in terms of observed characteristics.

Let us, for instance, consider the causal effect of a discount on customers' buying decision in an online marketplace. Then the selection-on-observables assumption implies that when only considering customers that are comparable in covariates, such as past buying behavior (which may jointly affect the current buying decision as well as the likelihood of being offered a discount), receiving a discount or not is as good as if it were randomly assigned. This appears plausible if the characteristics that importantly or exclusively determine a discount are known and observed: for instance, if the online marketplace offers discounts depending on a customer's past buying behavior, and if unobserved characteristics affecting the discount are unlikely to also affect the outcome after controlling for the covariates. The latter condition, for instance, is satisfied if the availability of discounts varies seemingly randomly over time among customers with the same buying history, due to ad hoc changes in a company's discount policy.

However, it is important to note that the selection-on-observables assumption is unlikely to be satisfied in many empirical contexts. Quite often, we do not plausibly observe all the factors that drive confounding of the treatment and outcome of interest. For instance, specific personal characteristics such as ability, intelligence, motivation, or personality traits like self-confidence and extrovertedness are rarely measured in data but might influence treatment decisions like training participation, while also having an impact on labor market outcomes like earnings or employment. Then, controlling for observed covariates alone will not account for all sources of bias when estimating causal effects and may in specific cases even increase biases coming from unobserved characteristics relative to not controlling for any variable, such as the results discussed in Brooks and Ohsfeldt (2013). The plausibility of the selection-on-observables assumption, therefore, needs to be scrutinized based on theoretical arguments, domain knowledge, or previous empirical findings, all of which may provide guidance on which variables likely affect both the treatment and the outcome in the causal evaluation problem at hand. In later sections, we will consider alternative causal strategies that may be applied when the selection-on-observables assumption would likely fail, given that other (but not necessarily weaker) assumptions are satisfied.

In addition to the selection-on-observables assumption, a second condition is common support, which requires that for any combination of covariate values occurring in the population, there are both treated and nontreated subjects. For instance, for any value that past buying behavior may take (measured by such statistics as the total volume of previous purchases), there must be individuals receiving and not receiving a discount as the treatment in the customer population of interest. Common support rules out that the covariates deterministically predict the treatment, which would imply that everyone or no one was treated for specific values of the covariates. As a final assumption, we stipulate that the covariates are not affected by the treatment,

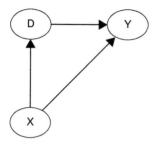

**Figure 4.1**
Selection on observables.

but measured at or prior to treatment assignment. Otherwise, controlling for them risks conditioning away part of the treatment effect or introducing collider bias, as we have already discussed in section 3.6.

To formally introduce these assumptions, let us focus on a binary treatment and, analogously to section 3.6, denote by $X = (X_1, X_2, \ldots, X_K)$ the observed covariates, whose number is $K$, and by $X(1), X(0)$ the potential covariate values with and without treatment:

$$\{Y(1), Y(0)\} \perp D | X, \quad 0 < p(X) < 1, \quad X(1) = X(0) = X, \tag{4.1}$$

where $p(X) = \Pr(D = 1 | X)$ is the conditional treatment probability, also known as propensity score. The first assumption in expression (4.1) states that the potential outcomes are conditionally independent of the treatment when controlling for (or conditioning on) covariates $X$. This implies that $D$ is as good as randomly assigned among subjects with the same values of $X$.

The causal graph in figure 4.1 displays a framework that satisfies this conditional independence assumption. The covariates $X$ jointly affect $D$ and $Y$, but conditional on $X$, there are no further unobserved variables that influence both $D$ and $Y$. This still allows for sets of unobserved variables affecting either $D$ or $Y$, which are omitted from the graph, so long as these sets are not statistically associated (e.g., correlated) with each other. The second assumption in equation (4.1) says that the propensity score is greater than zero and less than 1, such that $D$ is not deterministic in $X$ and common support holds. The third assumption requires that $X$ is not a function of $D$, and therefore must not contain posttreatment characteristics that are affected by the treatment. Under these assumptions, we can mimic the experimental context with the help of observed information. After creating groups with and without treatment that are comparable in the covariates, differences in the outcomes are assumed to be exclusively caused by the treatment.

We note that the first assumption in expression (4.1) is actually somewhat stronger than required for the evaluation of the ATE. The latter could also be identified under the somewhat weaker conditional independence of the potential outcome means, $E[Y(1)|D=1,X]=E[Y(1)|D=0,X]$ and $E[Y(0)|D=1,X]=E[Y(0)|D=0,X]$. In contrast, the conditional independence assumption in expression (4.1) also concerns higher moments of the potential outcome distributions, like the variance. In empirical applications, however, it might be hard to argue that conditional independence holds in means, but not in other distributional features, which would for instance rule out mean independence for nonlinear transformations of $Y$ that are common in empirical research. Labor market applications, for example, frequently consider the logarithm of the wage as the outcome rather than the wage itself. In this context, conditional mean independence would imply that only the potential average wages were conditionally independent of the treatment (e.g., some training), but not their logarithmic transformations, which might appear odd.

Furthermore, the stronger conditional independence assumption in expression (4.1) is required for the identification of causal effects that concern the distribution (rather than the mean) of the potential outcomes, like the quantile treatment effect (QTE). The latter corresponds to the effect at a particular rank of the potential outcome distribution, such as the impact of training on the median wage, which is the wage at the 0.5 rank such that 50 percent of the population earn more and 50 percent earn less than that wage. We will discuss the QTE in more detail in section 4.8.

A second remark concerns the identification of treatment parameters among the treated (rather than the total) population, like the average treatment effect on the treated (ATET), rather than the ATE. In this case, the conditional independence assumption in expression (4.1) may be relaxed to apply only to the potential outcomes under nontreatment: $Y(0) \perp D|X$. Furthermore, we may relax the common support restriction to $p(X) < 1$. The reason is that when assessing the ATET, for all treated observations, there must exist nontreated observations with comparable covariates $X$. Therefore, $p(X)$ must not be 1, which would imply that only treated existed for some values of the covariates $X$ (e.g., everyone older than 50 receives the treatment). However, as we are interested in the ATET rather than the ATE, we need not find comparable treated observations for each nontreated observation. This implies that $p(X)$ might be zero for some values of $X$, such that only nontreated observations exist with such values (e.g., no one younger than 20 receives the treatment). We will henceforth abstain from making such relaxations in our assumptions and stick to the conditions in expression (4.1), which permit assessing average and distributional causal effects in the total population.

To show how our assumptions permit identifying causal effects, let us use $\mu_d(x) = E[Y|D=d, X=x]$ to denote the conditional mean outcome, given that treatment $D$ is equal to value $d \in \{0, 1\}$ (i.e., $d$ is either 0 or 1) and $X$ is equal to value $x$. Because the treatment is as good as randomly assigned conditional on the covariates under expression (4.1), $\mu_1(x) - \mu_0(x)$ identifies the causal effect among subjects with the same values $x$ of observed covariates $X$, denoted by $\Delta_x$:

$$\Delta_x = E[Y(1)|X = x] - E[Y(0)|X = x] = \mu_1(x) - \mu_0(x). \tag{4.2}$$

Equation (4.2) is a conditional version (given $X$) of equation (3.2) in section 3.1 for the experimental case, for which controlling for covariates is not required. We will henceforth refer to $\Delta_x$ as the conditional average treatment effect (CATE), because it corresponds to the average effect under the condition that subjects share the same covariate values $X = x$.

The identification of the CATE also permits identifying the ATE—namely, by averaging CATEs across all values of $x$ (e.g., age measured in years), which the covariates $X$ take in the population:

$$\Delta = E[\mu_1(X) - \mu_0(X)]. \tag{4.3}$$

In the selection-on-observables framework, treated and nontreated groups may generally differ in terms of their covariates $X$, and therefore, also in terms of causal effects if the latter depend on the covariates. For instance, the effect of a discount could vary across the previous buying behavior. This is very much in contrast to the experimental framework where treated and nontreated groups are assumed to have comparable characteristics, including covariates $X$. It may be interesting, therefore, to consider the effects for subpopulations with a different distribution of $X$ than the total population.

One group that has received a lot of attention in empirical evaluations is the treated population. The causal effect on those receiving the treatment may in many contexts be more relevant than the effect on the total population, if it is not feasible or desirable to provide a treatment to everyone. For instance, a training program might deliberately only target individuals that satisfy specific criteria in terms of experience or education, such that the causal effect on exactly this group is of interest, rather than on someone else. To identify the ATET, the CATEs are averaged across the covariate values $x$ appearing among the treated (rather than the total) population, such that we condition on $D = 1$ when taking expectations of the CATE:

$$\Delta_{D=1} = E[\mu_1(X)|D = 1] - E[\mu_0(X)|D = 1] = E[Y|D = 1] - E[\mu_0(X)|D = 1]. \tag{4.4}$$

The second equality in equation (4.4) follows from the law of iterated expectations, implying that $E[\mu_1(X)|D = 1] = E[E[Y|D = 1, X]|D = 1] = E[Y|D = 1]$. We can apply an analogous approach for identifying the average treatment effect among the nontreated (ATENT):

$$\Delta_{D=0} = E[\mu_1(X)|D = 0] - E[\mu_0(X)|D = 0] = E[\mu_1(X)|D = 0] - E[Y|D = 0]. \tag{4.5}$$

## 4.2   Linear, Series, and Kernel Regression

By the selection-on-observables assumption in expression (4.1), it holds that the conditional potential outcome given $X$ corresponds to the conditional observed outcome given $D$ and $X$—that is, $E[Y(D)|X] = E[Y|D, X] = \mu_D(X)$. This permits identifying the CATE as outlined in equation (4.2) in the last section. Therefore, by defining and estimating a regression model for $E[Y|D, X]$, we may ultimately evaluate the CATE and any averages thereof as the ATE and the ATET. Analogously to equation (3.62) in section 3.6, where we assumed that $D$ was randomly assigned, we might be tempted to posutlate a linear regression model for $\mu_D(X)$; that is,

$$\mu_D(X) = \alpha + \beta_D D + \beta_{X_1} X_1 + \cdots + \beta_{X_K} X_K. \tag{4.6}$$

In analogy to equation (3.14) in section 3.3, but now controlling for $X$, the ordinary least squares (OLS) estimate of $\hat{\beta}_D$ can be shown to correspond to the conditional sample covariance of $Y$ and $D$ when controlling for $X$, divided by the conditional sample variance of $D$ when controlling for $X$:

$$\hat{\beta}_D = \frac{\widehat{Cov}(Y_i, D_i|X_i)}{\widehat{Var}(D_i|X_i)}. \tag{4.7}$$

However, there is an important conceptual difference when it comes to linear models in a selection-on-observables framework compared to social experiments. As $X$ may affect $D$, the treatment and covariates are generally correlated such that $Cov(D, X) \neq 0$. Such a correlation has implications for the properties of the OLS estimate $\hat{\beta}$. First, it entails a larger variance of $\hat{\beta}$ relative to the case of no correlation ($Cov(D, X) = 0$), as satisfied in the experimental context. Intuitively, this is because effect estimation only hinges on the variation in $D$, which is not associated with covariates $X$, as we control for (i.e., fix) the latter. Second, incorrectly assuming a linear regression model as in equation (4.6) generally implies that $\hat{\beta}$ is biased and inconsistent for estimating causal effects, very much in contrast to an experiment (see our previous discussion in section 3.6). In fact, if the linear specification in equation (4.6) does not reflect the true relationship between $Y$ and $X$, this misspecification generally spills over to the estimation of $\beta_D$ through the correlation of $D$ and $X$, such that treatment effects cannot be consistently estimated by $\hat{\beta}_D$.

One example for such a misspecification is the omission of multiplicative interactions between covariates, such as $X_1 \cdot X_2$. The latter implies that the association of $X_1$ (e.g., customer's past buying behavior) with $Y$ (e.g., a buying decision) differs across values of $X_2$ (e.g., customer's education), or, analogously, that the association of $X_2$ with $Y$ differs across values of $X_1$. A further example for misspecification is the omission of higher-order terms such as $X_1^2$, meaning that a customer's past buying behavior has a nonlinear association with the buying decision.

Finally, let us consider the omission of interactions between treatment $D$ and some or all variables in $X$ (or their respective higher-order terms), by which the treatment effect differs (i.e., is heterogeneous) across distinct values of the covariates. The omission generally causes $\beta_D$ in equation (4.6) to be different from $\Delta_x$, such that the estimator $\hat{\beta}_D$ is again biased and inconsistent for estimating the CATE. We also note that in the linear model postulated in equation (4.6), it holds that $E[\beta_D] = E[\Delta_x] = \Delta = \beta_D$. That is, by not including interactions between $D$ and $X$, average effects are implicitly assumed to be the same (or homogeneous) across values of $X$ such that CATE=ATE=ATET. In many empirical applications, it appears unrealistic to assume that causal effects are the same across groups with distinct observed characteristics. For instance, a discount could have a different effect on the buying decisions of customers previously buying little versus those spending a lot. Unless we have strong prior knowledge that effects are homogeneous, a method that allows heterogeneous effects across $X$ appears preferable to avoid model misspecification.

What can we do to improve the linear regression model in equation (4.6) to address such concerns of misspecification? One way to make the model more flexible in terms of its specification is to add such previously mentioned interaction terms between $D$ and $X$ or between elements in $X$, as well as higher-order terms in $X$, as additional regressors:

$$\mu_D(X) = \alpha + \beta_D D + \beta_{X_1} X_1 + \cdots + \beta_{X_K} X_K + \beta_{D,X_1} DX_1 + \ldots$$
$$+ \beta_{D,X_k} DX_K + \beta_{X_1^2} X_1^2 + \cdots + \beta_{X_1 X_2} X_1 X_2 + \ldots. \tag{4.8}$$

Estimating the model in the sample yields the coefficient estimates $(\hat{\alpha}, \hat{\beta}_D, \hat{\beta}_{X_1}, \ldots)$. The latter permit estimating $\hat{\mu}_1(X)$ by setting the treatment to 1 and $\hat{\mu}_0(X)$ by setting the treatment to zero in the estimated model of equation (4.8). We can then compute the ATE by averaging the difference $\hat{\mu}_1(X) - \hat{\mu}_0(X)$ in the sample:

$$\hat{\Delta} = \frac{1}{n} \sum_{i=1}^{n} [\hat{\mu}_1(X_i) - \hat{\mu}_0(X_i)]. \tag{4.9}$$

In analogy to equation (4.9), we may also estimate the ATET by averaging the CATEs in the subsample of treated observations as follows:

$$\hat{\Delta}_{D=1} = \frac{1}{n_1} \sum_{i:D_i=1} [\hat{\mu}_1(X_i) - \hat{\mu}_0(X_i)], \tag{4.10}$$

where $n_1 = \sum_{i=1}^{n} D_i$ is the number of treated observations. As noted in Imbens and Wooldridge (2009), the ATE estimate in equation (4.9) is numerically equivalent to the coefficient of $D_i$ obtained from a regression of $Y_i$ on a constant, $D_i$, $X_i$ (and possibly higher-order terms and interactions thereof), and the interactions

$D_i \cdot (X_i - \bar{X})$. $\bar{X}$ denotes the sample means of the covariates such that $(X_i - \bar{X})$ are demeaned covariate values in the data:

$$\hat{\mu}_D(X) = \hat{\alpha} + \underbrace{\hat{\beta}_D}_{\hat{\Delta}} D_i + \beta_{X_1} X_{i1} + \cdots + \beta_{X_K} X_{iK} + \beta_{D,X_1} D_i \cdot (X_{i1} - \bar{X}_1) + \cdots .$$

$$+ \beta_{D,X_k} D_i \cdot (X_{iK} - \bar{X}_K) + \beta_{X_1^2} X_1^2 + \cdots + \beta_{X_1 X_2} X_1 \cdot X_2 + \cdots . \tag{4.11}$$

By running this convenient regression, we directly estimate the ATE by $\hat{\beta}_D$ and avoid the two-step procedure of first computing the CATEs and then averaging over them.

It is worth noting that equation (4.8), which includes interactions between $D$ and $X$ and its higher-order terms, can equivalently be expressed by means of two separate equations for $D = 1$ and $D = 0$ without including interaction terms between $D$ and $X$:

$$\mu_1(X) = \alpha_1 + \beta_{X_1,1} X_1 + \cdots + \beta_{X_K,1} X_K + \cdots + \beta_{X_1^2,1} X_1^2 + \cdots + \beta_{X_1 X_2,1} X_1 \cdot X_2 + \cdots ,$$

$$\mu_0(X) = \alpha_0 + \beta_{X_1,0} X_1 + \cdots + \beta_{X_K,0} X_K + \cdots + \beta_{X_1^2,0} X_1^2 + \cdots + \beta_{X_1 X_2,0} X_1 \cdot X_2 + \cdots .$$

$$\tag{4.12}$$

This suggests estimating coefficients $(\hat{\alpha}_1, \hat{\beta}_{X_1,1}, \ldots)$ and $(\hat{\alpha}_0, \hat{\beta}_{X_1,0}, \ldots)$ by regressing $Y$ on a constant, $X$, and its interaction/higher-order terms only within the treated or nontreated observations. We can thus obtain $\hat{\mu}_1(X)$ and $\hat{\mu}_0(X)$ by estimating the first and second lines in equation (4.12), respectively, and computing the ATE or ATET based on averaging appropriately.

However, a practical issue in regression specifications like equations (4.8) and (4.12) is the question of how to optimally choose the number of interaction and higher-order terms. Similar to the discussion of a continuous treatment at the end of section 3.5, including too few higher-order or interaction terms may induce a bias in treatment effect estimation due to a poor approximation of the true model $\mu_D(X)$. On the other hand, including too many terms that have little or no influence on $\mu_D(X)$ (e.g., higher-order terms whose coefficients are zero)—that is, overfitting the true model of $\mu_D(X)$—may increase the variance. This is due to the problem that including irrelevant terms does not only approximate $\mu_D(X)$, but also captures part of the error terms (previously denoted by $\varepsilon$, which is a function of unobservable $U$ affecting the outcome) that are specific to the sample at hand. Therefore, the resulting estimates of the conditional means under treatment and nontreatment, denoted by $\hat{\mu}_1(X)$ and $\hat{\mu}_0(X)$, might not generalize well to other randomly drawn samples with a different distribution of error terms. It is this sensitivity of $\hat{\mu}_1(X)$ and $\hat{\mu}_0(X)$ with regard to the particularities of a data set that drives the estimation uncertainty, and therefore the variance of an estimator.

As already briefly discussed in section 3.5, one way to optimally balance the trade-off between the bias and the variance to minimize the overall estimation error in $\hat{\mu}_1(X)$ and $\hat{\mu}_0(X)$—namely, the MSE introduced in section 3.3—is leave-one-out cross-validation. Considering the estimation of $\hat{\mu}_1(X)$, it is based on finding the combination of terms of $X$ that minimize the sum of squared deviations (or residuals) between an observed treated outcome $Y_i$ and the estimated conditional mean among the treated observations (satisfying $i : D_i = 1$):

$$\sum_{i:D_i=1} [Y_i - \hat{\mu}_{1,-i}(X_i)]^2. \tag{4.13}$$

A key feature in the leave-one-out procedure of expression (4.13) is that any treated observation $i$ is itself not included (i.e., left out) when estimating the conditional mean outcome given the covariate values of $X_i$. The estimate of the conditional mean outcome for observation $i$ is therefore denoted by $\hat{\mu}_{1,-i}(X_i)$, with $-i$ indicating the exclusion of observation $i$. The aim of this approach is to appropriately account for overfitting, and thus an estimator's variance when searching for the optimal model of the conditional mean that minimizes the overall estimation error, which consists of both the bias and the variance. To see this, let us consider the contrary case of including observation $i$ when estimating the conditional mean and denote this estimate by $\hat{\mu}_{1,+i}(X_i)$. Then we can always further reduce the squared residual $[Y_i - \hat{\mu}_{1,+i}(X_i)]^2$ by including more higher-order or interaction terms, such that $\hat{\mu}_{1,+i}(X_i)$ better approximates $Y_i$ due to its increased flexibility. However, this approach will eventually overfit (i.e., fit away) the true error term $\varepsilon_i = Y_i - \mu_1(X_i)$, which represents that part of the outcome $Y_i$ that is not due to covariates $X_i$, but rather to unobserved characteristics.

For this reason, we leave observation $i$ out and use only the remaining sample when estimating the conditional mean outcome for observation $i$. The latter can then be regarded as a sample on its own of just one observation, as it is held out of the remaining sample used to compute $\hat{\mu}_{1,-i}(X_i)$. By estimating the conditional mean outcome in one sample, say among all observations but $i$, and assessing the estimation error in the other sample, say for observation $i$ based on the squared residual $[Y_i - \hat{\mu}_{1,-i}(X_i)]^2$, we properly consider the overfitting or variance problem. As we take an estimated model from one sample to assess its error measured by the squared residual of another sample (or out-of-sample), the variability of observations across samples that drives an estimator's variance is accounted for, very much in contrast to model estimation and error assessment in the very same sample. This implies that adding more interaction and higher-order terms does not necessarily decrease the squared residual when using the leave-one-out approach because both the variance and the bias enter this squared residual.

Leave-one-out estimation corresponds to reestimating $\mu_1(X)$ for all treated observations in the sample by alternately dropping the respective treated observation (for which $\mu_1(X)$ is estimated) from the sample to compute the squared residuals for all treated observations and add them up. Cross-validation then consists of recomputing the sum of squared residuals in equation (4.13) for different model specifications in terms of interaction and higher-order terms of $X$ and ultimately selecting the one that minimizes equation (4.13). Formally, let $p$ correspond to the number of higher-order and interaction terms included in an estimator for $\mu_1(X_i)$, which we denote by $\hat{\mu}_{1,-i,p}(X_i)$ to make the dependence on $p$ explicit. Then, the optimal specification of terms, denoted as $p_{\text{opt}}$, is obtained by minimizing the sum of squared residuals as a function of $p$ among a set of possible choices for higher-order and interaction terms, denoted as $P$:

$$p_{\text{opt}} = \arg \min_{p \in P} \sum_{i:D_i=1} [Y_i - \hat{\mu}_{1,-i,p}(X_i)]^2. \tag{4.14}$$

We then obtain the estimate of $\mu_1(X_i)$ for observation $i$ based on estimating the conditional mean with the optimal specification $p_{\text{opt}}$, now including observation $i$. Formally,

$$\hat{\mu}_1(X_i) = \hat{\mu}_{1,p_{\text{opt}}}(X_i). \tag{4.15}$$

That is, after excluding $i$ in order to optimally trade off bias versus variance when selecting $p_{\text{opt}}$, we include it again for the actual estimation of $\hat{\mu}_1(X_i)$ based on the selected $p_{\text{opt}}$. The leave-one-out cross-validation procedure can be analogously applied to the estimation of $\mu_0(X)$. In general, the number of terms in $p_{\text{opt}}$ depends on the number of observations and tends to grow as the sample size increases. This is because the variance tends to decrease in larger samples, thus increasing the scope of more interaction and higher-order terms for improving model flexibility and reducing the bias. However, the optimal number of terms grows at a slower pace than the sample size; otherwise, the bias would be reduced at a faster rate than the variance, which would not entail an optimal bias-variance trade-off for minimizing the overall error.

The estimation and cross-validation procedure that we just described is a form of so-called series or polynomial regression. While it is more flexible in terms of model specification than linear regression models, it shares the feature that it permits estimating the conditional means $\mu_D(X)$ and the CATEs for any thinkable covariate value $X$, even those with few or no observations in the data. Based on the coefficient estimates $(\hat{\alpha}_1, \hat{\beta}_{X_1,1}X_1, \ldots)$, one could, for instance, predict $\hat{\mu}_1(X)$ for values of $X$ that are far larger than the ones observed in our sample, such as for individuals that are presumably 150 years old. As our coefficients are obtained based on the $X$ values observed in the data, however, predictions based on these coefficients may be quite poor for $X$ values far beyond our sample. This issue needs to be kept in mind when using linear or series regression or any other so-called global estimation method based on estimating the coefficients in the entire sample.

As an alternative, there are local approaches to the estimation of $\mu_1(X)$ and $\mu_0(X)$, that, in contrast to series regression, do not permit predictions far beyond the observed data. The underlying idea is that we may estimate the mean conditional outcome under treatment for a value $X = x$, $\mu_1(x)$ as a local average of observations with values of $X$ that are close to $x$. The simplest approach is to average the outcome of treated subjects with covariate values that lie within a certain distance $h$ around $x$. As $X$ consists of $K$ covariates, $h$ contains $K$ values that define the maximum discrepancy that an observation may have in terms of each of the $K$ covariate values $x = (x_1, \ldots, x_K)$ for being considered in the average. Taking age as a covariate, for instance, we may set $h = 2$, implying that any observations no more than 2 years older or younger than $x{=}40$ years are included for the computation of the average treated outcome at the age of 40. Formally, such an average can be expressed as

$$\hat{\mu}_{1,h}(x) = \frac{\sum_{i:D_i=1} I\{|X_i - x| \le h\} \cdot Y_i}{\sum_{i:D_i=1} I\{|X_i - x| \le h\}}, \tag{4.16}$$

where $I\{|X_i - x| \le h\}$ is an indicator for all $K$ covariate values within the bandwidth and $\sum_{i:D_i=1} I\{|X_i - x| \le h\}$ is the number of treated observations satisfying this bandwidth-related condition. The notation $\hat{\mu}_{1,h}(x)$ makes explicit that the conditional mean estimate depends on the specific choice of the bandwidth $h$.

Equation (4.16) provides an unweighted average in the sense that it gives the same weight (or importance) to all observations within the bandwidth. However, even if within the bandwidth, we might want to give more weight to observations whose $X$ values are closer to $x$ than to those whose values are less comparable. For instance, we might judge someone who is only 1 year younger or older than the reference age of 40 to be more similar than someone who is 2 years older or younger. For this reason, a more sophisticated approach consists of weighting observations by a kernel function, as already briefly discussed in section 3.5, which gives a higher weight to observations with covariate values closer to $x$. Formally, the weighted average corresponds to

$$\hat{\mu}_{1,h}(x) = \frac{\sum_{i:D_i=1}^{n} \mathcal{K}\left(\frac{X_i - x}{h}\right) \cdot Y_i}{\sum_{i:D_i=1}^{n} \mathcal{K}\left(\frac{X_i - x}{h}\right)}, \tag{4.17}$$

which is known as *local constant kernel regression* or the Nadaraya Watson estimator (Nadaraya 1964; Watson 1964). $\mathcal{K}\left(\frac{X_i - x}{h}\right)$ denotes a kernel function that is assumed to satisfy the following conditions: it integrates to 1 (formally: $\int \mathcal{K}(a) da = 1$, where $a$

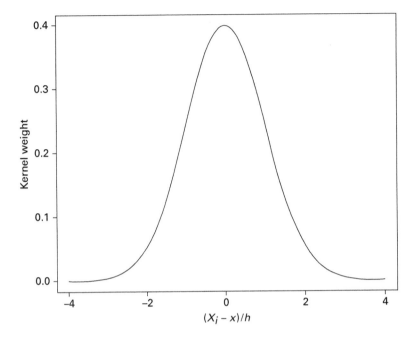

**Figure 4.2**
Standard normal kernel function.

is a specific value in the kernel function), is symmetric around zero ($\int a\mathcal{K}(a)da = 0$, such that positive and negative differences $X_i - x$ with the same absolute value obtain the same weight), and is of a bounded second order such that $\int a^2\mathcal{K}(a)da < \infty$, implying a finite variance of the weighting approach.

A kernel function satisfying these properties is, for instance, the standard normal density function as provided in figure 4.2, among other functions such as the Epanechnikov or triangular kernels. What these kernels have in common is that they assign a greater weight to observations with $\frac{X_i - x}{h}$ being close to zero and less (and, depending on the kernel function, possibly zero) weights for values further from zero. Quite intuitively, the weight is greater when $X_i$ is close to $x$, however, it is the bandwidth $h$ that determines by how much the kernel weight depends on the absolute difference of $X_i - x$. Under a larger bandwidth $h$, any $\frac{X_i - x}{h}$ is closer to zero (and thus obtains a greater weight) than under a smaller bandwidth. Therefore, weights are more uniform and less dependent on $X_i - x$ when $h$ is large. In the extreme case that $h \to \infty$ (i.e., the bandwidth approaches infinity), each observation gets exactly the same weight $\mathcal{K}(0)$, such that equation (4.17) simply corresponds to the average outcome among the treated observations in the sample. In contrast, for a bandwidth approaching zero ($h \to 0$), only observations with $X_i$ virtually identical to $x$ obtain a nonnegligible weight.

Similar to parameter $p$ in series estimation, the question is how to appropriately choose $h$. A rather large bandwidth may entail a substantial bias due to giving large weights to observations whose value $X_i$ is far from $x$. A rather small bandwidth may entail large variance due to giving nonnegligible weight to only very few observations, which may cause overfitting if the error terms $\varepsilon_i$ of these few observations do not properly average out to zero. Similar to determining $p_{opt}$ in series estimation, we may use leave-one-out cross-validation for finding the optimal bandwidth $h_{opt}$ among a range of candidate values denoted by $H$:

$$h_{opt} = \arg\min_{h \in H} \sum_{i:D_i=1} [Y_i - \hat{\mu}_{1,-i,h}(X_i)]^2, \tag{4.18}$$

where $\hat{\mu}_{1,-i,h}(X_i)$ denotes the estimate of the conditional mean outcome for observation $i$ when applying bandwidth $h$, when leaving observation $i$ out, and using the remaining sample for estimation.

In the next step, we estimate the conditional mean $\mu_1(X_i)$ for observation $i$ using the optimal bandwidth $h_{opt}$, now including observation $i$. Formally,

$$\hat{\mu}_1(X_i) = \hat{\mu}_{1,h_{opt}}(X_i). \tag{4.19}$$

As $p_{opt}$, the optimal bandwidth $h_{opt}$ generally depends on the sample size. It tends to decrease as the sample size grows to reduce the bias related to relying on observations with covariate values too far from $x$. However, $h$ should decrease at a slower pace than the growth of the sample size. Otherwise, the variance would be too large relative to the bias, which would not be optimal in terms of minimizing the MSE of $\hat{\mu}_1(X_i)$; for instance, see, Wand and Jones (1994). We can apply an analogous cross-validation procedure to the estimation of $\mu_0(X)$.

It is important to note that the cross-validation procedures outlined in this chapter yield a choice of $p$ or $h$, which is optimal for the estimation of the conditional mean outcomes, but not necessarily for the estimation of the CATE, ATE, or ATET—that is, the (average) differences of conditional mean outcomes. Nevertheless, leave-one-out cross-validation constitutes a feasible approach for picking adequate values for $p$ or $h$ in practice, particularly as theoretically optimal rules for selecting the kernel bandwidth in ATE evaluation might perform poorly in moderate samples; for instance, see the discussion in Frölich (2005). However, it seems advisable to investigate the sensitivity of the effect estimates with regard to various values of $p$ or $h$, such as by taking the cross-validation-based choices as default values and multiplying and dividing them by a certain factor.

We may even combine kernel-based averaging and regression to obtain local regression estimators of $\mu_1(X)$ and $\mu_0(X)$. Local linear regression, for instance, consists of running a weighted linear regression of $Y_i$ on $X_i$ within treatment groups, where the weight of each observation corresponds to the kernel function $\mathcal{K}\left(\frac{X_i-x}{h}\right)$.

While conventional linear regression (or OLS) assigns the same importance to all observations, this approach gives more weight to observations whose $X_i$ is close to $x$ when running the regression. Considering the estimation of $\mu_1(x)$, for instance, this approach permits estimating regression coefficients that are specific to the covariate value $x$ at which the CATE is to be computed—that is, $(\hat{\alpha}_1(x), \hat{\beta}_{X_1,1}(x), \ldots)$, to predict $\hat{\mu}_1(x)$. Local linear regression generally has a smaller bias than local constant regression at the boundaries of the data—that is, close to the maximum or minimum covariate values observed for $X$; for instance, see the discussion in Frölich and Sperlich (2019).

It is worth mentioning that series regression and kernel-based estimates of the conditional mean outcomes $\mu_1(X)$ and $\mu_0(X)$, as well as the CATE, converge to the true values in the population at a slower pace than the fasted possible convergence rate of $\frac{1}{\sqrt{n}}$, as discussed in section 3.3. Considering kernel regression as described in equation (4.17), this occurs because the method (for a sufficiently small bandwidth) strongly depends on a subset of observations with covariate values close to the point of interest $x$, while giving less weight to (and thus making less use of) observations further away. This is in contrast to the linear regression model in equation (4.6), which equally exploits observations in the entire sample, and therefore converges at a rate of $\frac{1}{\sqrt{n}}$ (with $n$ being the sample size), but at the price of imposing linearity. Even under the slower convergence rate of series- and kernel-based methods for estimation of the conditional mean outcomes and the CATE, the estimation of the ATE and ATET can nevertheless be shown to be $\sqrt{n}$-consistent under specific conditions. The intuition for this result is that the ATE or ATET estimates are obtained by averaging over many CATEs with different values of $x$. This may average out the estimation errors in the CATEs occurring at specific values $x$ and thus entail the desirable property of $\sqrt{n}$-consistency.

Let us conclude this section by discussing a regression in R for assessing the National Supported Work (NSW) Demonstration. The latter is a training program in the US providing work experience for a period of up to 18 months to individuals with economic and social problems, with the aim to increase their labor market performance. Our data, which are a subset of the experimental sample in LaLonde (1986) and have previously been analysed in Dehejia and Wahba (1999), consist of 185 treated and 260 nontreated individuals who were assigned or not assigned to NSW, respectively. We load the packages *Matching* by Sekhon (2011), which contains the data set of interest; *Jmisc* by Chan (2014), which contains a function for demeaning variables; *lmtest*; and *sandwich* using the *library* command. We then use the *load* command to load the *lalonde* data with 445 observations and the *attach* command to store each variable into an own R object.

Calling *?lalonde* opens the help file, with a more detailed documentation about the variables. This concerns the treatment (i.e., training participation), the covariates like age, education, marital status, ethnicity, and previous labor market performance; and the outcome, namely real earnings (i.e., earnings adjusted for inflation over time) after treatment in 1978. Accordingly, we define the treatment, outcome, and covariates by *D=treat*, *Y=re78*, and *X=cbind(age,educ, nodegr,married,black,hisp,re74,re75,u74,u75)*, respectively, where *cbind* appends the covariates columnwise. We thus assume that after making the treated and nontreated groups comparable in terms of these covariates, they do not systematically differ in their potential earnings with and without training.

We estimate the ATE based on the regression formulation in equation (4.11), but without including higher-order terms of or interactions between the covariates. This approach requires computing interactions between the treatment and the demeaned covariates $(X_i - \bar{X})$. To this end, we apply the *demean* command to $X$ and store the interaction with $D$ in an R object named *DXdemeaned=D\*demean(X)*. We then run an OLS regression of $Y$ on $D$, $X$, and *DXdemeaned* using the *lm* command, and store the output in an object named *ols*. Finally, we use the *coeftest* command to investigate the results. The R code for the various steps is provided in the box here.

```
library(Matching)              # load Matching package
library(Jmisc)                 # load Jmisc package
library(lmtest)                # load lmtest package
library(sandwich)              # load sandwich package
data(lalonde)                  # load lalonde data
attach(lalonde)                # store all variables in own objects
?lalonde                       # call documentation for lalonde data
D=treat                        # define treatment (training)
Y=re78                         # define outcome
X=cbind(age,educ,nodegr,married,black,hisp,re74,re75,u74,u75) # covariates
DXdemeaned=D*demean(X)         # interaction of D and demeaned X
ols=lm(Y~D+X+DXdemeaned)       # run OLS regression
coeftest(ols, vcov=vcovHC)     # output
```

Running the code yields the following output:

|             | Estimate    | Std. Error  | t value  | Pr(>\|t\|) |    |
|-------------|-------------|-------------|----------|-----------|----|
| (Intercept) | 7.1612e+03  | 3.9760e+03  | 1.8011   | 0.072397  | .  |
| D           | 1.5835e+03  | 7.1117e+02  | 2.2266   | 0.026502  | *  |
| Xage        | 4.0710e+01  | 4.8020e+01  | 0.8478   | 0.397043  |    |
| Xeduc       | 8.2151e+01  | 2.1745e+02  | 0.3778   | 0.705771  |    |
| Xnodegr     | −1.6882e+02 | 1.1115e+03  | −0.1519  | 0.879346  |    |
| . . .       |             |             |          |           |    |

The result suggests that training participation increases average real earnings by $1.5835e+03 = 1583.5$ US dollars (USD) in 1978 when running our specification, and the ATE estimate is statistically significant at the 5 percent level, with a p-value of 0.0265 (or roughly 2.7 percent). We note that this regression specification allows interaction effects of the treatment and the covariates, and thus for some level of effect heterogeneity. Yet, it is not fully flexible, as it contains neither higher-order terms (in particular of the almost continuous variable age), nor interactions between covariates. Such terms could be added as further regressors to verify whether the ATE estimate is sensitive to an increased level of model flexibility, which would seem to argue against our current specification.

## 4.3  Covariate Matching

A further class of methods that we may apply in the context of a selection-on-observables framework are matching estimators, such as those considered in Rosenbaum and Rubin (1983b, 1985); Heckman, Ichimura, and Todd (1998); Heckman, Ichimura, Smith, and Todd (1998); Dehejia and Wahba (1999); and Lechner, Miquel, and Wunsch (2011). The idea of matching is to find and match treated and nontreated observations with similar (or, ideally, identical) covariate values to create a sample of treated and nontreated groups that are comparable in terms of covariate distributions, just as it would be the case in a successful experiment. The most basic form is pair or nearest-neighbor matching, implying that for a specific observation in one treatment group, the most similar observation in the other treatment group in terms of $X$ is selected to form a match.

Let us, for instance, consider finding for each treated observation with $D = 1$ the respective best matches among the nontreated observations with $D = 0$, as illustrated in figure 4.3. For the sake of simplicity, $X$ consists of only a single covariate, like age. Taking the mean difference in the outcomes of the treated and matched nontreated samples then yields an estimate of the ATET. Assuming that for each treated unit, there is a single nontreated unit that is most similar in terms of $X$, the ATET estimate is formally defined as

$$\hat{\Delta}_{D=1} = \frac{1}{n_1} \sum_{i:D_i=1} \{Y_i - \sum_{j:D_j=0} I\{||X_j - X_i|| = \min_{l:D_l=0} ||X_l - X_i||\}Y_j\}, \tag{4.20}$$

where $||X_j - X_i||$ is a measure yet to be defined for the discrepancy or distance between covariate vectors $X_j$ and $X_i$ and accordingly, $\min_{l:D_l=0} ||X_l - X_i||$ is the minimum distance among all nontreated observations compared to the treated $X_i$. Here, $n_1 = \sum_{i=1}^{n} D_i$ denotes the number of treated observations in the sample. In an analogous way, we can estimate the ATENT, denoted by $\Delta_{D=0} = E[Y(1) - Y(0)|D = 0]$, by finding the closest match among the treated observations for each of the nontreated observations as follows:

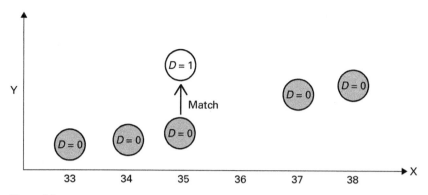

**Figure 4.3**
Pair matching.

$$\hat{\Delta}_{D=0} = \frac{1}{n_0} \sum_{i:D_i=0} \{ \sum_{j:D_j=1} I\{||X_j - X_i|| = \min_{l:D_l=1} ||X_l - X_i||\} Y_j - Y_i\}, \qquad (4.21)$$

where $n_0 = \sum_{i=1}^{n}(1 - D_i)$ is the number of nontreated observations.

The ATE can (by the law of total probability) be expressed as a weighted average of ATET and ATENT, with weights depending on the shares of treated and nontreated subjects in the population:

$$\Delta = \Pr(D=1) \cdot \Delta_{D=1} + \Pr(D=0) \cdot \Delta_{D=0}. \qquad (4.22)$$

This suggests estimating the ATE by taking a weighted average of equations (4.20) and (4.21) based on the shares of treated and nontreated observations in the sample:

$$\hat{\Delta} = \frac{n_1}{n} \cdot \hat{\Delta}_{D=1} + \frac{n_0}{n} \cdot \hat{\Delta}_{D=0}. \qquad (4.23)$$

The pair-matching algorithms described so far are examples of matching with replacement, implying that the same observation may serve multiple times as a match. In equation (4.20), for instance, a nontreated observation could be used as a match for several treated units, depending on how often it is most comparable in terms of X. In contrast, matching without replacement means that a nontreated unit serves as a match no more than once. This may reduce the variance by forcibly relying on a larger number of nontreated observations for matching rather than reusing the same nontreated observation several times. But it also may increase the bias because of not necessarily finding the most comparable match for each treated observation due to a previous use of that match. Our subsequent discussion will focus on matching with replacement. The latter may appear preferable in applications where the number of potential matches (in this example, the nontreated observations) is not substantially larger than the number of reference observations for which matches are to be found (in this example, the treated observations).

A crucial question for matching is how we should define the distance metric $||X_j - X_i||$—that is, based on which grounds we judge the covariate values of two observations $i$ and $j$ to be similar or dissimilar. An obvious (but naive) approach is to sum the squared differences in any of the covariates and select the observation $j$ with the smallest difference. This is the intuition underlying the Euclidean distance, which is defined as

$$||X_j - X_i||_{\text{Euclidean}} = \sqrt{\sum_{k=1}^{K} (X_{jk} - X_{ik})^2}, \tag{4.24}$$

where the subscript $k$ indexes a specific covariate and $K$ is the total number of covariates. The problem with this approach is that a specific difference (e.g., of 1) is considered equally important for each covariate $X_k$, independent of its distribution, particularly the range of values that the variable can take and how dispersed these values are. This implies that a difference of 1 in a binary variable like nationality (such as taking the value 1 for a native and 0 for a foreigner) contributes equally to the distance metric as a difference of 1 in a continuous variable like income (e.g., 5,000 versus 5,001 USD). Arguably, a difference of 1 in the binary variable for nationality is a qualitatively much bigger change than earning 1 USD more. For this reason, a standardized version of the Euclidean distance appears more appropriate, which takes into account the differences in covariate distributions by normalizing any covariate difference between observations $j$ and $i$ based on the inverse of the sample variance of the respective covariate:

$$||X_j - X_i||_{\text{Variance}} = \sqrt{\sum_{k=1}^{K} \frac{(X_{jk} - X_{ik})^2}{\widehat{Var}(X_k)}}. \tag{4.25}$$

In addition to the variance of $X_k$, one may normalize covariate differences by the inverse of $X_k$'s covariance with the remaining covariates in X. In this case, the difference $X_{j,k} - X_{i,k}$ gets a smaller importance (or less weight) if $X_k$ strongly correlates (i.e., shows a pattern that is strongly associated) with other covariates. Then, finding good matches in terms of the other covariates would generally imply a decent match in terms of $X_k$ too due to that correlation. Therefore, giving a smaller individual weight to differences in $X_k$ appears appropriate. For instance, it might be the case that someone's labor market experience strongly correlates with the level of education. If the values in $X_k$ are independent of (i.e., not associated with) those of other covariates, however, finding good matches in terms of the latter will generally not entail decent matching with respect to $X_k$. Then it makes sense to give a greater weight to $X_{j,k} - X_{i,k}$ to ensure a satisfactory match quality for this covariate too.

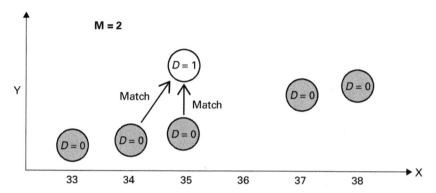

**Figure 4.4**
1:M matching.

The Mahalanobis distance metric incorporates inverse weighting by both the variance and covariance and is formally defined as follows:

$$\|X_j - X_i\|_{\text{Mahalanobis}} = \sqrt{\sum_{k=1}^{K}\sum_{l=1}^{K} \frac{(X_{jk} - X_{ik})(X_{jl} - X_{il})}{\widehat{Cov}(X_k, X_l)}}, \tag{4.26}$$

where for all cases in which $k = l$, it holds that $(X_{jk} - X_{ik})(X_{jl} - X_{il}) = (X_{jk} - X_{ik})^2$ and $\widehat{Cov}(X_k, X_l) = \widehat{Var}(X_k)$. Several other distance metrics have been proposed; see, for instance, the review in Zhao (2004), who also considers metrics that depend on how well a covariate $X_k$ predicts $D$, $Y$, or both to take the strength of confounding associated with that covariate into account. Furthermore, the genetic matching algorithm of Diamond and Sekhon (2013) uses a weighted distance metric in which the weights are such that predefined features (or moments) of the covariate distribution, like the covariate means, are as similar as possible across treated and nontreated matches.

A generalization of pair matching is one-to-many or 1:M matching, where $M$ is an integer larger than 1. It is based on matching to a treated reference observation $i$ several, namely $M$ nontreated observations that are closest in terms of $X$, in order to estimate $\mu_0(X_i)$ based on the average outcomes of these matches. Figure 4.4 provides an illustration of this approach for the simplistic case of a single covariate $X$ when $M = 2$. More formally, let $J(i)$ denote the set of $M$ closest nontreated observations matched to a treated reference observation $i$. Then, the 1:M matching estimator of the ATET corresponds to

$$\hat{\Delta}_{D=1} = \frac{1}{n_1} \sum_{i:D_i=1} [Y_i - \hat{\mu}_0(X_i)], \tag{4.27}$$

where

$$\hat{\mu}_0(X_i) = \frac{1}{M} \sum_{j \in J(i)} Y_j. \tag{4.28}$$

One can show that this ATET estimator may be equivalently expressed as

$$\hat{\Delta}_{D=1} = \frac{1}{n_1} \sum_{i=1}^{n} \left[ D_i - \frac{W_i}{M} \right] \cdot Y_i, \tag{4.29}$$

with $W_i$ denoting the number of times that a nontreated observation is matched to any treated observation. Similar to the choice of $p$ or $h$ for series estimation or kernel regression, the optimal choice of $M$ depends on the data, such as the availability of similar comparison observations, and the sample size. $M$ should increase as the sample size grows to reduce the bias, but at a slower pace than the growth of the sample size, which also reduces the variance. In many empirical studies, however, $M$ is chosen in an ad hoc manner, without using methods aiming at an optimal selection, even though cross-validation in principle could be applied for determining $M$. It appears advisable to investigate the sensitivity of the effect estimates with regard to various choices of $M$.

A further matching variant called *radius* or *caliper matching* does not fix the number of matches, but it does define a maximum admissible level of dissimilarity in $X$ between matched treated and nontreated observations. This level, for instance, may be based on thresholds for the distance metrics in equation (4.25) or (4.26), denoted by $\mathcal{B}$. For any treated reference observation $i$, we estimate $\mu_0(X_i)$ as the average of all nontreated observations whose distance metric $||X_j - X_i||$ is less than or equal to a positive number $\mathcal{B}$:

$$\hat{\mu}_0(X_i) = \frac{\sum_{j:D_j=0} I\{||X_j - X_i|| \leq \mathcal{B}\} \cdot Y_i}{\sum_{j:D_j=0} I\{||X_j - X_i|| \leq \mathcal{B}\}}. \tag{4.30}$$

The more similar potential matches are available in the data, the more comparison observations are actually matched. In contrast to 1:$M$ matching, radius matching does not determine the number of matches a priori, but it does make this choice data-dependent. This may decrease the variance if many similar comparison observations exist, without the large costs in terms of bias that would be induced by using noncomparable matches.

As a modification of this idea, we can also use a kernel function to make weights dependent on the magnitude of the distance metric:

$$\hat{\mu}_0(X_i) = \frac{\sum_{j:D_j=0} \mathcal{K}\left( \frac{||X_j - X_i||}{\mathcal{B}} \right) \cdot Y_i}{\sum_{j:D_j=0} \mathcal{K}\left( \frac{||X_j - X_i||}{\mathcal{B}} \right)}. \tag{4.31}$$

Radius matching is conceptually closely related to the kernel approach outlined in equation (4.17), which also permits computing $\hat{\mu}_0(X_i)$ in order to estimate, for instance, the ATET based on equation (4.27), an approach known as *kernel matching*. One difference between radius matching based on equation (4.30) or (4.31) and kernel matching based on equation (4.17), however, is that $\mathcal{B}$ is a scalar (i.e., a single number) that relates to an aggregated distance metric for all covariates, while $h$ in equation (4.17) is a vector containing kernel bandwidths for each covariate. For the same reason, $\mathcal{K}$ in equation (4.17) is typically a multiplicative (or product) kernel function tailored to a vector of covariates.

It is worth noting that the various matching estimators can differ in their large sample behavior. Abadie and Imbens (2006) show that pair or 1:$M$ matching do not necessarily converge with a rate of $\frac{1}{\sqrt{n}}$ to the true effect—that is, they are not $\sqrt{n}$-consistent—if $X$ contains more than one continuous element (such as past earnings). This is due to the use of a fixed number of matches, which does not optimally trade off the bias and variance of the estimator. In contrast, kernel matching as discussed in Heckman, Ichimura, and Todd (1998) may attain $\sqrt{n}$-consistency if the bandwidth $h$ is appropriately adapted to the sample size. Even in the case of $\sqrt{n}$-consistency, pair or 1:$M$ matching tends to have a higher asymptotic variance than the most precise treatment effect estimators, which attain the lowest possible variance under the same selection-on-observables assumptions; see the discussion in Hahn (1998).

Furthermore, Abadie and Imbens (2008) demonstrate that bootstrapping approaches as introduced in section 3.4 are inconsistent for estimating the standard error (e.g., by replacing $\hat{\beta}^b$ in equation (3.48) with bootstrap-sample-specific ATET estimates $\hat{\Delta}^b_{D=1}$ based on matching). This is due to the discontinuity of weights in pair and 1:$M$ matching, implying that either one or $M$ matches get a positive weight when computing $\hat{\mu}_0(X_i)$, while the weight (or importance) of any other observation in the sample is zero. Abadie and Imbens (2006), however, provide a consistent asymptotic approximation of the estimator's variance, which corresponds to the following expression for the ATET:

$$Var(\hat{\Delta}_{D=1}) = \frac{1}{n_1} \left\{ E[(\Delta_{X_i} - \Delta_{D=1})^2 | D_i = 1] \right.$$

$$\left. + E\left[ \frac{1}{n_1} \sum_{i=1}^{n} \left[ D_i - (1 - D_i) \cdot \frac{W_i}{M} \right]^2 \cdot \sigma^2(D_i, X_i) \right] \right\}, \tag{4.32}$$

where $\sigma^2(D_i, X_i) = Var(Y | D = D_i, X = X_i)$ is the conditional variance of the outcome, given the treatment and the covariates.

The variance formula in equation (4.32) may be estimated by

$$
\widehat{Var}(\hat{\Delta}_{D=1}) = \frac{1}{n_1} \left\{ \frac{1}{n_1} \sum_{i=1}^{n} D_i \cdot [Y_i - \hat{\mu}_0(X_i) - \hat{\Delta}_{D=1}]^2 + \frac{1}{n_1} \sum_{i=1}^{n} (1 - D_i) \right.
$$
$$
\left. \cdot \left[ \frac{W_i \cdot (W_i - 1)}{M^2} \right] \cdot \hat{\sigma}^2(D_i, X_i) \right\},
\tag{4.33}
$$

where we may obtain the conditional variance estimate $\hat{\sigma}^2(D_i, X_i)$ based on matching within the nontreated group (rather than across treated and nontreated groups, as for effect estimation). Let us to this end denote by $\mathcal{J}_i$ the set of $M$ nontreated observations that are closest to some nontreated reference observation $i$. Then a matching-based, within-group conditional variance estimator is given by

$$
\hat{\sigma}^2(D_i, X_i) = \frac{M}{M+1} \cdot \left( Y_i - \frac{1}{M} \sum_{j \in \mathcal{J}(i)} Y_j \right)^2.
\tag{4.34}
$$

Even though this conditional variance estimate is inconsistent (due to a fixed number of $M$), averaging this variance estimator over the $n_1$ treated observations in equation (4.33) averages out estimation errors and implies that equation (4.33) is a consistent variance estimator of $\hat{\Delta}_{D=1}$.

To improve the properties of pair or 1:$M$ matching, we may combine the estimators with a regression-based correction for the bias that comes from not finding fully comparable matches for a reference observation, as discussed in Rubin (1979) and Abadie and Imbens (2011). Taking the ATET estimator such as in equation (4.27), estimation bias is rooted in the fact that nontreated observations entering $J(i)$ do typically not have exactly the same $X$ values as observation $i$. However, let us assume that we can well approximate $\mu_0(X)$ by a regression of $Y$ on $X$ among the nontreated observations. In this case, we may correct for the bias due to $X_j - X_i \neq 0$ by the difference in the regression-based estimates for $\mu_0(X_j) - \mu_0(X_i)$, with $i$ denoting the treated reference observation and $j$ denoting a nontreated observation in $J(i)$. This suggests modifying equation (4.28) to

$$
\hat{\mu}_0(X_i) = \frac{1}{M} \sum_{j \in J(i)} [Y_j - (\tilde{\mu}_0(X_j) - \tilde{\mu}_0(X_i))],
\tag{4.35}
$$

with $\tilde{\mu}_0(X)$ being an estimate obtained from regressing $Y$ on $X$ among the nontreated.

The bias correction removes the bias without affecting the asymptotic variance, and under specific conditions entails a $\sqrt{n}$-consistent and asymptotically normal ATET estimator. This approach may nonnegligibly reduce the bias even if the regression model for $Y$ given $X$ and $D = 0$ is somewhat misspecified. The price is that the bias correction increases the variance of ATET estimation in small samples due to estimating $\tilde{\mu}_0(X)$. We note that in contrast to pair or 1:$M$ matching without bias correction, kernel matching or radius matching, based on nondiscontinuous, smooth kernel weights as in equations (4.31) and (4.17), are less problematic in the context of the application of bootstrap-based inference for computing standard errors and confidence intervals. Also in this case, however, bias correction may be beneficial in terms of bias reduction.

As an empirical illustration, let us implement pair matching in R by reconsidering the NSW data set already analyzed at the end of section 4.2. To this end, we load the *Matching* package using the *library* command and follow the exact same steps as before for defining training participation as treatment $D$, earnings outcome $Y$, and covariates $X$. We then run the *Match* command for pair matching, in which the first argument corresponds to the outcome, the second to the treatment, and the third to the covariates, and save the output in a variable named *pairmatching*. Finally, we wrap the latter using the *summary* command to investigate the results. The box here provides the R code for each of the steps.

```
library(Matching)                                   # load Matching package
data(lalonde)                                       # load lalonde data
attach(lalonde)                                     # store all variables in own objects
D=treat                                             # define treatment (training)
Y=re78                                              # define outcome (earnings in 1978)
X=cbind(age,educ,nodegr,married,black,hisp,re74,re75,u74,u75) # covariates
pairmatching=Match(Y=Y, Tr=D, X=X)                  # pair matching
summary(pairmatching)                               # matching output
```

Running the R code yields the following output:

```
Estimate...  1686.1
AI SE......  866.4
T-stat.....  1.9461
p.val......  0.051642
```

By default, the procedure provides an estimate (*Estimate*) of the ATET—that is, the average effect of the training among participants, which amounts to roughly 1,686 USD. By setting the *estimand* argument to *"ATE,"* such that the command becomes *Match(Y=Y, Tr=D, X=X, estimand = "ATE"),* we would instead obtain an estimate of the ATE. The second line of the output, *AI SE*, yields the standard error, which is estimated based on the formula in equation (4.33) and matching within the non-treated group as outlined in equation (4.34). The third line *T-stat* yields the t-statistic, and the fourth line *p.val*, the p-value. The ATET is statistically significantly different to zero at the 10 percent level but not at the 5 percent level, as the p-value of 0.0516 slightly exceeds 5 percent.

The *Match* command contains a range of options for matching estimation. For instance, setting the argument *BiasAdjust = TRUE* performs a bias correction as discussed in equation (4.35). Furthermore, setting *M=3* (rather than using the default value of 1) performs 1:*M* matching with three nearest neighbors. To see how these choices affect the estimated ATET, we rerun the procedure with the following settings:

```
matching=Match(Y=Y, Tr=D, X=X, M=3, BiasAdjust = TRUE)
summary(matching)
```

The ATET estimate is now moderately lower than before and again statistically significant at the 10 percent level, such that our findings remain quite robust when modifying the options for matching estimation:

```
Estimate...  1535
AI SE......  792.09
T-stat.....  1.9379
p.val......  0.052632
```

## 4.4  Propensity Score Matching

A caveat of covariate matching and kernel or series regression that aim at controlling for X nonparametrically—that is, in a flexible way without imposing parametric assumptions like linearity as in equation (4.8)—is the curse of dimensionality: As the number of covariates in X (i.e., the dimension of X) and the number of possible values of the covariates (i.e., the support of X) grows, the probability of finding good matches with similar values in all variables entering X decays rapidly in finite samples. Instead of directly controlling for X, an alternative approach is to control for the conditional treatment probability given the covariates, denoted by $p(X) = \Pr(D = 1|X)$ and commonly referred to as the *propensity score*. This follows from the propensity score's balancing property demonstrated in Rosenbaum and Rubin (1983b): Conditioning on $p(X)$ equalizes, or balances, the distribution of X across treatment groups such that the covariates are independent of the treatment conditional on the propensity score. Formally, $X \perp D|p(X)$. Figure 4.5 provides a graphical intuition for this result. As the propensity score is a function of X through which any effect of X on D operates, controlling for $p(X)$ blocks any impact of X on D and thus, any confounding by X, which no longer jointly affects D and Y.

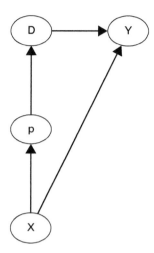

**Figure 4.5**
A causal graph including the propensity score (denoted by $p$).

For this reason, we can also identify the ATE and ATET when controlling for the propensity score $p(X)$ rather than $X$:

$$\Delta = E[\mu_1(p(X)) - \mu_0(p(X))], \tag{4.36}$$

$$\Delta_{D=1} = E[\mu_1(p(X)) - \mu_0(p(X))|D=1] = E[Y|D=1] - E[\mu_0(p(X))|D=1].$$

This suggests that we may estimate the treatment effects when substituting the covariates $X$ by an estimate of $p(X)$ in any of the matching and regression approaches previously discussed in sections 4.2 and 4.3. In practice, propensity score matching appears to be more popular than directly matching on covariates, as it implies matching on a single variable—namely, an estimate of $p(X)$—rather than a possibly high-dimensional $X$: that is, a large number of variables. Relatedly, propensity score matching does not require a distance metric because we need not aggregate the distances in several covariates. In pair matching, for instance, we simply match to a treated reference observation the nontreated subject with the most similar estimated propensity score, denoted by $\hat{p}(X)$, as illustrated in figure 4.6.

At first glance, an advantage of propensity score matching appears to be avoidance of the curse of dimensionality, as it is easier to match observations that are similar in $p(X)$ alone than in all elements of $X$. In fact, different combinations of values in $X$ could still yield similar propensity scores. In empirical applications, however, the true propensity score is typically unknown and needs to be estimated in the data prior to matching. Here, the curse of dimensionality kicks back in with regard to the estimation of $p(X)$ if we apply nonparametric methods like series or

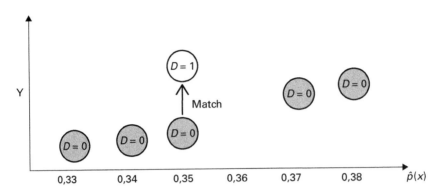

**Figure 4.6**
Propensity score matching.

kernel regression. In practice, however, $p(X)$ is frequently estimated by means of a parametric binary choice model, which circumvents the curse of dimensionality by imposing specific assumptions concerning the association of $X$ and $D$, rather than fully flexibly estimating it:

$$\Pr(D=1|X)=p(X)=\Lambda(\alpha_0+\alpha_{X_1}X_1+\cdots+\alpha_{X_K}X_K). \tag{4.37}$$

The model in equation (4.37) assumes that we may model $D$ based on a combination of a linear index $\alpha_0+\alpha_{X_1}X_1+\cdots+\alpha_{X_K}X_K$ and a nonlinear link function $\Lambda$ of that index. This implies that larger values of the index, due to a higher level of the covariates education or work experience, entail a larger probability (or propensity) to receive a treatment such as training. However, how much the propensity score changes as a reaction to a shift in the index depends on the initial value of the index.

Typically, $\Lambda$ is either a normal or logistic distribution function, implying a probit or logit model, respectively, for the propensity score. For either model, $p(X)$ is strictly between 0 and 1 (or 0 percent and 100 percent), as theoretically required for a probability. Furthermore, for extremely positive or negative indexes, a small change in the index only marginally affects the propensity score, while for intermediate levels of the index, a small change has a larger (and close to linear) impact on the propensity score. Intuitively, for someone with a large index that has a training probability of close to 1, an increase in a covariate like education can only marginally affect the already very high training probability, which cannot go beyond 1. In contrast, the impact of an increase in education is higher at intermediate index values and treatment probabilities.

It is important to note that the choice of $\Lambda$ directly restricts the distribution of unobserved characteristics affecting the treatment decision. Imposing the probit or logit model, unobservables are assumed to follow the normal or logistic distribution, respectively. This implies that propensity score estimation may be inconsistent when the treatment decision cannot be characterized by a linear index of the observed covariates $X$, if the unobserved characteristics do not follow a normal/logistic distribution, or both. The avoidance of the curse of dimensionality, therefore, comes with the cost of a smaller flexibility of the estimator due to stronger model assumptions when compared to directly matching on $X$ or using a nonparametric propensity score estimate. As a compromise, somewhat more flexible, semiparametric methods exist that maintain the linear index $\alpha_0 + \alpha_{X_1} X_1 + \cdots + \alpha_{X_K} X_K$, but put no distributional assumption on $\Lambda$; for instance, see the methods proposed by Klein and Spady (1993) or Ichimura (1993).

Parametric binary choice specifications like probit and logit models are conventionally estimated by maximum likelihood estimation. The latter determines the coefficient estimates $\hat{\alpha}_0, \ldots, \hat{\alpha}_{X_K}$ by solving the following maximization problem based on the log likelihood function:

$$\hat{\alpha}_0, \ldots, \hat{\alpha}_{X_K} = \arg \max_{\alpha_0^*, \ldots, \alpha_{X_K}^*} \sum_{i=1}^{n} D_i \ln(\Lambda(\alpha_0^* + \cdots + \alpha_{X_K}^* X_{iK}))$$

$$+ (1 - D_i) \ln(1 - \Lambda(\alpha_0^* + \cdots + \alpha_{X_K}^* X_{iK})), \tag{4.38}$$

with $\alpha_0^*, \ldots, \alpha_{X_K}^*$ denoting candidate values for the coefficients. Intuitively, maximum likelihood estimation aims at finding the coefficient values that maximize the joint probabilities (i.e., the ensemble of individual propensity scores) to obtain the treatment states actually observed in the sample. We then compute the propensity score estimate $\hat{p}(X)$ based on the following prediction:

$$\hat{p}(X) = \Lambda(\hat{\alpha}_0 + \hat{\alpha}_{X_1} X_1 + \cdots + \hat{\alpha}_{X_K} X_K). \tag{4.39}$$

Matching on the estimated propensity score has a different variance than matching directly on $X$, which in the case of the ATET can be either higher or lower, as pointed out in Heckman, Ichimura, and Todd (1998). Abadie and Imbens (2016) provide an asymptotic variance approximation for 1:$M$ propensity score matching, which appropriately accounts for uncertainty due to propensity score estimation. In the case of ATET estimation, it consists of the variance formula provided in equation (4.32) of section 4.3, with $X$ being replaced by $p(X)$, plus a correction term for propensity score estimation that involves the conditional covariance $Cov(X, Y|D, p(X))$. We may estimate the latter covariance by matching within treatment groups in analogy to the conditional variance estimator in equation (4.34). Ignoring the correction term and implicitly assuming that the propensity score is known generally entails a bias in the variance estimation of propensity score matching for the evaluation of the ATET. For the case of ATE estimation, it turns out that the bias is never negative, such that ignoring the correction generally overestimates the true variance (at least in large samples). An alternative inference method that takes the uncertainty due to propensity score estimation into account is to estimate the variance based on bootstrapping (see the discussion in section 3.4). In this case, we are required to reestimate both the propensity score and the ATET in each bootstrap sample to appropriately account for the variance related to either estimation step.

Let us reconsider the NSW data already analyzed at the end of sections 4.2 and 4.3 for demonstrating the use of propensity matching in R. We follow the same steps as before for loading the *Matching* package and the *lalonde* data, as well as defining the treatment $D$, outcome $Y$, and covariates $X$. However, an important difference is that we now estimate a propensity score model. To this end, we apply the *glm* command, whose first argument is (similar to the *lm* command) the regression formula (namely, $D\sim X$), as we estimate the treatment probability as a function of the covariates. A further argument is *family*, which defines the kind of nonlinear regression to be executed. Importantly, setting *family=binomial* assumes a logit model, while *family=binomial(probit)* imposes a probit model.

Furthermore, as we are predominantly interested in the predicted propensity scores rather than the coefficient estimates, we may directly append to the *glm(D∼X,family=binomial)* command the extension *$fitted*. The latter only retrieves the fitted values—that is, the estimated propensity scores—from the logit regression. We store the propensity score estimates in a variable named *ps* and use it as the sole covariate in the *Match* command to perform propensity score matching and switch on the bias correction: *Match(Y=Y, Tr=D, X=ps, BiasAdjust = TRUE)*. We save the output in a variable named *psmatching* and wrap it by the *summary* command to investigate the results. The box here provides the R code for each of the steps.

```
library(Matching)                              # load Matching package
data(lalonde)                                  # load lalonde data
attach(lalonde)                                # store all variables in own objects
D=treat                                        # define treatment (training)
Y=re78                                         # define outcome
X=cbind(age,educ,nodegr,married,black,hisp,re74,re75,u74,u75) # covariates
ps=glm(D~X,family=binomial)$fitted             # estimate the propensity score by logit
psmatching=Match(Y=Y, Tr=D, X=ps, BiasAdjust = TRUE) # propensity score matching
summary(psmatching)                            # matching output
```

Running the R code yields the following output:

```
Estimate...   2138.4
AI SE......   797.77
T-stat.....   2.6805
p.val......   0.0073518
```

The estimated ATET (*Estimate*) suggests an average increase in annual earnings of roughly 2,138 USD among training participants and is highly statistically significant, as the p-value (*p.val*) of 0.007 is very low. However, the standard error, t-statistic, and p-value are based on equation (4.33), and for this reason, they do not take into account the estimation of the propensity score. To improve this situation, we implement a bootstrap approach for inference, in analogy to the empirical example at the end of section 3.4. We therefore load the *boot* library and define a bootstrap function named *bs*, in which we draw the bootstrap data, estimate the propensity score using the *glm* command, and estimate the ATET using the *Match* command. By appending *$est* to the latter, we exclusively retrieve the ATET estimate, which we store in an R object named *effect* and return by our *bs* function. We then use the *data.frame* command to append Y, D, and X into a data matrix to be used for bootstrapping. We set a seed (*set.seed(1)*) for the replicability of our results, run the bootstrap procedure 999 times, store the output in an R object named *results*, and call the latter:

```
library(Matching)                              # load Matching package
library(boot)                                  # load boot package
data(lalonde)                                  # load lalonde data
attach(lalonde)                                # store all variables in own objects
D=treat                                        # define treatment (training)
Y=re78                                         # define outcome
X=cbind(age,educ,nodegr,married,black,hisp,re74,re75,u74,u75) # covariates
bs=function(data, indices) {                   # defines function bs for bootstrapping
  dat=data[indices,]                           # bootstrap sample according to indices
  ps=glm(dat[,2:ncol(dat)],data=dat,family=binomial)$fitted # propensity score
  effect=Match(Y=dat[,1], Tr=dat[,2], X=ps, BiasAdjust = TRUE)$est # ATET
  return(effect)                               # returns the estimated ATET
}                                              # closes the function bs
bootdata=data.frame(Y,D,X)                     # data frame for bootstrap procedure
set.seed(1)                                    # set seed
results = boot(data=bootdata, statistic=bs, R=999) # 999 bootstrap estimations
results                                        # displays the results
```

Running the R code yields the following output:

```
Bootstrap Statistics :
     original      bias    std. error
t1* 2138.396  −302.1217     885.6012
```

The bootstrap-based standard error (*std. error*) of roughly 886 USD, which accounts for propensity score estimation, is somewhat higher than the previous standard error that relies on equation (4.33) and ignores propensity score estimation.

To assess statistical significance using the bootstrap-based standard error, we compute the t-statistic and p-value (similar to the empirical example at the end of section 3.4):

```
tstat=results$t0/sd(results$t)        # compute the t-statistic
2*pnorm(-abs(tstat))                  # compute the p-value
```

This yields the following result:

```
              [,1]
[1,]  0.01575134
```

We find the ATET estimate to be statistically significant at the 5 percent level, as the p-value amounts to roughly 1.58 percent.

## 4.5   Inverse Probability Weighting, Empirical Likelihood, and Entropy Balancing

Matching and regression on the propensity score are not the only approaches that use $p(X)$ for treatment evaluation. We can alternatively assess treatment effects by inverse probability weighting (IPW) going back to Horvitz and Thompson (1952), which weights observations by the inverse of the propensity score. Intuitively, observations with propensity score values that are underrepresented or overrepresented in their treatment groups relative to some target population, like the total population if the ATE is of interest, are given more or less weight, respectively. After this weighting step, treated and nontreated groups are comparable in terms of the distribution of covariates $X$, such that we can properly assess treatment effects under the selection-on-observables assumptions in expression (4.1).

More formally, IPW identifies the ATE in the population due to the following relation:

$$\Delta = E[\mu_1(X) - \mu_0(X)] = E\left[\frac{E[Y \cdot D|X] \cdot D}{p(X)} - \frac{E[Y \cdot (1-D)|X] \cdot (1-D)}{1 - p(X)}\right]$$

$$= E\left[\frac{Y \cdot D}{p(X)} - \frac{Y \cdot (1-D)}{1 - p(X)}\right], \tag{4.40}$$

where the second equality follows from probability theory, implying that $\mu_1(X) = E[Y \cdot D|X]/p(X)$ and $\mu_0(X) = E[Y \cdot (1-D)|X]/(1 - p(X))$, and the third from the law of iterated expectations. Similarly, the ATET can be shown to be identified by

$$\Delta_{D=1} = E\left[\frac{Y \cdot D}{\Pr(D=1)} - \frac{Y \cdot (1-D) \cdot p(X)}{(1 - p(X)) \cdot \Pr(D=1)}\right]. \tag{4.41}$$

We may thus estimate treatment effects based on the sample versions (or analogs) of IPW equations (4.40) and (4.41). Hirano, Imbens, and Ridder (2003) consider this approach when using nonparametric series regression for estimating the propensity score and show that it can attain the lowest possible variance in large samples among all estimators relying on the selection-on-observables assumptions. This variance property is known as *semiparametric efficiency*. As an alternative approach, Ichimura and Linton (2005) and Li, Racine, and Wooldridge (2009) discuss IPW with kernel-based propensity score estimation. Practitioners, however, mostly rely on logit or probit specifications, which is generally not semiparametrically efficient; see Chen, Hong, and Tarozzi (2008). In any case, it is common and recommended to use normalized sample versions (or sample analogs) of the expressions in equations (4.40) and (4.41), which ensure that the weights of observations within treatment groups add up to 1. Particularly in smaller samples, this typically entails more accurate effect estimates; for instance, see the findings in Busso, DiNardo, and McCrary (2014). For the ATE, such a normalized estimator takes the form

$$\hat{\Delta} = \sum_{i=1}^{n} \frac{Y_i \cdot D_i}{\hat{p}(X_i)} \bigg/ \sum_{i=1}^{n} \frac{D_i}{\hat{p}(X_i)} - \sum_{i=1}^{n} \frac{Y_i \cdot (1 - D_i)}{1 - \hat{p}(X_i)} \bigg/ \sum_{i=1}^{n} \frac{1 - D_i}{1 - \hat{p}(X_i)}, \tag{4.42}$$

where $\sum_{i=1}^{n} D_i/\hat{p}(X_i)$ and $\sum_{i=1}^{n}(1 - D_i)/(1 - \hat{p}(X_i))$ normalize the weights such that they add up to 1 within the treatment groups.

Compared to matching, IPW has the advantages that it is computationally inexpensive (i.e., does not take a lot of computer time) and does not require choosing tuning parameters (other than for nonparametric propensity score estimation, if applied) such as the number of matches. On the negative side, IPW estimates might be more sensitive to errors in propensity scores that are very close to 1 or 0, which may drive up their variance, especially in small samples, as discussed in Frölich (2004) and Khan and Tamer (2010). Furthermore, IPW may be less robust (that is, more prone to estimation errors) when using an incorrect model for the propensity score than matching, which merely uses the score to match treated and nontreated observations; for instance, see the evidence provided in Waernbaum (2012).

A specific variant of IPW are empirical likelihood (EL) methods such as those proposed in Graham, Pinto, and Egel (2012) and Imai and Ratkovic (2014). These methods are based on modifying an initial propensity score estimate, such as by changing the coefficients of a logit model, in an iterative way until predefined features (or moments) of X like the covariate means are maximally balanced, that is, as similar as possible across treatment groups. We have encountered this balancing idea already in the context of genetic matching in section 4.3. EL methods are based on the intuition that after weighting by the inverse of the true propensity score $p(X)$ in the population, the covariates $X$, and thus any of their moments, are balanced across treated and nontreated groups. For the ATET, this implies

$$E\left[\frac{\tilde{X} \cdot D}{\Pr(D=1)} - \frac{\tilde{X} \cdot (1-D) \cdot p(X)}{(1-p(X)) \cdot \Pr(D=1)}\right] = 0$$

$$\Leftrightarrow E\left[\tilde{X} \cdot D - \frac{\tilde{X} \cdot (1-D) \cdot p(X)}{1-p(X)}\right] = 0, \tag{4.43}$$

where $\tilde{X}$ denotes a function of $X$. For instance, for $\tilde{X} = X$ or $\tilde{X} = [X - E[X]]^2$, equation (4.43) implies that the mean or variance of $X$, respectively, are balanced across treatment groups after reweighting.

EL methods aim at enforcing the condition in equation (4.43), which refers to the population, to hold in the sample:

$$\frac{1}{n}\sum_{i=1}^{n}\left[\tilde{X}_i \cdot D_i - \frac{\tilde{X}_i \cdot (1-D_i) \cdot \tilde{p}(X_i)}{1-\tilde{p}(X_i)}\right] = 0, \tag{4.44}$$

where $\tilde{p}(X_i)$ is an adjusted version (through tweaking the coefficients) of the initial propensity score estimate $\hat{p}(X_i)$, which may not fully balance $\tilde{X}$. Using $\tilde{p}(X_i)$ instead of $\hat{p}(X_i)$ in IPW thus guarantees that $\tilde{X}$ is perfectly balanced across treatment groups in the sample. By applying this approach, we may avoid manually searching for propensity score specifications (e.g., by including interaction or higher-order terms of specific covariates) that entail decent balancing. Besides IPW, we could use $\tilde{p}(X_i)$ in the context of other estimators like propensity score matching. An approach related to EL methods is entropy balancing (EB), as proposed in Hainmueller (2012). The method iteratively modifies an initial (e.g., uniform) default weights until the predefined balance criterion with regard to $X$ is maximized under the constraint that the weights must sum to 1 (and be nonnegative) in either treatment group. In contrast to the previously mentioned EL approaches, EB does not necessarily require an initial estimate of the propensity score for finding the final weights.

EL and EB approaches aim at perfect covariate balance across treatment groups to make treated and nontreated observations fully comparable, which avoids biases due to dissimilarities in $X$ across treatment groups. As typically nothing is for free in statistics, however, this may come at the cost of a higher variance, which also contributes to the overall error of the estimator; see the discussion of the MSE in equation (3.34) in chapter 3. In contrast to striving for a perfect covariate balance in $\tilde{X}$, we may also trade off balance (or bias) and variance in estimation when defining the weights. In this spirit, Zubizarreta (2015) suggests a method for computing the weights entailing the smallest estimation variance under the condition that covariate differences across treatment groups do not exceed a specific level, implying that covariate balance is approximately (rather than exactly) satisfied.

Let us apply IPW in R to investigate the effect of mothers' smoking behavior during pregnancy on the birth weight of newborn children. To this end, we consider a data set of 189 mothers and their newborns collected in 1986 at a medical center in the US and previously analyzed in Hosmer and Lemeshow (2000). First, we load the *causalweight* package (which contains IPW estimators) and the *COUNT* package by Hilbe (2016) (which contains the data of interest) using the *library* command. Second, we load the *lbw* data into the R workspace and store all variables in separate R objects using the *data* and *attach* commands, respectively. We define *smoke*, a binary treatment indicator for whether a mother smoked during pregnancy, as treatment $D$, and *bwt*, a child's birth weight in grams, as outcome $Y$. As smoking behavior is most likely not random, we control for several covariates that may jointly affect $Y$ and $D$, which are attached columnwise using the *cbind* command and stored in the R object $X$. The latter includes a binary indicator for nonwhite or white ethnicity, *race==1*, as well as the variables *age, lwt, ptl, ht, ui,* and *ftv*, which provide health-relevant information on mother's age, weight, and number of physician visits, among other characteristics.

For replicability purposes, we set a seed of 1 using *set.seed(1)*. We then estimate the ATE by normalized IPW as outlined in equation (4.42) using the *treatweight* command, whose first, second, and third arguments are the outcome, treatment, and covariates, respectively. Furthermore, we set the argument *boot=999* to compute the standard error of the ATE estimate based on 999 bootstrap samples. We note that the *treatweight* command estimates the propensity scores by means of a probit model unless we change it to a logit model by using the argument *logit=TRUE*. We store the estimation output in an R object named *ipw*. The box here presents the R code of each of the steps.

```
library(causalweight)                          # load causalweight package
library(COUNT)                                 # load COUNT package
data(lbw)                                       # load lbw data
attach(lbw)                                     # store all variables in own objects
D=smoke                                         # define treatment (mother smoking)
Y=bwt                                           # outcome (birthweight in grams)
X=cbind(race==1, age, lwt, ptl, ht, ui, ftv)   # covariates
set.seed(1)                                     # set seed
ipw=treatweight(y=Y,d=D,x=X, boot=999)         # run IPW with 999 bootstraps
```

After running the code, we can access any output stored in *ipw* using the $ sign, such as the ATE *ipw$effect*, standard error *ipw$se*, and p-value *ipw$pval*:

```
ipw$effect                           # show ATE
[1] -339.7966
ipw$se                               # show standard error
[1] 115.7358
ipw$pval                             # show p-value
[1] 0.003325092
```

Our results suggest that smoking during pregnancy on average reduces the birth weight of newborns by almost 340 grams. The effect is highly statistically significant, as the p-value (i.e., the error probability of incorrectly rejecting zero ATE), amounts to only 0.003 (or 0.3 percent).

In a next step, we apply the EL approach suggested in Imai and Ratkovic (2014) as an alternative method to estimate the ATE of smoking on birth weight. To this end, we load the *CBPS* package, as well as the previously seen *lmtest* and *sandwich* packages. We then apply the *CBPS* command. It contains as the first argument the regression specification $D{\sim}X$ for estimating the covariate-balancing weights based on iteratively modified propensity scores when considering the covariate means as the balancing criterion in equation (4.44). Furthermore, we set the argument *ATT = 0* to estimate the ATE rather than the ATET. We store the output in an R object named *cbps* and can address the *weights* using the $ sign. In the next step, we use the *lm* command to run a weighted regression of $Y$ on $D$, in which we weight observations by the previously estimated EL weights based on including the argument *weights=cbps$weights*. Such a weighted regression is in fact equivalent to IPW using EL-based propensity scores. We store the output in an R object called *results* and wrap it by the *coeftest* command to inspect the ATE estimate. The box here presents the R code of each of the steps.

```
library(CBPS)                        # load CBPS package
library(lmtest)                      # load lmtest package
library(sandwich)                    # load sandwich package
cbps=CBPS(D~X, ATT = 0)              # covariate balancing for ATE estimation
results=lm(Y~D, weights=cbps$weights) # weighted regression
coeftest(results, vcov = vcovHC)     # show results
```

Running the code yields the following output:

```
          Estimate  Std. Error  t value   Pr(>|t|)
(Intercept) 3096.624   71.987   43.0167  < 2.2e−16  ***
D           −332.043  110.245  −3.0119   0.002956   **
──────
Signif. codes:  0 '***' 0.001 '**' 0.01 '*' 0.05 '.' 0.1 ' ' 1
```

The EL-based ATE estimate amounts to an average reduction in the birth weight of roughly 332 grams, which is similar to the IPW-based result. Furthermore, the estimated effect is highly statistically significant, with a p-value of just 0.3 percent. As a word of caution, however, we need to point out that this p-value does not take any uncertainty into account that comes from the estimation of the modified propensity scores and the related weights. This could be improved, however, by running a bootstrap procedure in which we reestimate both the EL-based weights and the ATE in each bootstrap sample.

## 4.6  Doubly Robust Methods

While equations (4.3) and (4.4) demonstrate the identification of treatment effects based on conditional mean outcomes $\mu_1(X)$ and $\mu_0(X)$ and equations (4.40) and (4.41) based on weighting using the propensity score $p(X)$, we may combine both approaches. In fact, the ATE ($\Delta$) and ATET ($\Delta_{D=1}$) are identified based on the following doubly robust (DR) expressions, which use both propensity scores and conditional means, such as those discussed in Robins, Rotnitzky, and Zhao (1994); Robins and Rotnitzky (1995); and Hahn (1998):

$$\Delta = E\left[\phi(X)\right], \text{ with } \phi(X) = \mu_1(X) - \mu_0(X) + \frac{(Y - \mu_1(X)) \cdot D}{p(X)} - \frac{(Y - \mu_0(X)) \cdot (1-D)}{1-p(X)},$$

$$\Delta_{D=1} = E\left[\frac{(Y - \mu_0(X)) \cdot D}{\Pr(D=1)} - \frac{(Y - \mu_0(X)) \cdot (1-D) \cdot p(X)}{(1-p(X)) \cdot \Pr(D=1)}\right], \tag{4.45}$$

where $\phi(X)$ is the efficient influence function.

Even though the expressions in equation (4.45) may seem more complicated than the previously considered identification results based on regression and IPW, they coincide with equations (4.3) and (4.4) in section 4.1, as well as equations (4.40) and (4.41) in section 4.5. This follows from the fact that by the law of iterated expectations, some terms cancel out in the first and second lines of equation (4.45), which yield the ATE and ATET, respectively:

$$E\left[\frac{(Y - \mu_1(X)) \cdot D}{p(X)} - \frac{(Y - \mu_0(X)) \cdot (1-D)}{1-p(X)}\right] = E\left[\frac{\varepsilon \cdot D}{p(X)} - \frac{\varepsilon \cdot (1-D)}{1-p(X)}\right] = 0 \quad \text{and} \quad (4.46)$$

$$E\left[\frac{-\mu_0(X)\cdot D}{\Pr(D=1)} - \frac{-\mu_0(X)\cdot(1-D)\cdot p(X)}{(1-p(X))\cdot\Pr(D=1)}\right] = E\left[\mu_0(X)\cdot\left(\frac{p(X)}{\Pr(D=1)} - \frac{p(X)}{\Pr(D=1)}\right)\right] = 0,$$

with the error term $\varepsilon = Y - \mu_D(X)$ and $E[\varepsilon|D,X] = 0$.

Despite the equivalence of DR, IPW, and outcome regression for identifying causal effects in the population when assuming correct models for the conditional mean outcomes and the propensity score $\mu_1(X), \mu_0(X), p(X)$, DR bears an attractive property when it comes to estimation, which may be prone to misspecification. In fact, DR estimators based on the sample versions of equation (4.45) with estimates for $p(X), \mu_1(X), \mu_0(X)$, henceforth referred to as *plug-in parameters*, are consistent if either the conditional mean outcomes or the propensity scores are correctly specified. This property is discussed in Robins, Mark, and Newey (1992) and Robins, Rotnitzky, and Zhao (1995). Put differently, DR remains consistent if either the propensity score or the conditional mean outcome models are incorrect. In contrast, regression estimators solely rely on the correct specification of $\mu_1(X)$ and $\mu_0(X)$, while IPW solely relies on the correct specification of $p(X)$. This makes DR estimation more robust to model misspecification (reducing the threat of estimation error) than the other two methods, as suggested by its name, because we now have two shots at getting either $p(X)$ or $\mu_1(X)$ and $\mu_0(X)$ right.

If we even manage to correctly set up (or parameterize) the models for both the conditional mean outcomes (using a linear model, for instance) and the propensity score (using a probit model, for instance), then DR estimation is semiparametrically efficient: that is, it has the smallest possible asymptotic variance. This is also the case if the plug-in parameters $p(X), \mu_1(X), \mu_0(X)$ are nonparametrically estimated by kernel or series regression, as demonstrated by Cattaneo (2010). Furthermore, Rothe and Firpo (2013) show that in small samples, nonparametric DR has a lower bias and variance than either IPW, using a nonparametric propensity score, or nonparametric (e.g., kernel-based) outcome regression. This implies that for a limited sample size, the accuracy of the DR estimator is less dependent on the accuracy of the estimated propensity scores and conditional mean outcomes, as determined by the choice of the bandwidth $h$ in kernel regression. This better finite sample behavior makes DR estimation attractive from a practical perspective, even when IPW and outcome regression are consistent, meaning that they collapse to the true effect in infinitely large samples.

Estimation based on equation (4.45) is, however, not the only approach satisfying the DR property. Another method is targeted maximum likelihood estimation (TMLE), as suggested by van der Laan and Rubin (2006). It is based on first obtaining initial estimates of $\mu_1(X)$ and $\mu_0(X)$ by regression and updating (or robustifying) the estimates in a second step by regressing them on a function of an estimate of the propensity score $p(X)$. If the regression specifications for $\mu_1(X), \mu_0(X)$ are incorrect while those for $p(X)$ is correct, the second step will correct for the biases of the first step. If, however, the regression specifications for $\mu_1(X), \mu_0(X)$ are correct while those for $p(X)$ is incorrect, then the second step will do no harm. For this reason, TMLE is doubly robust. Yet another DR approach consists of running a weighted version of an outcome regression as in equation (4.8), with the weights corresponding to the IPW weights computed based on the inverse of the propensity score. Therefore, all DR methods have in common that they combine the estimation of conditional mean outcomes and propensity scores to obtain an estimate of the causal effect of interest, which fully exploits the information in the data concerning the associations of $X$ and $D$, as well as $D, X$ and $Y$.

To illustrate DR estimation in R, let us reconsider the birth-weight data already investigated at the end of section 4.5. To this end, we load the *drgee* package by Zetterqvist and Sjölander (2015), which contains a DR estimator of the ATE, and the *COUNT* package, which contains the *lbw* data of interest. We follow the same steps as in section 4.5 in terms of defining treatment $D$ as a dummy for smoking during pregnancy, the birth-weight outcome $Y$, and the covariates $X$. We then use the *drgee* command to estimate the ATE. The argument *oformula* requires specifying a regression model for the outcome as a function of the covariates, but without including the treatment (even though the latter will be included in the model to be estimated): *oformula=formula(Y~X)*. By default, a linear regression is performed, which appears appropriate for our continuously distributed birth-weight outcome, but it might be changed for other (e.g., binary) outcome variables. The argument *eformula* requires specifying a propensity score model for the treatment as a function of the covariates: *eformula=formula(D~X)*. We estimate the propensity scores based on a logit (rather than a linear) model by setting the argument *elink="logit"*. We save the output in an R object named *dr*, which we wrap by the *summary* command to inspect the results. The box here presents the R code of each of the steps.

```
library(drgee)                              # load drgee package
library(COUNT)                              # load COUNT package
data(lbw)                                   # load lbw data
attach(lbw)                                 # store all variables in own objects
D=smoke                                     # define treatment (mother smoking)
Y=bwt                                       # outcome (birthweight in grams)
X=cbind(race==1, age, lwt, ptl, ht, ui, ftv) # covariates
dr=drgee(oformula=formula(Y~X), eformula=formula(D~X), elink="logit") # DR reg
summary(dr)                                 # show results
```

Running the code yields the following output:

```
   Estimate  Std.  Error  z value  Pr(>|z|)
D   -358.0          102.7   -3.485  0.000492  ***

Signif. codes:  0 '***' 0.001 '**' 0.01 '*' 0.05 '.' 0.1 ' ' 1
```

The ATE estimate suggests that mothers' smoking during pregnancy decreases the weight of newborns by 358 grams on average and is highly statistically significant, even at the 0.1 percent level. This is qualitatively in line with the findings based on the IPW and EL methods discussed at the end of section 4.5.

## 4.7  Practical Issues: Common Support and Match Quality

After learning about several estimation methods in the selection-on-observables framework, let us consider some practical issues that are relevant for the accuracy of estimated causal effects—namely, common support in samples and match quality. Common support implies that the distributions of the propensity scores in the treated and nontreated groups of our sample overlap in terms of the range of values. Common support, however, does, not mean that the shape of the propensity score distribution is exactly the same across treatment groups. In fact, the distribution generally differs depending on the treatment state due to treatment selection based on X. Only in experiments, where the random treatment assignment does not depend on X, one would expect the propensity score distributions to be the same across treated and nontreated groups.

We note that in the population, common support holds by one of the assumptions in expression (4.1), requiring $0 < p(X) < 1$. In a sample with a limited (i.e., finite) number of observations used for treatment effect estimation, however, common support is not automatically guaranteed by this assumption. For the estimation of the ATET, common support implies that for each treated observation, nontreated matches with similar propensity scores exist in the data, while for the ATE, it also needs to hold that for each nontreated observation, treated matches with similar propensity scores exist. We can graphically check the satisfaction of common support based on histograms or density plots of the propensity scores separately for treated and nontreated observations in the sample to see whether the range of propensity score values overlaps.

Strictly speaking, common support is violated whenever for any reference observation, no observation in the other treatment group with exactly the same propensity score is available. In practice, propensity scores should be sufficiently similar, which requires defining a trimming criterion based on which dissimilar observations may be discarded from the data to enforce common support. However, if we discard observations, then effect estimation might not be fully representative of the initial target population (e.g., the treated). We thus sacrifice external validity, in the sense that we only estimate the effect for a subpopulation of the target population with common support in the data. On the other hand, doing so likely reduces estimation bias within this subpopulation satisfying common support, thus enhancing internal validity. Whenever we drop observations due to a lack of common support, we should report their number or share relative to the total sample in order to quantify how much we compromise on external validity.

The literature on causal analysis has suggested several trimming criteria for enforcing common support. One approach is to discard observations with propensity score values that have a likelihood (or density) of or close to zero in at least one treatment group—that is, which are either very rare or do not occur at all in one of the treatment groups; see Heckman, Ichimura, Smith, and Todd (1998). This requires estimating the density function of the propensity score (e.g., by kernel density estimation; see Rosenblatt (1956) and Parzen (1962)) and defining a threshold for the minimally required density to keep from being dropped from the data. Choosing the threshold may be data-driven—for instance, by defining it to be a specific quantile in the distribution of estimated propensity score densities. Other criteria are based on dropping observations with propensity score values that are higher than a specific threshold value such that nontreated observations with such high propensity scores unlikely occur in finite samples, or lower than a value such that treated observations unlikely occur. For ATET estimation, Dehejia and Wahba (1999) propose discarding all treated observations with an estimated propensity score higher than the highest value among the nontreated. For the ATE, one also discards nontreated observations with a propensity score lower than the lowest value among the treated.

Crump, Hotz, Imbens, and Mitnik (2009) suggest discarding observations with propensity scores close to zero or 1 in a way that minimizes the variance of ATE estimation in the remaining sample under specific assumptions. They find that dropping all observations with propensity scores outside the interval $[0.1, 0.9]$ (or $[10\ percent, 90\ percent]$) captures most of the variance reduction, which yields a simple rule of thumb for analysts and researchers. However, by minimizing the variance, we may compromise other goals such as ensuring external validity, which requires us to retain as many observations as possible in the sample. For this reason, practitioners typically also consider wider intervals, such as $[0.05, 0.95]$ or $[0.01, 0.99]$.

In general, any of such trimming rules should be sample size dependent to appropriately trade off bias (due to a loss of external validity) and variance (due to propensity scores close to 1 or zero). That is, the larger the sample size, the closer propensity scores may approach zero or 1 and still be retained in the sample, as the likelihood of the existence of such extreme propensity scores in the other treatment group increases with the sample size. Only then, trimming approaches zero in large samples such that external validity is maintained asymptotically: that is, as the sample size approaches infinity. A trimming rule that obeys these considerations consists of defining a maximum relative importance or weight (rather than a maximum value of the propensity score) any observation may have within its treatment group when estimating the treatment effect, such as discussed in Huber, Lechner, and Wunsch (2013). In the IPW-based estimator of the ATE provided in equation (4.42), for instance, the weight of a treated observation is given by $\frac{D_i}{\hat{p}(X_i)} \Big/ \sum_{i=1}^{n} \frac{D_i}{\hat{p}(X_i)}$. We can easily verify that the weight increases in $\hat{p}(X_i)$. Fixing the maximum weight to 0.05 (or 5 percent), for instance, implies that we drop observations with a higher contribution for the estimation of $E[Y(1)]$ than this maximum weight from the sample. Lechner and Strittmatter (2019) provide a comprehensive overview of alternative common support criteria.

A further practical issue for any propensity score–based method concerns the match quality: that is, the question of whether the estimated propensity score successfully balances $X$ across treatment groups, such as in matched samples or after weighting covariates (rather than outcomes) by IPW; see equation (4.44) in section 4.5. A poor balance (or match quality) implies that neither matching nor IPW succeeded in generating treated and nontreated groups that are comparable in $X$, entailing the threat of treatment selection bias when estimating causal effects. Poor match quality may either be due to a misspecified propensity score model, so we should consider a more flexible model, or a matching algorithm creating unsatisfactory matches, so we should use a different algorithm. A further possible reason for poor match quality is a lack of common support, in which case we might apply a trimming approach.

To check covariate balance, practitioners frequently consider hypothesis tests, such as two-sample t-tests (the Welch test for samples with unequal variances) to test for equality in means of each covariate $k$ across matched or IPW-weighted treated and nontreated groups (Welch 1947). The null hypothesis to be tested is thus $E[X_k^m | D=1] = E[X_k^m | D=0]$, with $m$ referring to matched units. Formally, the test statistic of the two-sample t-test is

$$\frac{\bar{X}_k^{m1} - \bar{X}_k^{m0}}{\sqrt{\frac{\widehat{Var}(X_k^{m1})}{n^{m1}} + \frac{\widehat{Var}(X_k^{m0})}{n^{m0}}}}, \tag{4.47}$$

where $\bar{X}_k^{m1}$ and $\bar{X}_k^{m0}$ denote the means of covariate $k$ in the groups of matched (or weighted) treated ($m1$) and nontreated ($m0$) observations, respectively. $\widehat{Var}(X_k^{m1})$ and $\widehat{Var}(X_k^{m0})$ denote the sample variances of $X_k$ in the matched treatment groups, while $n^{m1}$ and $n^{m0}$ denote the sample sizes of matched treated and matched nontreated observations, respectively. This corresponds to the heteroscedasticity-robust t-value obtained by a regression of $X$ on a constant and the treatment indicator among matched observations.

Rather than testing for mean differences, we can test for differences across the entire distribution of matched treated and nontreated observations, such as by means of a Kolmogorov-Smirnov test. In the latter case, the null hypothesis states that any covariate $X_k$ has the same distribution across matched treated and nontreated groups. Indeed, successful balancing implies that the entire distribution of a covariate, not just its mean, is equal across these groups, which is generally required for consistent treatment effect estimation. A further test consists of regressing $X_k$ on a constant, treatment $D$, the estimated propensity score $\hat{p}(X)$, higher-order terms of $\hat{p}(X)$, and interactions of $\hat{p}(X)$ and $D$ in the total (rather than the matched) sample; see Smith and Todd (2005). If $\hat{p}(X)$ balances the covariates well, such that $X_k$ and $D$ are independent conditional on $\hat{p}(X)$, then any coefficients on $D$ or its interaction with $\hat{p}(X)$ should be close to zero. This can be verified by running a joint statistical test for the null hypothesis that all those coefficients are equal to zero, such as a so-called F-test.

It is important to note that running hypothesis tests like two-sample t-tests for several covariates entails a multiple hypothesis testing problem. The more hypothesis tests we conduct, the higher is the likelihood that one of them spuriously rejects the null hypothesis in our sample, even when the hypothesis is satisfied for every covariate in the population. Quite intuitively, when running two sample t-tests for 100 covariates with a significance level of $\alpha = 0.05$, we would expect the tests to incorrectly reject the null hypothesis in five cases, as we are willing to accept an error probability of incorrect rejections of 5 percent. We need to keep this in mind when interpreting the results, implying that a few rejections among a large number of tests does not necessarily point to a violation of the balancing property of the propensity score.

An alternative strategy to avoid such issues related to multiple hypotheses consists of applying joint statistical tests for jointly checking imbalances in all elements of $X$, such as an F-test. We can implement this approach by estimating a system of equations consisting of several linear regressions in which each $X_k$ is regressed on a constant and $D$ among matched observations, and by testing whether the coefficients on $D$ are jointly zero. A different joint test consists of reestimating the propensity score among the matched observations, only to obtain an estimate of $\Pr(D = 1 | X_m)$, as suggested by Sianesi (2004). If the resulting pseudo-$R^2$, which analogously to $R^2$ for linear models (see section 3.4) is a measure of the goodness of fit for nonlinear models, is close to zero for the reestimated propensity score, then $X$ poorly predicts $D$ after matching, which points to a decent balance of $X$ across treated and nontreated matches.

One issue that all hypothesis tests have in common is that the test statistic underlying the rejection or nonrejection of the balancing hypothesis is sample size dependent—for instance, a function of the number of observations in the matched samples $n^{m1}$ and $n^{m0}$. A t-test will reject balance under the slightest mean difference in covariates across treated and nontreated matches if the sample grows to infinity, even if the imbalance is negligible in terms of its magnitude (e.g., amounts to just 0.01 years in the age variable). In contrast to classical hypotheses tests, the standardized difference test suggested by Rosenbaum and Rubin (1985) is insensitive to the sample size, at the cost of not providing a clear-cut rejection rule based on p-values and levels of significance. Formally, the test statistic is given by

$$100 \cdot \frac{\bar{X}_k^{m1} - \bar{X}_k^{m0}}{\sqrt{\frac{\widehat{Var}(X_k^1) + \widehat{Var}(X_k^0)}{2}}}, \tag{4.48}$$

where $\widehat{Var}(X_k^1)$ and $\widehat{Var}(X_k^0)$ denote the variances of $X$ in the original treated and nontreated samples, respectively: that is, prior to matching (in contrast to $\widehat{Var}(X_k^{m1})$ and $\widehat{Var}(X_k^{m0})$ in expression (4.47), which refer to the matched subsamples only).

Unlike the t-statistic, expression (4.48) considers mean differences relative to the variances in the original treated and untreated samples, but it is not a function of the sample sizes $n^{m1}$ and $n^{m0}$. This implies that the standardized difference is expectedly the same for small and large samples. Rather than judging balance based on a p-value as in hypothesis tests, a standardized absolute difference larger than a specific threshold such as 10 or 20 (or 0.1 and 0.2 when omitting multiplication by 100 in expression (4.48)) may be considered as an indication of imbalance. There is, however, admittedly some arbitrariness in the choice of the threshold, similar to choosing significance levels in classical hypothesis tests. We may also test the balance across all covariates jointly based on standardized differences by averaging the absolute standardized difference of each covariate over all covariates and verifying whether this average passes the threshold.

As a practical example in R, let us inspect the common support of the propensity score in the birth-weight data considered in sections 4.5 and 4.6. To this end, we load the *COUNT* package containing the *lbw* data, as well as the *kdensity* package by Moss and Tveten (2020) for the kernel-based estimation of a density function. The latter provides the distribution of a variable based on the likelihood (or density) with which the variable takes on specific values. We follow the same steps as in the last two sections in terms of defining treatment $D$ as smoking during pregnancy, the birth-weight outcome $Y$, and the covariates $X$. As in section 4.4, we use the *glm* command with the regression formula $D \sim X$ and the argument *family=binomial* to estimate the treatment propensity score by a logit model. We append the extension *$fitted* to the command and save the predicted propensity scores in an R object called *ps*.

In the next step, we apply the *kdensity* command to the propensity scores separately in the treated and nontreated groups (namely, *ps[D==1]* and *ps[D==0]*), to estimate the densities of the propensity scores under either treatment state. We save the estimated densities (i.e., the likelihoods that specific propensity score values occur among the treated and nontreated observations) in two R objects named *psdens1* and *psdens0*. We use the command *par(mfrow=c(2,2))* to specify a figure in R that consists of $2 \times 2 = 4$ empty graphs. Finally, we fill the empty graphs with plots of our propensity score densities under treatment and nontreatment, as well as histograms of the propensity scores, as yet another approach for showing the propensity score distributions. To this end, we first apply the *plot* command to the estimated densities *psdens1* and *psdens0*, respectively, and then the *hist* command to the propensity scores in the treated and nontreated groups, *ps[D==1]* and *ps[D==0]*. The box here provides the R code for each of the steps.

```
library(COUNT)                              # load COUNT package
library(kdensity)                           # load kdensity package
data(lbw)                                   # load lbw data
attach(lbw)                                 # store all variables in own objects
D=smoke                                     # define treatment (mother smoking)
Y=bwt                                       # outcome (birthweight in grams)
X=cbind(race==1, age, lwt, ptl, ht, ui, ftv) # covariates
ps=glm(D~X,family=binomial)$fitted          # estimate the propensity score by logit
psdens1=kdensity(ps[D==1])                  # density of propensity score among treated
psdens0=kdensity(ps[D==0])                  # density of propensity score among nontreated
par(mfrow=c(2,2))                           # specify a figure with four graphs (2X2)
plot(psdens1)                               # plot density for treated
plot(psdens0)                               # plot density for nontreated
hist(ps[D==1])                              # plot histogram of p-score for treated
hist(ps[D==0])                              # plot histogram of p-score for nontreated
```

Running the code yields the density plots and histograms of figure 4.7. An inspection of the density functions in the upper graphs and the histograms in the lower graphs reveals that the propensity scores are not equally distributed across treatment groups (as would be the case in an experiment with successful treatment randomization), but are on average larger among the treated than among the nontreated. This points to systematic differences in covariates across treated and nontreated groups. Even though the propensity score distributions differ for $D = 1$ and $D = 0$, common support appears close to being satisfied. Both among the treated and nontreated observations, the propensity score estimates seem to cover a similar range of values. An exception is the propensity score values around 0.8 (or 80 percent), which occur among the treated but are absent among the nontreated. As a further check,

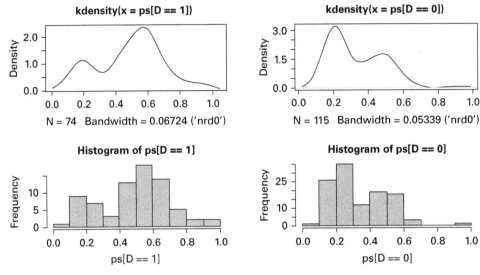

**Figure 4.7**
Propensity score distributions.

we can take a look at the summary statistics of the propensity score distributions by means of the *summary* command:

```
summary ( ps [D==1])
    Min. 1st Qu.  Median    Mean 3rd Qu.    Max.
  0.0897  0.3692  0.5295  0.4890  0.6060  0.9581
summary ( ps [D==0])
    Min. 1st Qu.   Median     Mean 3rd Qu.     Max.
 0.09918 0.20595 0.25930 0.32881 0.46018 0.90674
```

The minimum values of the propensity score are quite similar among treated and nontreated observations and amount to 8.97 percent and 9.92 percent, respectively. There is a somewhat larger gap in the respective maximum propensity scores of 95.81 percent versus 90.67 percent, providing some scope for propensity score trimming. Nevertheless, our common support check suggests that for most observations, we succeed in finding matches with comparable propensity score estimates in the other treatment group.

In our second example, we investigate the match quality after pair matching for ATET estimation by inspecting the propensity score distributions of matched treated and nontreated observations. As already mentioned, a matching algorithm that performs well in terms of match quality successfully balances the covariates across treatment states, and for this reason, the propensity score distributions (because the propensity score is a function of the covariates) as well. We load the *MatchIt* package by Ho, Imai, King, and Stuart (2011) and run the *matchit* command with the regression formula $D{\sim}X$ as the argument. The command first estimates the propensity scores based on a logit regression, which are then used in a second step to generate matched samples of treated and nontreated observations for ATET estimation. The ATET is not directly computed by the command, but it could be obtained by running a regression of the outcome on the treatment in the matched sample. We save the results of the *matchit* command in an R object named *output*. We wrap the latter using the *plot* command, where we also use *type="hist"* as the second argument, to plot the propensity score distributions across treatment states before and after matching by means of histograms. The box here contains the R code for each of the steps.

```
library(MatchIt)                    # load MatchIt package
output=matchit(D~X)                 # pair matching (ATET) on the propensity score
plot(output,type="hist")            # plot common support before/after matching
```

Running the code generates the graphs in figure 4.8, where the left and right histograms provide the propensity score distributions by the treatment state before and after matching, respectively. These histograms show that matching makes the propensity score distribution among the nontreated substantially more similar to that of the treated when compared to the raw sample prior to matching. Yet the match quality is not perfect, as some distributional differences remain, such as for propensity scores close to 0.8 (or 80 percent).

As an alternative to a graphical inspection, we next consider standardized mean difference and hypothesis tests for assessing the covariate balance before and after matching on the propensity score. To this end, we load the *Matching* package and use the *Match* command (similar to the example at the end of section 4.4) for pair matching on the propensity score using the previously defined *Y*, *D*, and *ps* as the outcome, treatment, and propensity score to match on, respectively. We store the results in an R object called *output1*. Finally, we use the *MatchBalance* command to investigate the balance of one of our covariates, *ptl*, which is the number of false premature labors. To this end, we use the formula $D{\sim}ptl$ as the first argument and *match.out=output1* to feed in the output of the pair-matching procedure:

```
library(Matching)                           # load Matching package
output1=Match(Y=Y, Tr=D, X=ps)              # pair matching (ATET) on p-score
MatchBalance(D~ptl, match.out=output1)      # covariate balance before/after matching
```

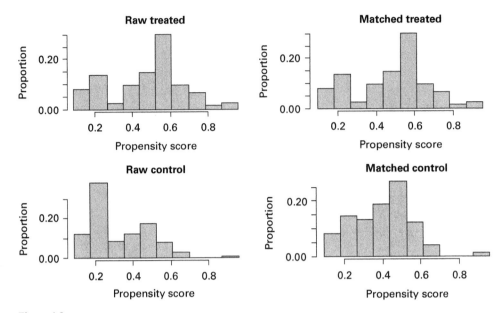

**Figure 4.8**
Propensity score distributions before and after matching.

Running the code yields the following output:

```
***** (V1) ptl *****
                              Before Matching        After Matching
    mean treatment.......        0.31081            0.31081
    mean control.........        0.12174            0.44595
    std mean diff........          30.61             -21.878

    mean raw eQQ diff.....        0.18919            0.15385
    med  raw eQQ diff.....              0                  0
    max  raw eQQ diff.....              1                  1

    mean eCDF diff........       0.047268           0.038462
    med  eCDF diff........       0.025088           0.019231
    max  eCDF diff........         0.1389            0.11538

    var ratio (Tr/Co).....          2.669            0.62893
    T-test p-value........       0.019878            0.16422
    KS Bootstrap p-value..          0.004              0.138
    KS Naive p-value......        0.35008            0.67676
    KS Statistic..........         0.1389            0.11538
```

Even though the standardized mean difference (*std mean diff*) of *ptl* across treatment groups is smaller in absolute terms in the matched sample (*After Matching*) than in the total sample (*Before Matching*), the value of $-21.878$ nevertheless points to an unsatisfactorily high imbalance. And even though neither the two-sample t-test (*T-test*) nor the Kolmogorov-Smirnov test (*KS Bootstrap p-value*) rejects the null hypotheses of equality in covariate means or distributions in the matched sample at conventional levels of significance, the p-values are not much higher than 10 percent. Let us see whether we can improve upon this rather unconvincing match quality by applying the trimming rule of Dehejia and Wahba (1999) and discarding all treated observations with an estimated propensity score greater than the highest value among the nontreated. To this end, we set the argument *CommonSupport* to *TRUE* in the *Match* command and rerun the previous analysis:

```
output2=Match(Y=Y, Tr=D, X=ps, CommonSupport=TRUE) # pair matching (ATET)
MatchBalance(D~ptl , match.out=output2) # covariate balance before/after matching
```

This gives the following results:

```
***** (V1) ptl *****
```

| | Before Matching | After Matching |
|---|---|---|
| mean treatment........ | 0.31081 | 0.30986 |
| mean control.......... | 0.12174 | 0.35915 |
| std mean diff........ | 30.61 | $-8.5659$ |
| | | |
| mean raw eQQ diff..... | 0.18919 | 0.13333 |
| med raw eQQ diff..... | 0 | 0 |
| max raw eQQ diff..... | 1 | 1 |
| | | |
| mean eCDF diff........ | 0.047268 | 0.033333 |
| med eCDF diff........ | 0.025088 | 0.033333 |
| max eCDF diff........ | 0.1389 | 0.066667 |
| | | |
| var ratio (Tr/Co)..... | 2.669 | 1.0377 |
| T-test p-value........ | 0.019878 | 0.6048 |
| KS Bootstrap p-value.. | 0.008 | 0.468 |
| KS Naive p-value...... | 0.35008 | 0.99626 |
| KS Statistic.......... | 0.1389 | 0.066667 |

We see that the common support restriction indeed improves the balance of *ptl* across treatment groups in the matched sample, as the standardized difference (*std mean diff*) now only amounts to $-8.566$, which is below the threshold of 10 in absolute terms. Furthermore, the p-values of the t-test (*T-test p-value*) and the

Kolmogorov-Smirnov test (*KS Bootstrap p-value*) are now considerably higher than before. All in all, the match balance looks pretty convincing for *ptl* after propensity score trimming. However, for judging how much to compromise in terms of external validity, we should check how many observations have been dropped due to the trimming rule. To this end, we wrap out previous matching outputs with and without trimming, *output1* and *output2*, by the *summary* command:

```
summary(output1)                    # ATET without common support

Estimate...  -134.55
AI SE......  179.1
T-stat.....  -0.75124
p.val......  0.45251

Original number of observations.............  189
Original number of treated obs..............  74
Matched number of observations..............  74
Matched number of observations (unweighted).  78


summary(output2)                    # ATET with common support

Estimate...  -135.16
AI SE......  181.59
T-stat.....  -0.74433
p.val......  0.45668

Original number of observations.............  186
Original number of treated obs..............  71
Matched number of observations..............  71
Matched number of observations (unweighted).  75
```

We see that the ATETs (*Estimate*) both with and without trimming propensity scores are quite similar, amounting to $-134.55$ and $-135.16$, respectively. By comparing the number of observations (*Original number of observations*) without trimming (189) to those with trimming (186) we learn that the trimming rule discards only 3 observations from the sample. This nevertheless considerably improves the match quality. We note that we can repeat the previous procedures for any of the other covariates of interest.

Let us now also consider a method for checking covariate balance in the context of IPW-based estimation by reconsidering the *treatweight* command in the *causalweight* package, as previously used in section 4.5. In fact, we can investigate the mean difference in a covariate like *ptl* across treatment groups after weighting by considering it as an outcome variable in the IPW procedure. If the propensity score underlying the IPW estimator successfully balances the covariates, then this mean difference should be close to zero. We set a seed (*set.seed(1)*) prior to running the *treatweight* command with *ptl* as the outcome, *D* as the treatment, and *X* as covariates for probit-based propensity score estimation. We set the number of bootstraps to 999 and save the results in an R object named *ipw*. Finally, we investigate the estimated mean difference, *ipw$effect*, along with its p-value, *ipw$pval*:

```
library(causalweight)                      # load causalweight package
set.seed(1)                                # set seed to 1
ipw=treatweight(y=ptl,d=D,x=X, boot=999)   # run IPW with 999 bootstraps
ipw$effect                                 # show mean difference in X
ipw$pval                                   # show p-value
```

Running the code yields a mean difference of roughly −0.049, with a p-value of 0.284. We therefore do not find statistical evidence for a violation of covariate balance across reweighted treatment groups. It is also worth noting that the *treatweight* command contains a trimming rule (see the argument *trim*). By default, the latter is set to 0.05, implying that observations with rather extreme propensity scores of less than 5 percent or more than 95 percent are discarded when estimating the ATE. Let us now change this value to 10 percent, as considered by Crump, Hotz, Imbens, and Mitnik (2009), by setting *trim=0.1* in the *treatweight* command and saving the output in the R object *ipw*. In addition to the estimated mean difference in *ptl* across the reweighted treatment groups and its p-value, we now inspect the number of discarded observations, which is provided in *ipw$ntrimmed*:

```
set.seed(1)                                        # set seed to 1
ipw=treatweight(y=ptl,d=D,x=X, trim=0.1, boot=299) # run IPW with 999 bootstraps
ipw$effect                                         # show mean difference in X
ipw$pval                                           # show p-value
ipw$ntrimmed                                       # number of trimmed units
```

This gives a mean difference of 0.044 and a p-value of 0.31, which again does not point to a violation of covariate balance. Furthermore, only three observations are discarded, such that external validity is hardly compromised by this trimming rule.

## 4.8 Multivalued or Continuous Treatments and Distributional Effects

The selection-on-observables framework straightforwardly extends to multivalued, discrete treatments, which may either reflect distinct treatments, such as an information technology (IT) course or a sales training, or an ordered amount of a specific treatment, such as one, two, or three weeks of training. As in section 3.5, let $D \in \{0, 1, 2, \ldots, J\}$, where $J$ denotes the number of treatments that are nonzero. Under appropriate selection-on-observable assumptions, we can identify treatment effects by pairwise comparisons of any nonzero treatment with no treatment, or of two nonzero treatment values, if the effect of one nonzero treatment relative to another is of interest. More formally, let $d$ and $d'$ denote the treatment values to be compared (e.g., $d = 2$ and $d' = 0$ or $d = 2$ and $d' = 1$). Adapting the assumptions in expression (4.1) for a binary $D$ to hold for the pair of values $D = d$ and $D = d'$ of the multivalued treatment, one obtains the following identifying assumptions (see Imbens (2000)):

$$\{Y(d), Y(d')\} \perp D|X, \quad \Pr(D = d|X) > 0, \quad \Pr(D = d'|X) > 0, \quad X(d) = X(d') = X.$$

$$(4.49)$$

Under the assumptions in expression (4.49), we can identify the ATE in the total population when comparing $D = d$ versus $D = d'$, as well as the ATET when considering those with $D = d$ as the treated population. To this end, we simply have to adapt equations (4.3), (4.4), (4.36), (4.40), (4.41), and (4.45), which permit assessing treatment effects based on regression, matching, IPW, and DR, respectively, to the case of a nonbinary treatment. Everywhere in these expressions, we replace $D$ with the indicator function $I\{D = d\}$ and $1 - D$ with $I\{D = d'\}$, as well as the propensity score $p(X) = \Pr(D = 1|X)$ with $\Pr(D = d|X)$ (i.e., the conditional probability of receiving treatment value $d$), and $1 - p(X)$ with $\Pr(D = d'|X)$.

As in the binary treatment case, regression-, IPW-, or DR-based treatment effect estimation with multivalued discrete treatments can be $\sqrt{n}$-consistent and semi-parametrically efficient if we estimate the plug-in parameters (the conditional mean outcomes and propensity scores) nonparametrically, such as by series or kernel regression, as discussed in Cattaneo (2010). We may also apply matching approaches to assess the causal effect of assigning treatment $d$ versus $d'$, such as by matching observations with $D = d$ and $D = d'$ with similar estimates of both propensity scores $\Pr(D = d|X)$ and $\Pr(D = d'|X)$; see Lechner (2001). We note that if the conditions in expression (4.49) are satisfied for all possible pairwise comparisons of treatment values, then the selection-on-observables assumption holds for all potential outcomes: $\{Y(0), Y(1), Y(2), \ldots, Y(J)\} \perp D|X$. Depending on the context, this may appear to be a stronger restriction than imposing it only for a specific pair of treatment values; therefore, we need to scrutinize its plausibility in the empirical application at hand.

When $D$ does not have discrete values like $0,1,2,\ldots$ (or probability masses) but is continuously distributed, such that it may take infinitely many values like time spent training, the previously considered conditional treatment probability $\Pr(D = d|X)$

turns into a conditional density function, denoted by $f(D = d|X)$, which is known as the *generalized propensity score*. Replacing $\Pr(D = d|X) > 0$ and $\Pr(D = d'|X) > 0$ by $f(D = d|X) > 0$ and $f(D = d'|X) > 0$ in expression (4.49) thus yields the set of conditions required for assessing the effect of treatment dose $d$ (e.g., 50 hours of training) versus $d'$ (e.g., 20 hours and 30 minutes of training). If these conditions hold for any feasible treatment values, we may assess the full range of treatment doses. In the spirit of equation (4.3) in section 4.1 for binary treatments, we can estimate the causal effects of a continuous treatment based on a parametric or nonparametric (e.g., kernel) regression of $Y$ on $D$ and $X$, as considered in Flores (2007). This permits estimating the mean potential outcomes $\mu_d(x)$ and $\mu_{d'}(x)$ to assess the average effects of discrete changes in the treatment: that is, $E[\mu_d(x) - \mu_{d'}(x)]$. We may also estimate the derivatives $\frac{\partial \mu_d(x)}{\partial d}$ to assess the marginal treatment effect $\frac{\partial E[Y(d)]}{\partial d}$ previously considered in section 3.5.

In analogy to equation (4.36) in section 4.4, we may alternatively regress $Y$ on $D$ and an estimate of $f(D|X)$ (Hirano and Imbens (2005)) or apply the stratification approach of Imai and van Dyk (2004). We can also use IPW-based methods in the spirit of equation (4.40) in section 4.5, which requires replacing any indicator functions for discrete treatments, such as $I\{D = d\}$, with continuous weighting functions, as considered by Flores, Flores-Lagunes, Gonzalez, and Neumann (2012) and Galvao and Wang (2015). Let us to this end define the kernel weight $\mathcal{K}\left((D - d)/h\right)/h$, where $\mathcal{K}$ is a symmetric, second-order kernel function (e.g., the standard normal density function provided in figure 4.2), which assigns more weight to values of $D$ the closer they are to $d$; and $h$ is the bandwidth. Then, for instance, the ATE in the population is identified by the following IPW expression:

$$\Delta = \lim_{h \to 0} E\left[ \frac{Y \cdot \mathcal{K}\left((D - d)/h\right)/h}{f(D = d|X)} - \frac{Y \cdot \mathcal{K}\left((D - d')/h\right)/h}{f(D = d'|X))} \right], \qquad (4.50)$$

where $\lim_{h \to 0}$ means "as $h$ goes to zero." Related to equation (4.45) in section 4.6, we can also estimate the effects of continuous treatments based on DR approaches, as suggested by Kennedy, Ma, McHugh, and Small (2017).

So far, our discussion of causal effects has focused on average effects in a target population, particularly the ATE or the ATET of a binary or multivalued treatment. However, the selection-on-observables framework also extends to the evaluation of treatment effects on distributional features of the outcome other than the mean, as already briefly acknowledged in section 4.1. This follows from the fact that under our assumptions, we may replace $Y$ in the various identification results with a function of $Y$, so long as the function satisfies some mild properties (like having a finite variance). Replacing, for example, $Y$ in equation (4.50) with $I\{Y \leq y\}$, with $y$ being one value of the outcome of interest, identifies a distributional treatment effect. The latter tells us how the share of individuals whose outcome is less than or equal to $y$ changes in reaction to the treatment. More formally, let us return to the case of a binary treatment and denote by $F_{Y(1)}(y) = \Pr(Y(1) < y)$ and $F_{Y(0)}(y) = \Pr(Y(0) < y)$ the cumulative distribution functions of the potential outcomes under treatment and nontreatment, respectively, assessed at outcome value $y$.

Considering a monthly wage outcome of $y = 4{,}000$ euros (EUR), $F_{Y(1)}(y)$ and $F_{Y(0)}(y)$ correspond to the share of individuals earning no more than 4,000 EUR per month when receiving and not receiving the treatment, respectively. Put differently, the cumulative distribution functions give the ranks of a wage of $y = 4{,}000$ in the potential outcome distributions under treatment and nontreatment, respectively. For instance, $F_{Y(0)}(4{,}000) = 0.5$ implies that under nontreatment, 50 percent of the population of interest earn 4,000 EUR or less per month and therefore, individuals earning exactly 4,000 EUR obtain the median wage in the population. Accordingly, the distributional treatment effect is defined as the difference in the potential outcome distributions, $F_{Y(1)}(y) - F_{Y(0)}(y)$. It corresponds to the effect of the treatment on the share of individuals who obtain a wage smaller than or equal to $y = 4{,}000$ or, equivalently, on the rank of individuals earning $y = 4{,}000$. The estimation of potential outcome distributions is discussed in such sources as Donald and Hsu (2014); DiNardo, Fortin, and Lemieux (1996); and Chernozhukov, Fernández-Val, and Melly (2013).

A related causal parameter is the quantile treatment effect (QTE). It corresponds to the effect on subjects situated at a specific rank of the potential outcome distributions, such as at the median or the first or third quartile of the potential wage outcomes under treatment and nontreatment. QTEs are useful to investigate effect heterogeneity across ranks of outcome, especially when effects for specific subgroups in terms of outcome ranks are particularly policy relevant. For instance, if we are interested in the effect of an income support policy on household income, then the effect on poorer households situated at lower ranks of the income distribution may appear more relevant than the effect on high-income households. Furthermore, investigating QTEs permits assessing whether a treatment increases or decreases income inequality across outcome ranks, which is not revealed when merely looking at the ATE.

The evaluation of QTEs generally requires continuously distributed outcomes (e.g., a wage outcome taking many values) with a strictly increasing distribution function for the outcome ranks of interest. The latter condition rules out that there are ranges of outcome values for which no observations exist (e.g., no observed wages between 2,000 and 2,500 EUR); otherwise, quantiles are not uniquely defined, meaning that specific outcome values cannot be uniquely attributed to a specific rank. If, for instance, 20 percent of the population earn less than 2,000 EUR and there are no individuals earning between 2,000 and 2,500 EUR, then the rank for any of these values is equal to 0.2 (or 20 percent). Therefore, there is no one-to-one correspondence between quantiles (i.e., outcome values) and ranks, as required for a unique definition of QTEs.

To discuss the QTEs more formally, we note that the quantile function of a potential outcome is defined as the inverse of its cumulative distribution function assessed at a particular rank $\tau$ larger than zero and smaller than 1, $F_{Y(d)}^{-1}(\tau)$ with the rank $\tau \in (0, 1)$ (or 0 percent and 100 percent). The value $d$ in $F_{Y(d)}^{-1}(\tau)$ is zero for the potential outcome under nontreatment and 1 for that under treatment. Under selection-on-observables assumptions, for instance, we may identify quantiles (i.e., values of potential outcomes at specific ranks of interest) based on IPW when solving the following minimization problem, as outlined in Firpo (2007):

$$F_{Y(d)}^{-1}(\tau) = \min_{y} E\left[ \frac{D}{\Pr(D = d|X)} \cdot (Y - y) \cdot (\tau - I\{Y - y < 0\}) \right]. \tag{4.51}$$

Here, $(Y - y) \cdot (\tau - I\{Y - y < 0\})$ is a loss function tailored to the assessment of quantiles (rather than means, as in the squared-error-based loss function in equation (3.12) of section 3.2) and was first suggested by Koenker and Bassett (1978). Then the QTE, denoted by

$$\Delta(\tau) = F_{Y(1)}^{-1}(\tau) - F_{Y(0)}^{-1}(\tau), \tag{4.52}$$

is obtained as the difference between the quantiles in equation (4.51) under treatment ($d = 1$) and nontreatment ($d = 0$), assessed at the same rank $\tau$.

Figure 4.9 provides a graphical illustration of the identification of the QTE, based on plotting the ranks of the potential outcome distributions under treatment and nontreatment on the $y$-axis and the respective quantiles of an outcome (income) on the $x$-axis. A particular rank $\tau$ is associated with specific outcome values $y = F_{Y(0)}^{-1}(\tau)$ and $y' = F_{Y(1)}^{-1}(\tau)$. Therefore, the difference $y' - y$ corresponds to the QTE.

To illustrate the evaluation of a continuously distributed treatment in R, let us investigate the effect of expert ratings of video games on the sales of those games in a data set originally collected and analyzed by Wittwer (2020). To this end, we load the *causalweight* package, which contains the *games* data of interest, as well as the *npcausal* package by Kennedy (2021), which contains a DR method applicable to continuous treatments. The package is available on GitHub, an online platform for building and maintaining software. Accessing GitHub requires first installing and loading the *devtools* package developed by Wickham, Hester, and Chang (2021), and then running *install_github("ehkennedy/npcausal")* to install the *npcausal* package before loading it using the *library* command. We then use the *data* command to load

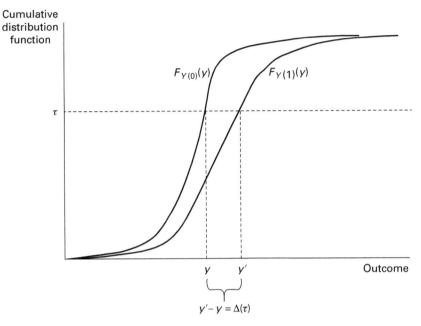

Cumulative
distribution
function

$F_{Y(0)}(y)$     $F_{Y(1)}(y)$

$\tau$

$y$     $y'$     Outcome

$y' - y = \Delta(\tau)$

**Figure 4.9**
Illustration of the QTE.

the *games* data, which contains missing values in some variables for some observations. Even though this is generally not the most appropriate way to deal with missing
data problems, we apply the *na.omit* command to the *games* data, which simply
drops observations with any missing information. We save the remaining sample as an
R object named *gamesnomis* and apply the *attach* command to store all variables in
own R objects.

Next, we define the variable *metascore*, a weighted average rating of a video game
by professional critics that ranges from 0 to 100, as continuous treatment $D$ and
*sales*, which corresponds to the global sales in millions of a video game until 2018,
as outcome $Y$. Furthermore, we use the *cbind* command to define the *year* of a
game's release, the *userscore* providing the average user rating, and *genre=="Action"*,
a dummy variable for the action genre, as covariates $X$. We then feed the outcome, treatment, and control variables into the *ctseff* command, a DR approach
that estimates outcome and generalized propensity score models nonparametrically
by an ensemble (i.e., combination) of machine learning methods; see section (5.2) in
chapter 5 for further details on causal machine learning. We also set the argument
*bw.seq=seq(from=1,to=5,by=0.5)* to find the optimal treatment kernel bandwidth to
be used in the *ctseff* command by cross-validation based on iterating over bandwidth
values $1, 1.5, \ldots, 4.5, 5$. We store the output in an object named *results* and wrap the

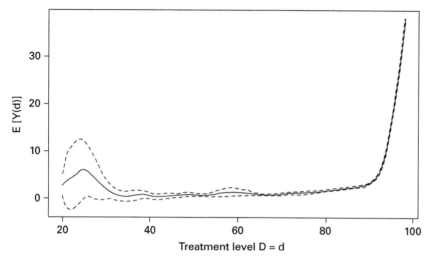

**Figure 4.10**
Mean potential outcome under a continuous treatment.

latter by the *plot.ctseff* command to plot the regression function: that is, the estimate of the mean potential outcome $E[Y(d)]$ as a function of treatment value $d$. The box here provides the R code for each of the steps.

```
library(causalweight)                       # load causalweight package
library(devtools)                           # load devtools package
install_github("ehkennedy/npcausal")        # install npcausal package
library(npcausal)                           # load npcausal package
data(games)                                 # load games data
games_nomis=na.omit(games)                  # drop observations with missings
attach(games_nomis)                         # attach data
X=cbind(year, userscore, genre=="Action")   # define covariates
D=metascore                                 # define treatment
Y=sales                                     # define outcome
results=ctseff(y=Y, a=D, x=X, bw.seq=seq(from=1,to=5,by=0.5)) # DR estimation
plot.ctseff(results)                        # potential outcome-treatment relation
```

Running the code yields the graph in figure 4.10, which suggests that the association between the average sales of video games and the rating of professional critics is highly nonlinear. The initially mostly flat regression line implies that marginal increases in the rating have a limited effect on sales up to a treatment value of $d = 90$. For larger treatment values, however, the outcome-treatment-association becomes much steeper, pointing to a substantial marginal effect of the rating on sales. Of course, these conclusions are conditional on the satisfaction of the selection-on-observables assumptions, given our covariates $X$. The graph also shows the 95 percent confidence intervals (dashed lines), which are mostly quite narrow, suggesting that the mean potential outcome as a function of the treatment is quite precisely estimated.

As a second empirical example based on the same data, let us consider the estimation of the QTE using the *qte* package. To this end, we create a binary treatment indicator $D$ for a high average rating of a video game by professional critics, which takes the value 1 if the continuous treatment *metascore* is greater than 75 and zero otherwise. Using the *data.frame* command, we generate a data set named *dat*, which contains outcome $Y$, treatment $D$, and covariates $X$. In the next step, we apply the *ci.qte* command, an IPW estimator of QTEs based on equation (4.51). Its first argument is a regression like formula of the outcome and the treatment, $Y{\sim}D$, followed by the covariates, $x{=}X$, and the data set, *data=dat*. We store the output in an object named $QTE$ and wrap the latter by the *ggqte* command to plot the QTEs of our high rating indicator on sales across various ranks $\tau$ in the outcome distribution along with confidence intervals. The R code is provided in the box here.

```
library(qte)                      # load qte package
D=metascore>75                    # define binary treatment (score>75)
dat=data.frame(Y,D,X)             # create data frame
QTE=ci.qte(Y~D, x=X, data=dat)    # estimate QTE across different ranks
ggqte(QTE)                        # plot QTEs across ranks (tau)
```

Running the code gives the graph in figure 4.11. The effects of obtaining a high as opposed to low expert rating on sales are generally positive and statistically significant (as the dashed confidence intervals do not include zero), but substantially larger at higher ranks of the sales distribution than at lower ones.

## 4.9   Dynamic Treatment Effects

A further conceptual extension of the standard treatment framework is the analysis of sequences of treatments or dynamic treatment effects, as discussed in Robins (1986) and Robins, Hernan, and Brumback (2000). In this case, we are interested in the evaluation of several discrete treatments assigned over several time periods, like consecutive training programs (e.g., a language course followed by an IT course), medical interventions (e.g., a surgery followed by rehabilitation), or sales promotions (e.g., a marketing campaign followed by a price discount). Such a causal analysis of distinct sequences of treatments requires us to control for confounder that jointly affect the outcome and the various treatments assessed in multiple periods. One approach is to sequentially impose selection-on-observables assumptions

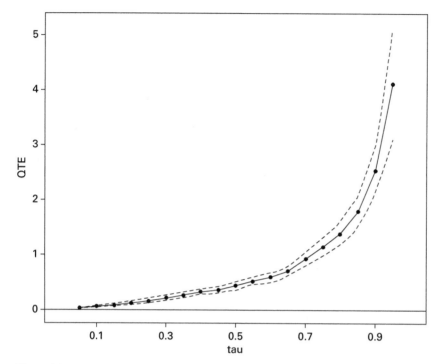

**Figure 4.11**
QTEs.

across all treatment periods, which implies that the treatment in each period is as good as randomly assigned, conditional on past treatment assignments, past outcomes, and the history of observed covariates up to the respective treatment assignment.

To discuss the dynamic treatment framework more formally, let us for the sake of simplicity consider just two sequential treatments and denote by $D_t$ and $Y_t$ the treatment and the outcome, respectively, in time period $T = t$. Therefore, $D_1$ and $D_2$ are the treatments in the first and second periods, respectively, and may take values $d_1, d_2 \in \{0, 1, \ldots, J\}$, with 0 indicating no treatment and $1, \ldots, J$ the various treatment choices (e.g., the various training programs), as in section 4.8. Let us denote by $Y_2$ the outcome (e.g., wages), which is measured in the second period after the realization of treatment sequence $D_1$ and $D_2$. Furthermore, $\underline{d}_2$ denotes a specific treatment sequence $(d_1, d_2)$, with $d_1, d_2 \in \{0, 1, \ldots, J\}$ (e.g., a language course followed by IT training) of the treatment variables $\underline{D}_2 = (D_1, D_2)$. Accordingly, $Y_2(\underline{d}_2)$ denotes the potential outcome hypothetically realized when the treatment values are set to sequence $\underline{d}_2$. We may then define the average treatment effect of two distinct treatment sequences $\underline{d}_2$ versus $\underline{d}_2'$ as

$$\Delta(\underline{d}_2, \underline{d}_2') = E[Y_2(\underline{d}_2) - Y_2(\underline{d}_2')].\tag{4.53}$$

An example is the ATE of a sequence of two distinct treatments (e.g., a language course followed by IT training) versus no training in either period. In this case, $\underline{d}_2 = (1,2)$ (with 1 =language course and 2 =IT training) and $\underline{d}_2' = (0,0)$.

To formalize the sequential selection-on-observables assumptions, let us denote by $X_t$ the observed covariates in period $T = t$. Here, $t$ equals 0 during the pretreatment period, 1 during the period of the first treatment, and 2 during the period of the second treatment. Accordingly, $X_0$ consists of pretreatment characteristics measured (at least shortly) prior to the first treatment participation $D_1$. $X_1$ consists of covariates measured during or even at the end of the first treatment period but (at least shortly) prior to the second treatment $D_2$ and may include intermediate outcomes observed in period 1, denoted by $Y_1$ (e.g., the wages just after the first treatment). Furthermore, $X_1$ may be influenced by both $D_1$ and $X_0$, such as the baseline wages prior to the first training, as well as training participation in the first period may affect the wages at the end of the first period. Covariates in a particular period, therefore, may be affected by previous covariates and treatments, implying that confounding may be dynamic in the sense that the assessment of causal effects requires controlling for time-varying covariates rather than for baseline characteristics (measured prior to the first treatment) alone.

In terms of identifying conditions, we impose a sequential version of the selection-on-observables assumptions as follows for any treatment sequence $\underline{d}_2$ of interest:

$$Y_2(\underline{d}_2) \perp D_1 | X_0 \text{ and } Y_2(\underline{d}_2) \perp D_2 | D_1, X_0, X_1 \text{ for } \underline{d}_2 = (d_1, d_2) \text{ and } d_1, d_2 \in \{0, 1, \dots, J\},$$
$$\Pr(D_1 = d_1 | X_0) > 0 \text{ and } \Pr(D_2 = d_2 | D_1, X_0, X_1) > 0 \text{ for } d_1, d_2 \in \{0, 1 \dots, J\}.\tag{4.54}$$

The first condition in the first line of expression (4.54) invokes the conditional independence of the treatment in the first period $D_1$ and the potential outcomes $Y_2(\underline{d}_2)$, given $X_0$. It rules out unobserved confounders jointly affecting $D_1$ and $Y_2(\underline{d}_2)$ conditional on $X_0$ in the same spirit as expression (4.1). The second condition in the first line invokes conditional independence of the second treatment $D_2$ given the first treatment $D_1$ and the (history of) covariates $X_0$ and $X_1$. It rules out unobserved confounders jointly affecting $D_2$ and $Y_2(\underline{d}_2)$, conditional on $D_1$, $X_0$, and $X_1$. The second line of expression (4.54) imposes common support, meaning that the treatment in each period is not a deterministic function of the observables in the conditioning set, which rules out conditional treatment probabilities (or propensity scores) of 0 or 1. This implies that conditional on each value of the observables occuring in the population, subjects with distinct treatment assignments $\{0, 1, \dots, J\}$ exist.

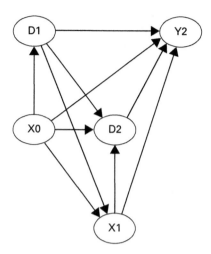

**Figure 4.12**
Sequential conditional independence with posttreatment confounders.

Figure 4.12 provides a causal graph that illustrates the implications of equation (4.54), with arrows representing causal effects. Each of $D_1$, $D_2$, and $Y_2$ might be causally affected by different sets of unobservables not displayed in figure 4.12, but none of these sets of unobservables may jointly affect $D_1$ and $Y_2$, given $X_0$, or $D_2$ and $Y_2$, given $D_1$, $X_0$, and $X_1$. We also note that the second treatment, $D_2$, is allowed to depend on the first treatment, $D_1$. The likelihood of participating in an IT course, for instance, might differ across individuals attending and not attending a language course in the first period.

In the spirit of equation (4.3) for treatment evaluation with a single treatment period, we can under the conditions in expression (4.54) assess dynamic treatment effects based on nested conditional means, which may be estimated by nested regressions or matching estimations; for example, see Lechner and Miquel (2010) and Blackwell and Strezhnev (2020):

$$\Delta(\underline{d}_2, \underline{d}_2') = E[E[E[Y_2|\underline{D}_2 = \underline{d}_2, X_0, X_1]|D_1 = d_1, X_0] - E[E[Y_2|\underline{D}_2$$
$$= \underline{d}_2', X_0, X_1]|D_1 = d_1', X_0]], \tag{4.55}$$

where $\underline{d}_2 = (d_1, d_2)$ and $\underline{d}_2' = (d_1', d_2')$ denote distinct treatment sequences. Alternatively, we may assess the effects based on IPW (see section 4.5) based on sequential propensity scores estimated for each treatment period, as in Lechner (2009)):

$$\Delta(\underline{d}_2, \underline{d}_2') = E\left[\frac{Y \cdot I\{D_1 = d_1\} \cdot I\{D_2 = d_2\}}{p^{d_1}(X_0) \cdot p^{d_2}(D_1, X_0, X_1)} - \frac{Y \cdot I\{D_1 = d_1'\} \cdot I\{D_2 = d_2'\}}{p^{d_1'}(X_0) \cdot p^{d_2'}(D_1, X_0, X_1)}\right], \tag{4.56}$$

where $p^{d_1}(X_0) = \Pr(D_1 = d_1 | X_0)$ and $p^{d_2}(D_1, X_0, X_1) = \Pr(D_2 = d_2 | D_1, X_0, X_1)$ are shorthand notations for the propensity scores in the two periods.

Finally, and in the spirit of section 4.6, we can use a DR approach, which is based on both outcome regression and propensity scores for evaluating the ATE of treatment sequences, as discussed by Robins (2000) and Tran et al. (2019):

$$\Delta(\underline{d}_2, \underline{d}_2') = E[\psi^{\underline{d}_2} - \psi^{\underline{d}_2'}],$$

$$\text{where } \psi^{\underline{d}_2} = \frac{I\{D_1 = d_1\} \cdot I\{D_2 = d_2\} \cdot [Y_2 - \mu^{Y_2}(\underline{d}_2, \underline{X}_1)]}{p^{d_1}(X_0) \cdot p^{d_2}(d_1, X_0, X_1)}$$

$$+ \frac{I\{D_1 = d_1\} \cdot [\mu^{Y_2}(\underline{d}_2, X_0, X_1) - \nu^{Y_2}(\underline{d}_2, X_0)]}{p^{d_1}(X_0)} + \nu^{Y_2}(\underline{d}_2, X_0), \qquad (4.57)$$

with $\mu^{Y_2}(\underline{d}_2, X_0, X_1) = E[Y_2 | \underline{D}_2 = \underline{d}_2, X_0, X_1]$ and $\nu^{Y_2}(\underline{d}_2, X_0) = E[E[Y_2 | \underline{D}_2 = \underline{d}_2, X_0, X_1] | D_1 = d_1, X_0]$ being shorthand notations for the (nested) conditional mean outcomes.

It is worth noting that our evaluation problem simplifies somewhat if the second treatment, $D_2$, is conditionally independent of the potential outcomes given the pretreatment variables $X_0$ and treatment $D_1$, implying that $X_1$ is not required to control for confounders jointly affecting the second treatment and the outcome. In this case, assumption $Y_2(\underline{d}_2) \perp D_2 | D_1, X_0, X_1$ in expression (4.54) may be replaced by the stronger condition $Y_2(\underline{d}_2) \perp D_2 | D_1, X_0$, which eases data requirements as it does not rely on information on posttreatment covariates $X_1$. Figure 4.13 provides a graphical illustration of the causal associations that satisfy this set of assumptions.

Accordingly, controlling for $X_1$ is not required in the propensity scores and conditional mean outcomes of any expression identifying the ATE $\Delta(\underline{d}_2, \underline{d}_2')$. For instance, $\psi^{\underline{d}_2}$ in equation (4.57) then simplifies to

$$\psi^{\underline{d}_2} = \frac{I\{D_1 = d_1\} \cdot I\{D_2 = d_2\} \cdot [Y_2 - \mu^{Y_2}(\underline{d}_2, X_0)]}{p^{d_1}(X_0) \cdot p^{d_2}(d_1, X_0)} + \mu^{Y_2}(\underline{d}_2, X_0). \qquad (4.58)$$

Intuitively, this implies that the second treatment is as good as being randomly assigned, given the first treatment and the baseline covariates. The evaluation problem in this case is equivalent to the evaluation of multivalued discrete treatments as discussed in section 4.8, with specific combinations of $D_1$ and $D_2$ defining the multivalued treatments.

However, this framework might appear unrealistic if there is a large time lag between the first and second treatments, which favors dynamic confounding. If the second treatment corresponds to a training program that takes place several years after a first training program, it appears likely that individual characteristics like health and labor market behavior that also affect the outcome (e.g., wages) change

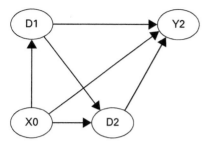

**Figure 4.13**
Sequential conditional independence without posttreatment confounders.

between the first and second treatments and therefore need to be controlled for. For this reason, we need to scrutinize the plausibility of dynamic, time-varying confounders when deciding whether we control only for baseline covariates or also take into account covariates in later time periods.

To illustrate the estimation of dynamic treatment effects in R, let us reconsider the Job Corps sample introduced at the end of section 3.1. To this end, we load the *causalweight* package and the *JC* data. The latter contains binary indicators for training participation in the first and second years after the start of the Job Corps program in columns 37 and 38, respectively. Accordingly, we define the treatment in the first and second periods as $D1=JC[,37]$ and $D2=JC[,38]$, respectively. That is, we use the second argument (right of the comma) in the square brackets to select the column in the data that contains the treatment of interest, while leaving the first argument (left of the comma) blank, such that all rows (or observations) are included. Columns 2 to 29 contain a range of characteristics measured prior to the first treatment (including gender, age, ethnicity, education, family background, and health), based on which we generate the baseline covariates $X0=JC[,2:29]$. Using the colon in the second argument of the square brackets permits us to choose the range of columns (2:29) to be included in the covariates. Accordingly, columns 30 to 36 contain covariates (on labor market participation and health) that are measured after the assignment of the first treatment but prior to the second treatment, defined as $X1=JC[,30:36]$. Finally, our outcome of interest is the weekly earnings in the fourth year, provided in column 44, and for this reason, we generate the R object $Y2=JC[,44]$.

After defining the covariates, treatments, and outcomes in the various time periods, we set a seed (*set.seed(1)*) and feed the variables into the *dyntreatDML* command, a DR approach for estimating the causal effects of treatment sequences based on equation (4.57) under the assumptions in expression (4.54). By default, the method estimates the ATE of being treated in both periods $(D_1 = 1, D_2 = 1)$ versus no treatment in either period $(D_1 = 0, D_2 = 0)$. It is also worth mentioning that the conditional mean outcomes and treatment propensity scores are based on lasso regression, a machine learning approach discussed in section 5.2 in chapter 5. We store the results in an object named *output* and inspect the ATE estimate, the standard error, and the p-value by calling *output$effect*, *output$se*, and *output$pval*, respectively. The box here provides the R code for each of the steps.

```
library(causalweight)        # load causalweight package
data(JC)                     # load JC data
X0=JC[,2:29]                 # define pretreatment covariates X0
X1=JC[,30:36]                # define posttreatment covariates X1
D1=JC[,37]                   # define treatment (training) in first year D1
D2=JC[,38]                   # define treatment (training) in second year D2
Y2=JC[,44]                   # define outcome (earnings in fourth year) Y2
set.seed(1)                  # set seed
output=dyntreatDML(y2=Y2,d1=D1,d2=D2,x0=X0,x1=X1)  # doubly robust estimation
output$effect; output$se; output$pval # effect, standard error, p-valu
```

The estimate suggests that the ATE $E[Y(1,1) - Y(0,0)]$ on weekly earnings in the fourth year amounts to 42.84 USD. The standard error of 5.07 is relatively small and the p-value is very close to zero, such that we can safely reject the null hypothesis of a zero ATE at any conventional level of statistical significance. We could easily change the default setting of comparing the sequences of being treated in both periods versus no treatment in either period. Setting, for instance, the arguments *d1treat=0*, *d2treat=1*, *d1control=1*, and *d2control=0* in the *dyntreatDML* command would estimate the ATE $E[Y(0,1) - Y(1,0)]$: that is, the effect of exclusive treatment in the later period versus exclusive treatment in the earlier period.

## 4.10   Causal Mechanisms (Mediation Analysis)

Causal mediation analysis assesses the causal mechanisms through which the treatment affects the outcome, and it has a causal structure that appears quite related to that of dynamic treatment evaluation, as discussed in section 4.9. More concisely, mediation analysis aims at disentangling a total treatment effect into one or several indirect effects, operating through one or several intermediate variables that are commonly referred to as *mediators*, as well as a direct effect, which includes any causal mechanisms not operating through the mediators of interest. As an example, let us reconsider the effect on employment or earnings of a sequence of training programs such as job application training and an IT course. Here, we could be interested in the direct effect of the job application training net of participation in the IT course by setting the latter to zero. This implies assessing the treatment effect $\Delta(\underline{d}_2, \underline{d}_2')$ with sequences $\underline{d}_2 = (1, 0)$ and $\underline{d}_2' = (0, 0)$, a parameter known as the *controlled direct effect*; see, for instance, Pearl (2001). That is, the net effect of $D_1$ is obtained by controlling for the mediator $D_2$, namely by setting it to the same value for everyone in the population (in our example, to zero).

However, the direct controlled effect can also be assessed when setting $D_2$ to a nonzero value (e.g., participation in an IT course). The sizes of the direct effects might generally differ across the values of $D_2$ if there are interaction effects between $D_1$ and $D_2$, such as if the job application training is particularly effective for individuals who also take an IT course. We can evaluate such controlled direct effects based on the same sets of assumptions as discussed in section 4.9.

Besides controlled direct effects, causal mediation analysis is also concerned with the evaluation of natural direct or indirect effects; for instance, see Robins and Greenland (1992). In this case, the direct effect of $D_1$, conditional on that value of the mediator $D_2$ that is naturally chosen as a reaction to $D_1$ (rather than being set to the same value for all, as for the controlled direct effect), is of interest, as well as the indirect effect operating through the choice of $D_2$ in reaction to $D_1$. For instance, some individuals might participate in an IT course after participating in job application training, while others might not. Therefore, the naturally chosen value of $D_2$ under a specific value of $D_1$ may vary across individuals (e.g., as a function of their preferences or other characteristics). Depending on whether specific values of $D_2$ can and should be imposed for everyone or subjects are permitted to choose $D_2$ in response to $D_1$, either controlled or natural effects may appear more relevant in a given empirical context.

To formalize the evaluation of natural direct and indirect effects, let us for the moment assume a binary $D_1$ and extend the potential outcome notation to the mediator $D_2$, such that $D_2(d_1)$ denotes the potential mediator as a function of the value $d_1 \in \{0, 1\}$. Furthermore, we denote by $E[Y_2(d_1, D_2(d_1'))]$ the potential outcome conditional on $D_1 = d_1$ and the potential mediator under $D_1 = d_1'$, with treatment values $d_1, d_1' \in \{0, 1\}$. Using this notation, the total ATE of $D_1$ on $Y_2$ corresponds to

$$\Delta(D_1) = E[Y_2(1, D_2(1)) - Y_2(0, D_2(0))]. \tag{4.59}$$

To separate the natural direct effect of $D_1$ from its indirect effect operating via $D_2$, we add and subtract either $E[Y_2(0, D_2(1))]$ or $E[Y_2(1, D_2(0))]$ on the right side of equation (4.59):

$$\Delta(D_1) = \underbrace{E[Y_2(1, D_2(1)) - Y_2(0, D_2(1))]}_{=\theta(1)} + \underbrace{E[Y_2(0, D_2(1)) - Y_2(0, D_2(0))]}_{=\delta(0)}$$

$$= \underbrace{E[Y_2(1, D_2(0)) - Y_2(0, D_2(0))]}_{=\theta(0)} + \underbrace{E[Y_2(1, D_2(1)) - Y_2(1, D_2(0))]}_{=\delta(1)}. \tag{4.60}$$

Here, $\theta(1)$ and $\theta(0)$ correspond to the natural direct effects of $D_1$ (i.e., net of indirect effect of $D_1$ on $Y_2$ through its impact on $D_2$) when setting the potential mediator to its value under either $d_1 = 1$ or $d_1 = 0$, respectively. Further, $\delta(1)$ and $\delta(0)$ are the natural indirect effects: that is, the difference in the mean potential outcomes due to a $D_1$-induced shift of the mediator from $D_2(0)$ to $D_2(1)$, while keeping $D_1$ fixed at either 1 or 0, respectively. As for the controlled direct effect, $\theta(1)$ and $\theta(0)$ might differ if there are interaction effects between $D_1$ and $D_2$. The same argument applies to $\delta(1)$ and $\delta(0)$. Equation (4.60) also highlights that the total effect of $D_1$ is the sum of the natural direct and indirect effects defined based on opposite treatment states: that is, either $\theta(1) + \delta(0)$ or $\theta(0) + \delta(1)$. This implies that interaction effects between $D_1$ and $D_2$ are either accounted for in the direct or the indirect effect, but not both at the same time.

Furthermore, it is worth noting that $Y_2(1, D_2(0))$ and $Y_2(0, D_2(1))$ are fundamentally counterfactual, in the sense that they are never observed for any subject, because mediator and outcome values can only be observed for the same (f)actual treatment, rather than for opposite treatment states. For this reason, the identification of natural direct and indirect effects hinges on stronger assumptions than what is required for controlled or dynamic treatment effects. A first additional assumption is that $D_1$ must be conditionally independent of the potential mediators (rather than of the potential outcomes only) such that no unobservables jointly affect $D_1$ and $D_2$, given $X_0$: $D_2(d_1) \perp D_1 | X_0$. Second, Avin, Shpitser, and Pearl (2005) show that the nonparametric identification of natural direct and indirect effects requires that the baseline covariates $X_0$ and the treatment $D_1$ be sufficient to control for confounders of $D_2$ and $Y_2$. This implies that conditioning on covariates $X_1$, which are possibly affected by $D_1$, is not necessary, a scenario also discussed at the end of section 4.9.

For the sake of generality, let us now allow for a multivalued, discrete treatment (e.g., alternative training programs) taking values $\{0, 1, \dots, J\}$ as already considered for the evaluation of dynamic treatment effects. Incorporating our previous considerations entails the following set of assumptions for evaluating natural direct and indirect effects:

$$\{Y_2(\underline{d_2}), D_2(d_1')\} \perp D_1 | X_0 \text{ and } Y_2(\underline{d_2}) \perp D_2 | D_1, X_0 \text{ for } \underline{d_2} = (d_1, d_2)$$

$$\text{and } d_1, d_1', d_2 \in \{0, 1, \dots, J\},$$

$$\Pr(D_1 = d_1 | X_0) > 0 \text{ and } \Pr(D_2 = d_2 | D_1, X_0) > 0 \text{ for } d_1, d_2 \in \{0, 1 \dots, J\}. \tag{4.61}$$

The causal graph provided in figure 4.13 satisfies the assumptions in expression (4.61) under the condition that any unobserved characteristics omitted from the graph do not jointly affect two out of the three variables, $D_1, D_2$, and $Y_2$.

Similar to previously considered causal effects, we may identify the mean potential outcomes that are required for the computation of natural direct and indirect effects in equation (4.60) based on alternative strategies. A first approach relies on nested conditional mean outcomes, which might be estimated by matching or regression estimators; for instance, see Imai, Keele, and Yamamoto (2010):

$$E[Y_2(d_1, D_2(d_1'))] = E[E[\mu^{Y_2}(d_1, D_2, X_0)|D_1 = d_1', X_0]], \qquad (4.62)$$

where $\mu^{Y_2}(D_1, D_2, X_0) = E[Y_2|D_1, D_2, X_0]$ denotes the conditional mean outcome, and $d_1$, $d_1'$ denote specific values of the first treatment. A second strategy assesses the mean potential outcomes by IPW (see, e.g., Hong (2010)):

$$E[Y_2(d_1, D_2(d_1'))] = E\left[\frac{I\{D_1 = d_1\} \cdot p^{D_2}(d_1', X_0) \cdot Y_2}{p^{d_1}(X_0) \cdot p^{D_2}(d_1, X_0)}\right] \qquad (4.63)$$

where $p^{d_2}(D_1, X_0) = \Pr(D_2 = d_2|D_1, X_0)$ denotes the propensity score of the mediator, which is a conditional density in the case of a continuous mediator.

Finally, DR identification combines IPW with conditional mean outcomes, as discussed by Tchetgen Tchetgen and Shpitser (2012):

$$E[Y_2(d_1, D_2(d_1'))] = E[\psi^{d_1, d_1'}],$$

$$\text{with } \psi^{d_1, d_1'} = \frac{I\{D_1 = d_1\} \cdot p^{D_2}(d_1', X_0)}{p^{d_1}(X_0) \cdot p^{D_2}(d_1, X_0)} \cdot [Y_2 - \mu^{Y_2}(d_1, D_2, X_0)]$$

$$+ \frac{I\{D_1 = d_1'\}}{p^{d_1'}(X_0)} \cdot [\mu^{Y_2}(d_1, D_2, X_0) - E[\mu^{Y_2}(d_1, D_2, X_0)|D_1 = d_1', X_0]]$$

$$+ E[\mu^{Y_2}(d_1, D_2, X_0)|D_1 = d_1', X_0]. \qquad (4.64)$$

It is also worth noting that by an application of Bayes' law, the mediator probability/density $p^{d_2}(D_1, X_0)$ appearing in equations (4.63) and (4.64) can be avoided by instead including another treatment propensity score $p^{d_1}(D_2, X_0) = \Pr(D_1 = d_1|D_2, X_0)$. As discussed in Huber (2014a), equation (4.63) is equivalent to

$$E[Y_2(d_1, D_2(d_1'))] = E\left[\frac{I\{D_1 = d_1\} \cdot p^{d_1'}(D_2, X_0) \cdot Y_2}{p^{d_1}(D_2, X_0) \cdot p^{d_1'}(X_0)}\right] \qquad (4.65)$$

From a practical perspective, relying on the estimation of $p^{d_1}(D_2, X_0)$ rather than $p^{D_2}(d_1, X_0)$ might be preferred if the mediator is continuously distributed and/or consists of several variables, such that the estimation of $p^{D_2}(d_1, X_0)$ (e.g., by kernel methods) might be cumbersome.

Assuming that the mediator is as good as being randomly assigned when controlling for treatment $D_1$ and baseline covariates $X_0$ alone appears to be a strong assumption. In many applications, it may seem more plausible that we also need to control for posttreatment covariates $X_1$, as discussed in section 4.9. This implies that assumption $Y_2(\underline{d_2}) \perp D_2|D_1, X_0$ in expression (4.61) is to be replaced by $Y_2(\underline{d_2}) \perp D_2|D_1, X_0, X_1$. However, the latter is not sufficient for the nonparametric identification of natural direct and indirect effects, which therefore requires additional assumptions, very much in contrast to controlled direct or dynamic treatment effects. An additional restriction is ruling out confounders that jointly affect (i) $D_1$ and $X_1$, given $X_0$, or (ii) $X_1$ and $D_2$ or $Y_2$, given $D_1, X_0$. This assumption is satisfied in figure 4.12, where assumably no unobservables jointly affect $D_1$ and $X_1$, $X_1$ and $D_2$, or $X_1$ and $Y_2$. Under this condition, we can identify the path-wise (or partial indirect) effect of $D_1$ on $Y_2$ directly operating via $D_2$, i.e., $D_1 \rightarrow D_2 \rightarrow Y_2$, for instance based on IPW, see Huber (2014a), which for a binary treatment corresponds to the following expression:

$$\delta^p(d_1) = E\left[ \frac{Y_2 \cdot I\{D_1 = d_1\}}{\Pr(D_1 = d_1 | D_2, X_0, X_1)} \cdot \frac{\Pr(D_1 = d_1 | X_0, X_1)}{\Pr(D_1 = d_1 | X_0)} \right.$$
$$\left. \times \left( \frac{\Pr(D_1 = 1 | D_2, X_0, X_1)}{\Pr(D_1 = 1 | X_0, X_1)} - \frac{1 - \Pr(D_1 = 1 | D_2, X_0, X_1)}{1 - \Pr(D_1 = 1 | X_0, X_1)} \right) \right], \quad (4.66)$$

with $\delta^p(d_1)$ denoting the pathwise effect of $D_1 \rightarrow D_2 \rightarrow Y_2$.

However, $\delta^p(d_1)$ represents only a partial indirect effect because it omits any indirect impact that operates via $X_1$ (namely, the path $D_1 \rightarrow X_1 \rightarrow D_2 \rightarrow Y_2$) and for this reason, it does not coincide with the natural indirect effect $\delta(d_1)$. To identify the arguably more interesting full natural indirect effect (along with the corresponding natural direct effect, such that both add up to the ATE), we would need to impose even further assumptions. One possibility is to rule out interaction effects between $D_1$ and $D_2$, such that the effect of the treatment does not depend on that of the mediator and vice versa; see, for instance, Robins (2003). For a binary treatment, this implies that $Y(1, m) - Y(0, m) = Y(1, m') - Y(0, m')$ for any distinct mediator values $m \neq m'$. Such an approach, however, may appear unattractive in many empirical contexts, as it severely restricts effect heterogeneity. Improving on the latter situation, Imai and Yamamoto (2013) demonstrate that natural effects can also be identified under a treatment-mediator interaction effect, so long as the latter is homogeneous (i.e., the same) for different subjects, which is admittedly not an uncontroversial restriction either.

Alternatively, Tchetgen Tchetgen and VanderWeele (2014) show that we can assess $\delta(d)$ and $\theta(d)$ if the average interaction effects of $X_1$ and $D_2$ on $Y_2$ amount to zero, which cannot be taken for granted either. Yet another, arguably very strong, restriction is that potential values of $X_1$ under treatment and nontreatment are statistically independent, or that the form of their statistical association is known; see Robins and Richardson (2010) and Albert and Nelson (2011). Finally, Xia and Chan (2021) provide a DR approach under the assumption that conditional on $X_0$, the average effects operating via the causal paths $D_1 \rightarrow Y_2$ and $D_1 \rightarrow X_1 \rightarrow Y_2$ are homogeneous across values of the potential mediator under nontreatment $M(0)$. While some of these additional assumptions required for assessing full rather than partial natural direct and indirect effects may appear more or less attractive than others, they all share the caveat that they impose rather specific constraints that may appear questionable in many empirical contexts, a price that we have to pay in the presence of posttreatment confounders $X_1$.

To illustrate the evaluation of the controlled direct effect in R under the assumptions in expression (4.54), let us use the same data set, variable definitions, and DR-based estimation approach as outlined at the end of section 4.9. The only modification that we make is setting the argument *d2treat=0* in the *dyntreatDML* command. The procedure then yields an estimate of the direct controlled effect of the sequence $E[Y(1,0) - Y(0,0)]$ (rather than of $E[Y(1,1) - Y(0,0)]$ as in section 4.9): that is, the net effect of the first training when switching off the second training for everyone:

```
output=dyntreatDML(y2=Y2,d1=D1,d2=D2,x0=X0,x1=X1, d2treat=0) # estimation
output$effect; output$se; output$pval # effect, standard error, p-value
```

The controlled direct effect amounts to 21.29 and is highly statistically significant, with a standard error of 5.30 and a p-value that is very close to zero.

As a further example for causal mediation analysis, we reconsider the *wexpect* data in the *causalweight* package introduced at the end of section 3.5 to estimate the natural direct and indirect effects of gender on the wage expectations of Swiss students. More concisely, we are interested in whether male and female students differ in their wage expectations because of choosing different study programs, which would imply an indirect effect via the mediator study program, or because of other reasons. The latter would point to a direct effect of gender on wage expectations, i.e., that is, net of the choice of the study program. After loading the *causalweight* package and the *wexpect* data, we wrap the latter by the *attach* command to save all variables as own R objects. Female and male students going to college might differ in terms of their family background, which might also affect their choice of study program and their wage expectations. For this reason, we aim at controlling for baseline characteristics that are jointly associated with gender, on the one hand, and the mediator or outcome, on the other hand.

To this end, we use the *cbind* command to define a set of covariates $X$ consisting of parents' eduction (*motherhighedu* and *fatherhighedu*), student's *age*, and an indicator for being *swiss*. Furthermore, we define a binary indicator for being male as treatment, $D=male$, a set of indicators for study program choice (business, economics, communication, or business informatics) as mediators,

*M=cbind(business,econ,communi,businform)*, and the expectations about the monthly gross wages three years after studying as the outcome: *Y=wexpect2*. Assuming that the conditions in expression (4.61) are satisfied (which may admittedly be challenged in our toy example with only a few control variables), we use *Y*, *D*, *M*, and *X* to run the *medDML* command. The latter is a DR procedure based on equation (4.64), which estimates any propensity scores and conditional mean outcomes based on lasso regression, a machine learning approach discussed in section 5.2. The box here provides the R code for each of the steps.

```
library(causalweight)                              # load causalweight package
data(wexpect)                                      # load wexpect data
attach(wexpect)                                    # attach data
X=cbind(age,swiss,motherhighedu,fatherhighedu)     # define covariates
D=male                                             # define treatment
M=cbind(business,econ,communi,businform)           # define mediator
Y=wexpect2                                          # define outcome
medDML(y=Y, d=D, m=M, x=X)                          # estimate causal mechanisms
```

This gives the following output:

```
$results
              total     dir.treat   dir.control  indir.treat indir.control  Y(0,M(0))
effect  1.355195e+00  1.088359e+00 1.146031e+00   0.20916428   0.26683632   8.691524
se      1.932545e-01  2.263925e-01 2.125098e-01   0.08880141   0.11041265   0.136727
p-val   2.341151e-12  1.529081e-06 6.935447e-08   0.01850198   0.01566112   0.000000

$ntrimmed
[1] 0
```

The first column provides the estimate of the total ATE, $\Delta$, along with its standard error and p-value; the second and third show those of the direct effects under treatment, $\theta(1)$, and nontreatment, $\theta(0)$; the fourth and fifth show those of the indirect effects under treatment, $\delta(1)$, and nontreatment, $\delta(0)$; and the final column contains an estimate of the mean potential outcome under nontreatment, $E[Y(0, M(0))]$. Keeping in mind that the outcome variable is measured in steps of 500 CHF, the ATE estimate suggests that males expect on average $1.355 \times 500 = 678$ CHF higher monthly gross wages than females who are comparable in terms of *X*.

This overall effect is mostly driven by the direct effect of gender, reflecting causal mechanisms other than the choice of the study program, as the estimates of $\theta(1)$ and $\theta(0)$ amount to $1.088 \times 500 = 544$ CHF and $1.146 \times 500 = 573$ CHF, respectively. The indirect effect, which is due to gender-specific differences in the choice of study program, is considerably smaller, with the estimates of $\delta(1)$ and $\delta(0)$ corresponding to roughly 105 CHF and 133 CHF, respectively. Any of the total, direct, and indirect effect estimates are statistically significant at the 5 percent level. Our results, therefore, suggest that differential wage expectations of female and male students can be explained only to some extent by the choice of study program, and to a much larger extent by other factors entering the direct effect (which might include personality traits or job expectations).

## 4.11   Outcome Attrition and Posttreatment Sample Selection

The evaluation of causal effects is frequently complicated by the issue that the outcome of interest is observed only for a nonrandom subsample in the data. One example is nonrandom outcome attrition, which occurs if the outcome is measured in a follow-up survey (e.g., by means of a questionnaire or interview several days, months, or years after treatment assignment) and some of the initial study participants cannot be interviewed anymore. This might be due to their relocation or reluctance to participate in the interview. Another example is posttreatment sample selection, implying that the outcome is observed only conditional on some other posttreatment variable. This applies to the evaluation of the effect of education on hourly wages, as the wages are observed only conditional on being employed, or to the assessment of educational interventions like school vouchers on college admissions tests, as test results are observed only conditional on participating in the test. In general, sample selection and outcome attrition create a bias when estimating causal effects, even in the case that the treatment is randomized. But are there scenarios or conditions that permit us to fix this problem?

Similar to for the treatment assignment in section 4.1, one approach is to impose a selection-on-observables assumption, now with respect to outcome attrition/sample selection, implying that the latter is as good as random, conditional on observed information like the covariates and the treatment. This is also known as a *missing at random (MAR) assumption*; see, for instance, Rubin (1976). The evaluation of the ATE or other causal effects then relies on a sequential selection-on-observables assumption with regard to treatment and the attrition problem, which appears somewhat related to the dynamic treatment context. Building on the notation of section 4.9, let $D_1$ denote the (possibly multivalued) treatment of interest, while $D_2$ is (contrary to the previous sections 4.9 and 4.10) not a treatment or mediator that might have an effect on outcome $Y$, but a binary indicator of whether the outcome is observed. That is, $Y$ is known only for observations with $D_2 = 1$, but unknown if $D_2 = 0$. Accordingly, the potential outcome $Y_2(D_1)$ is a function of $D_1$ only, but not of the indicator for its observability, $D_2$. We can assess the ATE of $D_1$ under the

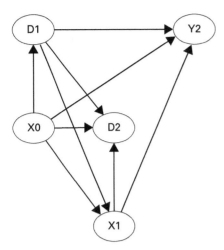

**Figure 4.14**
Causal paths under sequential conditional independence.

following assumptions, such as discussed in Bia, Huber, and Lafférs (2021):

$$Y_2(d_1) \perp D_1 | X_0 \text{ and } Y_2 \perp D_2 | D_1, X_0, X_1 \text{ for } d_1 \in \{0, 1, \dots, J\} \text{ and } d_2 \in \{0, 1\},$$

$$\Pr(D_1 = d_1 | X_0) > 0 \text{ and } \Pr(D_2 = d_2 | D_1, X_0, X_1) > 0 \text{ for } d_1 \in \{0, 1, \dots, J\} \text{ and } d_2 \in \{0, 1\}.$$

$$(4.67)$$

Figure 4.14 provides a causal graph that satisfies the conditional independence assumptions in equation (4.67), given that there are no omitted unobservables that jointly affect $D_1$ and $Y_2$ or $D_2$ and $Y_2$.

The conditions in expression (4.67) appear somewhat similar to those suggested in expression (4.54) of section 4.9. The key difference is that we now assume that $Y_2 \perp D_2 | D_1, X_0, X_1$, which implies that the selection indicator $D_2$ does not affect $Y$, as it is not a treatment. A further implication is that $D_2$ is not associated with unobserved characteristics affecting $Y$, conditional on covariates $X_0, X_1$ and treatment $D_1$. Under these conditions, we can assess the ATE $\Delta(d_1, d'_1) = E[Y(d_1) - Y(d'_1)]$ for two distinct values $d_1 \neq d'_1$ of treatment $D_1$ by equations (4.55), (4.56), and (4.57) when setting $D_2 = 1$ in any conditional mean outcome and propensity score. Therefore, we may apply the very same identification results of section 4.9 for assessing treatment effects under outcome attrition or sample selection when imposing the assumptions in expression (4.67) rather than expression (4.54).

In analogy to the discussion at the end of section 4.9, the evaluation framework simplifies somewhat if conditioning on $X_1$ is not required such that outcome attrition/sample selection is as good as random, given treatment $D_1$ and baseline covariates $X_0$ alone. Then, we can drop $X_2$ from the assumptions in expression (4.67) and any conditional mean outcomes or propensity scores entering the expressions for the identification of the ATE. For instance, the use of $\psi^{d_2}$ in equation (4.58) rather than that in equation (4.57) permits identifying $\Delta(d_1, d_1')$. Similar to the discussion of dynamic treatment effects, this approach may appear unrealistic in scenarios when there is a substantial time lag between the treatment and the measurement of the outcome and outcome attrition, such that posttreatment confounders affecting both $D_2$ and $Y_2$ likely exist.

# 5

# Causal Machine Learning

## 5.1 Motivation for Machine Learning and Fields of Application

In chapter 4, we encountered a range of methods for assessing causal effects based on using treated and nontreated observations that are comparable in terms of observed covariates $X$. We have so far (maybe more implicitly than explicitly) assumed that the researcher or analyst preselects the covariates to be controlled for before estimating any causal effect. This requires substantial, if not exact, contextual knowledge about which covariates are to be included to satisfy the selection-on-observables assumption. Even though reasoning based on theory, intuition, or previous empirical findings may and should guide the selection of covariates when aiming at making the selection-on-observables assumption plausible, the exact set of variables that is sufficient for satisfying this assumption (if such a set exists at all) is typically unknown. This applies in particular to scenarios where the number of available control variables is large. For instance, in big data contexts in which the set of covariates is so rich that the selection-on-observables assumption appears likely satisfied, an a priori unknown subset of those covariates might be sufficient to tackle confounding.

As researchers or analysts, we would like to learn a model specification that includes such a sufficient set of covariates. However, by considering $X$ to be preselected, statistically speaking, we assume away any model or estimation uncertainty coming from the selection of $X$. This stands in stark contrast to the common practice of selecting covariates based on how well they predict the treatment, the outcome, or both in the data. Without appropriately accounting for this selection step when estimating the treatment effect and determining its statistical significance, such an approach may entail misleading causal conclusions. In fact, treatment effect estimators like inverse probability weighting (IPW), matching, and regression are generally not robust to human-made ad hoc rules for selecting covariates in the sense that their p-values and confidence intervals might be incorrect. Is there a way to properly conduct and account for covariate selection in causal analysis under specific conditions? Indeed, causal machine learning (CML) approaches aim at (1) avoiding the ad hoc selection of covariates by controlling for the latter in a data-driven way and (2) providing valid inference (e.g., p-values and confidence intervals) under such a data-driven covariate selection.

CML, therefore, appears particularly useful in big, and more specifically in wide (or high-dimensional) data with a vast number of covariates that could potentially serve as control variables, which can render covariate selection by the researcher or analyst complicated, if not infeasible. However, it is important to stress that data-driven covariate selection cannot do away with fundamental assumptions required for the identification of causal effects. Just as for the methods outlined in chapter 4, the data must contain sufficiently rich covariate information to satisfy the selection-on-observables assumption in expression (4.1). Under this precondition, CML may be applied if there is a subset of covariate information that suffices to effectively tackle confounding, but it is a priori unknown. Under the assumption that a limited subset of covariate information (limited relative to the sample size) permits controlling for the most important confounders, CML can be approximately unbiased, meaning that the bias is negligibly close to zero and $\sqrt{n}$-consistent, even when confounding is not perfectly controlled for.

In addition to the estimation of average (or even distributional) effects in the total population, we can apply CML to the data-driven detection of important hetero-geneities in causal effects across subpopulations that are defined in terms of covariates $X$. For instance, it might be the case that a specific pricing policy (e.g., a discount on a product or service) is more successful in boosting sales among younger than among older customers, or among the less educated rather than the more educated. CML may reveal such a priori unknown effect heterogeneities. This also permits learning the optimal policy in terms of treatment assignment (e.g., granting discounts) across various subgroups depending on their observed characteristics, while at the same time taking into account treatment costs (e.g., the cost of granting discounts) in order to maximize the net benefits of the assignment. A further strand of CML that is in the spirit of optimal policy learning is reinforcement learning. It is based on repeatedly allocating alternative treatments (e.g., distinct advertisements) in different time periods to (1) iteratively learn from the data which treatment is most effective and (2) ultimately focus on the provision of the most effective treatment to maximize the causal impact.

## 5.2   Double Machine Learning and Partialling out with Lasso Regression

One CML approach for estimating the average treatment effect (ATE) or other causal effects is double machine learning (DML), as discussed in Chernozhukov et al. (2018), which relies on Neyman-orthogonal functions for treatment effect estimation (Neyman 1959). Considering a binary treatment, Neyman-orthogonality implies that treatment effect estimation is relatively robust (i.e., first-order insensitive) to approximation errors in the estimation of the treatment propensity score $p(X)$ and the conditional mean outcomes $\mu_1(X)$, $\mu_0(X)$ introduced in chapter 4. It turns out that the doubly robust (DR) estimator based on the sample analog of equation (4.45), whose desirable properties were discussed in section 4.6, satisfies this robustness property, as well as other DR approaches like targeted maximum likelihood estimation (TMLE). In contrast, estimation based on equation (4.3) or (4.4) is not robust to approximation errors of $\mu_1(X)$ and $\mu_0(X)$, while estimation based on equation (4.40) or (4.41) is not robust to errors in $p(X)$. Because DR incorporates both propensity score and conditional mean outcome estimation, the approximation errors can be shown to enter multiplicatively into the estimation problem, such that small errors in either become negligible when multiplied. This is key for the robustness property, as discussed in Farrell (2015).

CML and DML owe their name to the fact that the model parameters (e.g., regression coefficients) of $p(X)$, $\mu_1(X)$, and $\mu_0(X)$ are estimated by machine learning, a subfield of artificial intelligence. We will discuss some machine learning approaches later in this chapter, as well as in section 5.3. However, it needs to be stressed that CML is conceptually different to conventional (predictive) machine learning. The latter aims at accurately predicting an outcome by predictor variables based on minimizing the prediction error, such as the mean squared error (MSE), through optimally trading off prediction bias and variance.

This mere forecasting approach generally does not allow us to learn the causal effects of any of the predictors. One reason is that any predictor might be assigned less importance or weight (e.g., through a regression coefficient) in the forecasting process than implied by its true causal effect. This is the case, for instance, if the predictor is strongly correlated with other predictors (e.g., work experience might be strongly correlated with education), such that constraining the predictor's weight hardly affects the prediction bias (as the correlated predictor contains little additional information for prediction), while reducing the variance. Therefore, predictive machine learning with $Y$ as the outcome and treatment $D$ and covariates $X$ as predictors generally gives a biased estimate of the causal effect of $D$, due to correlations between the treatment and covariates. But even without such a correlation between $D$ and $X$, the estimate may be biased if the causal effect of $D$ on $Y$ is rather small (in absolute terms) relative to the importance of $X$ for predicting $Y$.

In DML, however, machine learning is not directly applied to treatment effect estimation, but rather merely for predicting the plug-in parameters $p(X)$, $\mu_1(X)$, and $\mu_0(X)$ which enter expression (4.45) in section 4.6 of chapter 4. To this end, we conduct separate machine learning predictions of $D$ as a function of $X$, $Y$ among the treated as a function of $X$, and $Y$ among the nontreated as a function of $X$. This is motivated by the fact that covariates $X$ merely serve the purpose of tackling confounding, while their causal effects are (very much in contrast to the effect of $D$) not of interest. This makes the estimation of $p(X)$, $\mu_1(X)$, and $\mu_0(X)$ a prediction problem to which we can apply machine learning, while we use the sample version (or sample analog) of equation (4.45) to estimate the causal effect of $D$.

To consider an example of machine learning, let us assume that in the spirit of equation (4.12), $\mu_1(X)$ and $\mu_0(X)$ are estimated by regressing $Y$ on $X$, as well as higher-order and interaction terms of $X$ in separate subsamples with $D = 1$ and $D = 0$ using lasso regression; see Tibshirani (1996). While both ordinary least squares (OLS) and lasso regression aim at finding the coefficient values that minimize the sum of squared residuals, only lasso regression includes a penalty term on the sum of the absolute values of the slope coefficients in the minimization problem. The aim of this penalization is to constrain (or regularize) the overall influence or importance of regressors when predicting the outcome. Intuitively, including too many regressors with low predictive power (as it would be the case in series estimation with irrelevant, nonpredictive, higher-order terms) likely increases the variance of prediction, with little gain in terms of bias reduction. On the other hand, omitting very important regressors implies a large increase in prediction bias relative to the gain in variance reduction. For this reason, lasso regression aims to optimally balance bias and variance through penalization, which is performed by shrinking the absolute coefficients that would be obtained in a standard OLS regression toward zero for less important regressors. The algorithm may even shrink coefficients exactly to zero, implying that the respective regressors are dropped from the model.

Taking the prediction of $\mu_1(X)$ as an example, lasso regression solves the following penalized minimization problem for obtaining the coefficients:

$$(\hat{\alpha}, \hat{\beta}_1, \ldots) = \arg \min_{\alpha^*, \beta_1^*, \beta_2^*, \ldots} \sum_{i: D_i = 1} (Y_i - \alpha^* - \beta_1^* X_{i1} - \beta_2^* X_{i2} - \ldots)^2 + \lambda \sum_{j=1}^{p} |\beta_j^*|, \qquad (5.1)$$

where $p$ denotes the number of regressors (consisting of $X$ and its higher-order/interaction terms), the nonnegative value $\lambda$ is the penalization term on the sum of absolute slope coefficients, and $|\cdot|$ stands for the absolute value. We note that for $\lambda = 0$, solving equation (5.1) corresponds to standard OLS regression without penalization of the coefficients and thus, the variance of predicting $\mu_1(X)$.

We may choose $\lambda$ by a cross-validation procedure for determining the optimal amount of shrinkage that minimizes the MSE for outcome prediction among a range of candidate values for $\lambda$. One possible approach is leave-one-out cross-validation in the spirit of equation (4.13) in section 4.2. In practice, a coarser version known as *k-fold cross-validation* is frequently applied, where the observations are randomly divided into $K$ nonoverlapping subsets. To this end, let us denote by $k \in \{1, \ldots, K\}$ a specific subset of observations, and let $k_i$ be the subset in which observation $i$

is situated. Then, we may select $\lambda$ such that it minimizes the squared residuals of predicted and observed treated outcomes:

$$\sum_{i:D_i=1} [Y_i - \hat{\mu}_{1,-k_i}(X_i)]^2, \tag{5.2}$$

where $\hat{\mu}_{1,-k_i}(X_i)$ is the prediction of $\mu_1(X)$ based on coefficients that have been estimated when exclusively using observations that are not in observation $i$'s subset $k_i$.

It is easy to see that this approach coincides with leave-one-out cross-validation when $k_i$ only contains a single observation (namely, $i$). However, particularly when samples are large, considering subsets with more than just one observation (and thus reducing the number of subsets, $K$) reduces the computational burden for selecting $\lambda$. As for $\mu_1(X)$ and $\mu_0(X)$, we can also estimate the propensity score $p(X)$ by a penalized (e.g., lasso logit) regression, which includes a penalty term in logit-based maximum likelihood estimation. As an alternative CML approach, the lasso-based estimation of $\mu_1(X)$ and $\mu_0(X)$ can be combined with approximate covariate balancing of Zubizarreta (2015) (see the discussion in section 4.5) instead of estimating a propensity score model for $p(X)$; see Athey, Imbens, and Wager (2018). Similar to the arguments made in section 4.2, it is worth mentioning that cross-validation yields a value of $\lambda$ that is optimal for estimating the plug-in parameter, but not necessarily for the ATE, the conditional average treatment effect (CATE), or the average treatment effect on the treated (ATET). Nevertheless, it constitutes a feasible approach for picking an adequate penalization in practice.

A further element of several CML approaches, including DML, is the use of independent samples for estimating the model parameters of the plug-ins $p(X)$, $\mu_1(X)$, and $\mu_0(X)$ on the one hand and of the treatment effects, such as $\Delta$ or $\Delta_{D=1}$, on the other hand. To this end, we randomly divide the total sample into two (nonoverlapping) parts or folds. We then estimate the model parameters for $p(X), \mu_1(X), \mu_0(X)$ (i.e., the coefficients in lasso regressions of the treatment and outcome) in the first fold. Based on these lasso coefficients, we predict the plug-ins $p(X), \mu_1(X), \mu_0(X)$ in the second fold and estimate the treatment effect of interest in the second fold, e.g., by the sample analog of equation (4.45). Figure 5.1 provides a graphical illustration of the workflow in the sample-splitting procedure.

Sample splitting avoids correlations between the two estimation steps (namely, estimating the model parameters of the plug-ins and the treatment effect). For this reason, sample splitting prevents overfitting bias related to fitting the respective models too much to the data points in a sample, which entails an underestimation of the error terms. However, sample splitting apparently comes with the cost that only part of the data is used to estimate the causal effect, thus increasing the variance. We can tackle this issue by cross-fitting, which consists of swapping the roles of the folds used for estimating the plug-in models and the treatment effect. That is, in a second

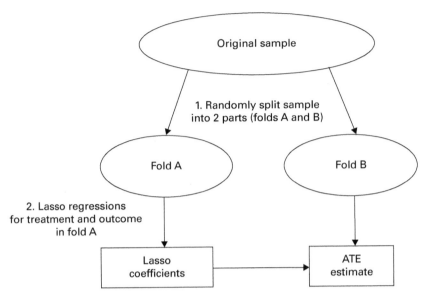

**Figure 5.1**
Sample splitting when estimating the ATE.

estimation round, we use the second fold for obtaining the lasso coefficients for the plug-in models and the first fold for treatment effect estimation. We ultimately estimate, for instance, the ATE by taking the average of the ATE estimates in either fold.

Under sufficiently well behaved plug-in estimators, treatment effect estimation based on DML with cross-fitting is $\sqrt{n}$-consistent and asymptotically normal. More concisely, $\sqrt{n}$-consistency of DML is satisfied if (among some other conditions) the plug-in estimates of $p(X), \mu_1(X), \mu_0(X)$ converge to their respective true values at a faster rate than $n^{-1/4}$; see Chernozhukov et al. (2018). A rate faster than $n^{-1/4}$ can obviously still be considerably slower than the conventional $\sqrt{n}$-rate and is attained by many machine learning or deep learning algorithms under specific conditions; for instance, see the discussions in Kueck, Luo, Spindler, and Wang (2022) and Farrell, Liang, and Misra (2021). A further property worth noting is that under these conditions, the asymptotic variance of DML is unaffected by the machine learning and cross-fitting steps: that is, it is not higher than if the covariates $X$ to be controlled for had been known a priori rather than learned in the data by machine learning. Due to this oracle property, we may thus compute standard errors by conventional asymptotic approximations without adjustment for the cross-fitting and machine learning steps in large enough samples (even though these steps may affect the variance of DML in data with a limited sample size).

As discussed in Chernozhukov et al. (2018), lasso regression attains the $n^{-1/4}$-rate requirement under approximate sparsity. The latter implies that the number of important covariates or interaction and higher-order terms required for obtaining a sufficiently decent (albeit not perfect) approximation of the plug-in parameters is small relative to the sample size $n$. To see the merits of cross-fitting, we note that when disregarding the latter and instead conducting the lasso and treatment estimation steps in the same total sample, the number of important predictors is required to be small relative to $\sqrt{n}$ (which is a stronger condition than being small relative to $n$); see Belloni, Chernozhukov, and Hansen (2014).

Also, TMLE, another doubly robust approach briefly introduced in section 4.6, can be combined with the machine learning–based estimation of the plug-in parameters $p(X)$, $\mu_1(X)$, and $\mu_0(X)$ and cross-fitting or related sample-splitting approaches, as discussed in Zheng and van der Laan (2011). Similar to DML, this implies that the estimation of the model parameters (e.g., lasso regression coefficients) of the plug-in parameters and the TMLE-based ATE estimation take place in different folds of the data, with the roles of the folds being swapped. In this case, ATE estimation also can attain $\sqrt{n}$-consistency if particular regularity conditions like $n^{-1/4}$-consistency of the plug-in estimators are met.

Yet another CML approach is based on purging, or partialling out, the influence of covariates $X$ on outcome $Y$ and treatment $D$ prior to assessing the treatment effect as discussed in Belloni, Chernozhukov, and Hansen (2014), which may be combined with cross-fitting, too. In this case, we randomly split the data into two folds and use the first fold to run lasso regressions of $Y$ on a constant and $X$, as well as of $D$ on a constant and $X$, respectively, to obtain coefficient estimates for the models $E[Y|X]$ and $E[D|X]$. The coefficients might either correspond to those directly obtained from the lasso regressions or to those coming from OLS regressions of $Y$ or $D$ on a constant and the lasso-selected variables in $X$. The latter procedure is known as *postlasso OLS* and therefore uses lasso regression only to select important covariates, while the association of the latter with $D$ or $Y$ is ultimately estimated by OLS regression. In the second fold, we then use these lasso or postlasso coefficient estimates to predict the outcome residuals $Y - E[Y|X]$ and treatment residuals $D - E[D|X]$ and regress the former on the latter using OLS to obtain an estimate for the ATE. As for other CML approaches, we may swap the roles of the data sets and average the ATE over both folds of the data.

The partialling out approach is inspired by the work of Robinson (1988) on partial linear models. It is based on the idea that after purging the confounding influence of the covariates on the outcome and treatment by only considering the residuals of $Y$ and $D$, the causal effect of the binary treatment is identified by linear regression.

This is in analogy to section 3.2 in chapter 3, where the treatment-outcome relation is assumed to be unconfounded due to random treatment assignment, such that partialling out is not even necessary. We also can apply partialling out to continuous treatments. In this case, however, the treatment effect is generally not permitted to vary across covariate values, which amounts to assuming a constant or homogeneous CATE across $X$ such that the CATE coincides with the ATE, which is very much related to the discussion in section 4.2. Similar to DML, the partialling out strategy can be shown to be $\sqrt{n}$-consistent if the postlasso-based estimators of $E[Y|X]$ and $E[D|X]$ converge at least with rate $n^{-1/4}$ to the respective true models. A more in depth review of various machine learning algorithms, CML, and DML is provided in Athey and Imbens (2019); Kreif and DiazOrdaz (2019); Knaus (2021); Shah, Kreif, and Jones (2021); Chernozhukov, Hansen, Spindler, and Syrgkanis (2022); and Lieli, Hsu, and Reguly (2022). Approximate sparsity or related assumptions on X are popular in CML. It is worth mentioning, however, that even under a large (rather than sparse) set of control variables, treatment effect estimation may under specific conditions be asymptotically normal, even though it generally has a larger variance; see the discussion by Cattaneo, Jansson, and Newey (2018) and Jiang et al (2022).

To illustrate the implementation of DML in R, we load the *causalweight* package and reconsider the *JC* data for the evaluation of the Job Corps program already analyzed in section 4.9. This time, we are interested in the effect of training participation in the first year after program assignment on the health state four years after assignment when controlling for the baseline covariates measured prior to training. After loading the *JC* data using the *data* command, we define the set of covariates (stored in columns 2 to 29), X=JC[,2:29], the treatment (in column 37), D=JC[,37], and the health outcome (in column 46), Y=JC[,46], which is measured from 1 (excellent) to 4 (poor). We then run the *treatDML* command using these variables, a DML procedure that by default applies lasso regression for the estimation of the propensity scores and conditional mean outcomes. We store the results in a variable called *output* and inspect the ATE estimate, its standard error, and the p-value by calling *output$effect*, *output$se*, and *output$pval*. The box here provides the R code for each step.

```
library(causalweight)                    # load causalweight package
data(JC)                                 # load JC data
X=JC[,2:29]                              # define covariates
D=JC[,37]                                # define treatment (training) in first year
Y=JC[,46]                                # define outcome (health state after 4 years)
output=treatDML(y=Y, d=D, x=X)          # double machine learning
output$effect; output$se; output$pval   # effect, standard error, p-value
```

The DML estimator yields an ATE of $-0.052$, which points to a very moderate improvement in general health because smaller values imply a better health state. This rather small impact is nevertheless statistically significant at the 1 percent level, as the standard error amounts to 0.017 and the p-value to 0.0025 (or 0.25 percent).

## 5.3  A Survey of Further Machine Learning Algorithms

Lasso regression, as introduced in section 5.2, is arguably close to conventional regression, with the crucial difference being the penalization of the sum of absolute slope coefficients in the optimization problem (see equation (5.1)) to take the variance into account. However, there are many more machine learning methods, which are different in terms of estimation but share the idea of optimally trading off the bias and variance in predicting the plug-in parameters: that is, the propensity score and conditional mean outcomes. Any of them may in principle be applied in the DML, TMLE, or partialling out approaches described in section 5.2 (or even other CML methods) if they satisfy specific regularity conditions like $n^{-1/4}$-convergence. This section provides a brief (and selective) introduction to several other machine learners, but a much more comprehensive discussion is provided in James, Witten, Hastie, and Tibshirani (2013) and Hastie, Tibshirani, and Friedman (2008). To ease notation, the fact that the plug-in models are only estimated in part of the data under cross-fitting will be omitted in any formal discussion in this section.

**Ridge Regression and Elastic Nets:** Ridge regression or Tihonov regularization, as discussed in Tihonov (1963) and Hoerl and Kennard (1970), is very much related to lasso regression, but it penalizes the sum of squared (rather than absolute) coefficients on the regressors. Considering equation (5.1), this means that $\lambda \sum_{j=1}^{p} |\beta_j|$ is to be replaced by $\lambda \sum_{j=1}^{p} \beta_j^2$. A noticeable difference between both methods is that ridge regression (whose penalty is based on an $L^2$ norm) cannot shrink coefficients of relatively unimportant regressors exactly to zero, while lasso regression can (as it is based on an $L^1$ norm). Therefore, lasso regression is able to perform variable selection (e.g., for defining parsimonious predictive models for the treatment and the outcome) based on dropping regressors with zero coefficients from the model, while ridge regression is not. However, neither method uniformly dominates the other with regard to predictive performance. Depending on the data, either ridge or lasso regression might do better for estimating $p(X), \mu_1(X), \mu_0(X)$. In fact, using a weighted average of both penalization approaches, which is known as an *elastic net*, might even outperform any single method. Similar to the choice of the penalty term $\lambda$, we may apply cross-validation to determine the optimal weights in an elastic net (e.g., 60 percent lasso and 40 percent ridge penalization) that minimizes the MSE when estimating the plug-in parameters.

**Decision Trees:** Decision trees, as suggested in Morgan and Sonquist (1963) and Breiman, Friedman, Olshen, and Stone (1984), are based on recursively splitting the covariate space—i.e., the set of possible values of $X$, for instance, when predicting treatment $D$—into a number of nonoverlapping subsets. Recursive splitting is performed such that after each split, a statistical goodness-of-fit criterion based on the differences between the actual treatments and the subset-specific average treatment (like the sum of squared residuals) is minimized across the newly created subsets. Let us, for instance, consider the case that migrant status is the most predictive element in $X$ for treatment assignment because almost all migrants but hardly any natives receive some form of treatment, like a language course. This implies that the treatment states within migrant status are more homogeneous than in the total sample. Therefore, splitting the data into migrant and native subsets implies that the squared deviations (or residuals) of the migrants' treatment states from the average treatment among migrants are on average lower than the squared residuals between the observed and the average treatment in the total sample. The same applies to the subset of natives. For this reason, the sum of the migrant status-specific sum of squared residuals is lower than the sum of squared residuals in the total sample prior to splitting. Formally,

$$\sum_{i:\text{migrant}=1} (D_i - \bar{D}_{\text{migrant}=1})^2 + \sum_{i:\text{migrant}=0} (D_i - \bar{D}_{\text{migrant}=0})^2 < \sum_{i=1}^{n} (D_i - \bar{D})^2, \quad (5.3)$$

where $\bar{D}_{\text{migrant}=1}$, $\bar{D}_{\text{migrant}=0}$, and $\bar{D}$ denote the sample averages of the treatment $D$ (i.e., the shares of treated) in the subsets of migrants, natives, and in the total sample, respectively.

Decision trees aim at finding the split that entails the highest reduction in the summed sums of squared residuals across subsets in a greedy manner: that is, the split that is optimal at the current stage without assessing performance several splits ahead. This approach is applied recursively, such that subsets are split into further subsets. For instance, in another step, the migrant subset might be split by age into migrants younger than 50 and 50 plus, if this split entails the largest additional reduction in the summed sum of squared residuals across all subsets. Decision trees owe their name to the fact that we can summarize the set of rules for splitting the covariate space by means of a tree structure. The latter contains nodes, which represent the covariate values at which the sample is split (e.g., "migrant $= 1$" versus "migrant $= 0$"), and leaves, which are the terminal subsets beyond which no further splitting occurs.

Interestingly, we also can represent such tree structures by regression equations in which the variable to be predicted (e.g., the treatment) is the dependent variable and indicator functions for the various leaves or terminal subsets (e.g., $I\{\text{migrant} = 1, \text{age} < 50\}$) serve as regressors, such as

$$D_i = \hat{\alpha} + \hat{\beta}_1 I\{\text{migrant} = 1, \text{age} < 50\} + \hat{\beta}_2 I\{\text{migrant} = 1, \text{age} \geq 50\} + \hat{V}_i. \quad (5.4)$$

Here, $\hat{\alpha}$ is the constant term reflecting the average treatment in a subset that serves as reference category (e.g., natives), $\hat{\beta}_1$ and $\hat{\beta}_2$ provide the differences in the treatment averages between the respective other subset and the reference category, and $\hat{V}_i$ is the estimated residual of the treatment equation. Based on the coefficient estimates, we can compute the estimated propensity score for each of the subsets, such as $\hat{p}(\text{migrant} = 1, \text{age} < 50) = \hat{\alpha} + \hat{\beta}_1$.

We repeat the splitting process until a specific stopping rule is reached, like a predefined maximum number of subsets or minimum number of observations in a subset. As for choosing the number of covariates to be included in lasso regression, we face a variance-bias trade-off concerning the number of splits in decision trees. More splits imply a finer grid of subsets, such that the observations within a subset are more similar in terms of $X$. This reduces the bias, as $\hat{p}(X)$ is estimated based on observations that are more homogeneous in terms of $X$. However, similar to a reduction of the bandwidth in radius matching, more splits imply that there are less observations within each subset to be used to estimate $\hat{p}(X)$ (or, similarly, the conditional mean outcomes), which increases the variance. As for lasso regression, we may use cross-validation to determine the optimal number of splits that minimizes the MSE by optimally trading off bias and variance.

Decision trees are a nonparametric method in the sense that splitting (or the use of indicator functions) does not impose functional form (e.g., linearity) assumptions about how $D$ and $X$ are associated. We might judge this to be an advantage over lasso regression, which is in principle a parametric approach (e.g., based on linear or logit models) and whose degree of model flexibility (in contrast to decision trees) depends on the inclusion of interaction and higher-order terms. However, a disadvantage of decision trees is that the estimated propensity score changes discontinuously across subsets: that is, it is not smooth in $X$ as would be the case in kernel regression. This follows from the fact that $\hat{p}(x)$ corresponds to an unweighted average of the outcomes of all observations with $X$ values in the same subset as value $x$:

$$\hat{p}(x) = \frac{\sum_{i=1}^{n} D_i \cdot I\{X_i \in L_x\}}{\sum_{i=1}^{n} I\{X_i \in L_x\}}, \tag{5.5}$$

where $L_x$ denotes the subset (or leaf) in which value $x$ is situated. Put differently, the discontinuity stems from the nonsmooth indicator functions in equation (5.5). Furthermore, the variance of tree structures is typically high. A small change in the data, therefore, can entail substantially different splitting rules, and thus definitions of the indicator functions in equation (5.4).

**Bagged Trees and Random Forests:** We can mitigate the issue that a single decision tree with many leaves likely suffers from a high variance by bootstrap aggregation, or "bagging," as discussed in Breiman (1996). The idea of bagged trees is to repeatedly draw bootstrap samples (see the discussion in section 3.4) from the original data with replacement (such that an observation might be drawn several times—or not at all—in a newly created bootstrap sample) and estimate the trees in each bootstrap sample. Then, the treatment or outcome is predicted based on averaging the predictions in the individual trees. Formally,

$$\hat{p}(x) = \frac{1}{B} \sum_{b=1}^{B} \hat{p}^b(x) = \frac{1}{B} \sum_{b=1}^{B} \frac{\sum_{i=1}^{n} D_i^b \cdot I\{X_i^b \in L_x^b\}}{\sum_{i=1}^{n} I\{X_i^b \in L_x^b\}}, \tag{5.6}$$

where $B$ denotes the number of bootstrap samples and $b$ indexes the various parameters (like treatments, covariates, or leaves) in a specific bootstrap sample $b$. This procedure not only has a smaller variance than basing propensity score estimation on a single tree, but it also implies that $\hat{p}(x)$ is a smooth function of $X$, which bears some similarly to kernel regression, as discussed in section 4.2. To better see this, note that we may rewrite equation (5.6) as

$$\hat{p}(x) = \sum_{i=1}^{n} w_i(x)^{\text{bagged}} \cdot D_i, \quad \text{where } w_i(x)^{\text{bagged}} = \frac{1}{B} \sum_{b=1}^{B} \frac{I\{X_i \in L_x^b\}}{\sum_{j=1}^{n} I\{X_j^b \in L_x^b\}}. \tag{5.7}$$

Due to bagging (i.e., averaging over the indicator functions of individual trees), the weights $w_i(x)^{\text{bagged}}$ are (in contrast to single trees) smooth in $X$, in the sense that they can take many values, given that the number of trees grows large. In bagged trees, the weights depend on the predictive power of the regressors, which can be a practically relevant advantage over kernel regression as described in equation (4.17), where weak predictors in $X$ may importantly affect the weight and thus exacerbate the curse-of-dimensionality problem mentioned in section 4.4.

Random forests, as discussed in Ho (1995) and Breiman (2001), are a further variation of tree-based methods and are similar to bagged trees in the sense that they rely on repeatedly drawing samples from the original data for estimating many trees and aggregating (or averaging over) predictions. They are, however, different in that only a random subset of (rather than all) covariates is chosen as potential variables for splitting at each split of a specific tree. Such a random selection of covariates aims at reducing the correlation of tree structures across samples (which are correlated because they are drawn from the same original data) to further reduce the variance in the estimation of the plug-in parameters. Similarly to bagged trees, we can represent random forest–based predictions by smooth weighting functions.

**Boosting and BARTs:** Boosting, as described in Freund and Schapire (1997), is yet another way to improve less sophisticated or weak machine learners by aggregation, but based on sequential application of such a weak learner (rather than averaging over many samples, as in the case of bagging and random forests). As an example, consider a simple decision tree with just a few splits for making predictions, which is likely to perform poorly. However, after a first application of the tree, we may compute the residuals, such as the difference between the treatment and the average treatment in the respective leaf, in order to apply the simple tree to those residuals again, which can substantially increase the predictive performance. Indeed, boosting consists of repeating these steps many times to sequentially apply the simple tree to the respective residuals of the previous prediction. This ultimately permits approximating the association of the covariates and the variable to be predicted in a very flexible way. A related method involves Bayesian additive regression trees (BARTs), as suggested in Chipman, George, and McCulloch (2010), which include a so-called regularization prior in the boosting process that penalizes too many splits in the tree structure to prevent excessive variance due to overfitting.

**Neural Networks:** Neural networks, as discussed in McCulloch and Pitts (1943) and Ripley (1996), aim at fitting a system of nonlinear regression functions that flexibly models the influence of a set of regressors (like covariates) on a variable to be predicted (like the treatment or outcome). Specifically, the regressors serve as inputs for specific nonlinear intermediate functions (e.g., logistic or rectifier functions) called hidden nodes, which themselves serve as inputs for the output layer: that is, the model of the variable to be predicted. The hidden nodes bear some similarity to the baseline functions in series regression (with the difference that they are learned from the data rather than predetermined), and with principal component analysis, which is based on dimension-reducing linear (rather than nonlinear) functions of the regressors. Indeed, when replacing the nonlinear functions by linear ones, neural networks collapse to a linear regression model.

Depending on the model complexity, hidden nodes may affect the outcome either directly or through other hidden nodes, such that several layers of hidden nodes allow modeling interactions among the functions. The number of hidden nodes and layers thus gauges the flexibility of the model, with more parameters reducing the bias but increasing the variance. Figure 5.2 provides an example of a neural network with five covariates and two hidden layers with four and three hidden nodes (e.g., logistic or rectifier functions, denoted by $\Lambda$), respectively, for estimating the treatment propensity score.

The basic approach of neural networks has been extended and refined in various dimensions, which is commonly referred to as *deep learning*. One refinement involves convolutional neural networks (CNNs); for instance, see LeCun, Bottou, Bengio, and Haffner (1998), which do not rely on providing a list of regressors but may autonomously learn to create relevant predictors from objects like images that a priori do not have a clear data structure. To this end, filters are applied, which slide over a prespecified amount of pixels in images to map them into numeric values based on specific functions, which permits representing particular spatial patterns (such as edges) by numeric features. This is typically followed by a pooling step that aggregates these features (e.g., by taking average or maximum values over a prespecified

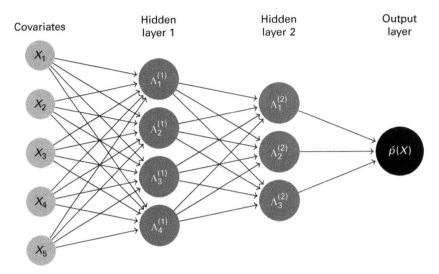

**Figure 5.2**
A neural network for treatment prediction.

number of adjacent numeric features). We may actually repeatedly apply filtering (of already pooled features) and pooling steps for the purpose of further transforming and aggregating the features. Finally, the refined features are used as regressors in a standard (i.e., feed-forward) neural network as described before, such as in figure 5.2. A further interesting development among many others are recurrent neural networks (RNNs) that allow feedback processes (rather than sequential effects) between hidden nodes situated in distinct hidden layers when optimizing prediction.

**Support Vector Machines:** Support vector machines, as discussed in Boser, Guyon, and Vapnik (1992) and Cortes and Vapnik (1995), can be best described by considering a binary variable to be predicted, like the treatment indicator $D$, even though a version for continuous outcomes exists as well. Put simply, the method aims at nonlinearly transforming the covariates in a way that permits fitting a linear hyperplane across the transformed covariate space, such that the hyperplane accurately separates the treatment values into two subsets. This implies that in one subset, there are mostly treated units, while in the other subset, there are mostly nontreated units. The hyperplane thus serves as a frontier between transformed covariate values with mostly treated or nontreated observations. This frontier is fitted in a way that maximizes the distance to the closest observation from either subset with $D = 1$ and $D = 0$, respectively, to maximize confidence in the classification of the transformed covariate space into predominantly treated and nontreated subsets.

**Ensemble Methods:** Finally, an ensemble method relies on a combination of several machine learning algorithms for making predictions by taking a simple or weighted average of the individual predictions of those algorithms; for instance, see Zhou (2012) and van der Laan, Polley, and Hubbard (2007). As for other tuning parameters in machine learning, we may determine the optimal weight of each algorithm by cross-validation to maximize the predictive accuracy. Based on this approach, an ensemble method may outperform each of the individual algorithms in terms of prediction.

Let us reconsider our DML application in R discussed at the end of section 5.2, but now use the random forest rather than lasso regression for estimating the propensity scores and conditional mean outcomes. To this end, we set the argument *MLmethod* in the *treatDML* command to *"randomforest"* and otherwise run the same commands as before, as shown in the box here.

```
output=treatDML(y=Y,d=D,x=X,MLmethod="randomforest") # double machine learning
output$effect; output$se; output$pval # effect, standard error, p-value
```

DML now yields an ATE of $-0.041$, which is in absolute magnitude slightly lower than the lasso-based estimate presented in section 5.2. As before, this points to a very moderate health-improving effect that is statistically significant at the 5 percent level (with a standard error of 0.018 and a p-value of 0.027). As already mentioned, the random forest might be preferred over lasso regression because being an aggregated tree-based method, it permits for arbitrary nonlinearities in the associations of $X$ and $Y$ or $D$, respectively. When using lasso regression, we would need to include interaction and higher-order terms of $X$ to allow such nonlinearities.

## 5.4   Effect Heterogeneity

The discussion in chapters 3 and 4, as well as section 5.2, predominantly focused on the evaluation of aggregated effects like the ATE or the ATET. In many empirical contexts, however, researchers and analysts are interested in whether causal effects differ importantly (i.e., are heterogeneous) across specific subgroups that can be described in terms of observed characteristics (i.e., covariates $X$). For instance, we might want to know whether the effectiveness of a training program or a marketing campaign differs across characteristics like gender, age, income, or other variables. Learning which groups a specific treatment is particularly effective for can be helpful for improving treatment allocation, such as targeting those customer groups among which a marketing campaign entails particularly high effects on sales.

CML, combined with sample splitting, can be fruitfully applied to investigate treatment effect heterogeneity across $X$, while at the same time avoiding inferential issues of multiple hypothesis testing when searching for subgroups with significantly different effects. Such issues arise when a researcher or analyst investigates effect heterogeneities across a large set of subgroups and then only reports the treatment effects for those subgroups across which statistically significant differences in causal effects occur. This approach will tend to find too many significant differences that are spurious in the sense that they do not occur in the population. This problem is related to the fact that conventional t-statistics and the related p-values are only valid for testing a single hypothesis, but they need to be adjusted when conducting several hypothesis tests at the same time. The problem of finding too many subgroups with effect heterogeneities due to multiple hypothesis testing is closely related to the issue of overfitting: that is, finding too many predictors associated with an outcome of interest. This motivates the application of appropriately designed machine learning approaches for analyzing effect heterogeneity.

Let us first consider the case that treatment $D$ is randomly assigned, as in successful experiments. Then, covariates $X$ are not required for making subjects comparable across treatment states when assessing causal effects but may nevertheless be exploited to assess effect heterogeneity. In this context, Athey and Imbens (2016) suggest a causal tree method that builds on two key modifications of conventional decision trees for prediction, as discussed in section 5.3. First, instead of $Y$, the difference in $Y$ across treatment groups serves as the outcome to be predicted when recursively splitting the covariate space into subgroups. The algorithm thus aims at generating covariate value-specific subgroups in a way that minimizes the sum of squared residuals in effect estimation (rather than outcome prediction). This corresponds to finding the splits that maximize effect homogeneity within, or put differently, effect heterogeneity across subgroups defined based on values of $X$. Therefore, the structure of a causal tree yields a definition of subgroups with the most heterogeneous treatment effects up to the number of splits considered.

As a second modification, a causal tree applies sample splitting to use distinct parts (or folds) of the data for estimating the tree's structure and the treatment effects within subsets. This avoids spuriously large effect heterogeneities due to overfitting, and thus the previously mentioned inference issues related to multiple hypothesis testing. We also can apply the causal tree method when the treatment is not randomly assigned, but the selection-on-observables assumptions in expression (4.1) hold for a preselected (rather than machine learning–selected) set of covariates. In this case, we may reweight outcomes by the inverse of the propensity score in analogy to equation (4.40) in section 4.5 prior to taking differences in treated and nontreated outcomes and applying the causal tree approach.

A further and related method for investigating effect heterogeneity is the causal random forest; see Wager and Athey (2018) and Athey, Tibshirani, and Wager (2019). As the name suggests, it is a modified version of the random forest and also can (in contrast to the causal tree) be applied when important control variables are not preselected but adjusted for by machine learning, as discussed in section 5.2, given that the selection-on-observables assumption in expression (4.1) is satisfied.

The causal forest consists of several steps. First, we predict both $Y$ and $D$ as a function of $X$ using random forests and leave-one-out cross-validation. The latter implies that the outcome or treatment of each observation is predicted based on all observations in the data but its own, in order to prevent overfitting when conditioning on $X$. Second, we use the predictions for computing residuals of the outcomes and treatments, which corresponds to the partialling out strategy discussed at the end of section 5.2. Third, we predict the effect of the residuals of $D$ on the residuals of $Y$ as a function of $X$ by yet another random forest, which averages over a large number of causal trees that use different folds of the respective tree-specific samples for modeling effect heterogeneity by a tree-structure and estimating treatment effects within the subsets. Put simply, this method combines the idea of sample splitting and partialling out to control for important confounders (as discussed in section 5.2) with the causal tree approach for finding effect heterogeneity, but based on averaging over many trees.

Even though our discussion focuses on the random forest, it is worth noting that we may also use other machine learning algorithms for this two-step approach of partialling out and detecting effect heterogeneity, which Nie and Wager (2020) refer to as "R-learning" (in recognition of Robinson (1988)).

When comparing a single causal tree to a causal forest, an advantage of the former is that it directly yields easy-to-interpret splitting rules for defining subgroups based on the most predictive covariates in terms of effect heterogeneity. On the negative side, tree structures tend to have a higher variance, such that a small change in the data may entail very different splitting rules. The causal forest is more attractive in terms of variance, but on the other hand, it does not provide straightforward guidance on how to define subgroups due to averaging over many trees. It, however, yields an estimate of the CATE $\Delta_x = E[Y(1) - Y(0)|X = x]$; see equation (4.2). Therefore, we can investigate the heterogeneity of the CATE as a function of the covariates $X$. We may, for instance, split the sample into several categories with higher and lower CATE estimates based on specific quantiles (like the median) of the CATE distribution, and investigate whether the averages of certain or all covariates differ importantly across categories. It is also worth mentioning that appropriately averaging over the CATE estimates in the total sample or among the treated provides consistent estimates of the ATE and ATET, respectively.

Another interesting feature of the causal forest is that we can also apply it to a continuously distributed treatment $D$, as discussed in section 4.8. In this case, the method yields the conditional average partial effect (CAPE) of marginally increasing the currently given treatment intensity when keeping the covariates fixed at some value $x$. This formally corresponds to the derivative of the conditional mean of $Y$

given $X$ and $D$ with regard to $D$, in analogy to the average marginal effect discussed in section 3.5 in chapter 3 where we did not control for $X$:

$$\Delta_x = E\left[\frac{\partial E[Y(d)|X=x]}{\partial d}\bigg|X=x\right] = E\left[\frac{\partial E[Y|D=d, X=x]}{\partial d}\bigg|X=x\right]. \tag{5.8}$$

A further approach for learning about the heterogeneity of the CATE is based on the DML approach outlined in section 5.2, and more specifically on the machine learning–based estimate of the efficient influence function $\phi(X)$ in equation (4.45) of section 4.6. Just as the unconditional average of this influence function yields the ATE of a binary treatment, $\Delta = E[\phi(X)]$, its conditional average given $X$ yields the CATE: $\Delta_x = E[\phi(X)|X=x]$. This suggests that we may investigate effect heterogeneity by regressing the estimated efficient influence function, denoted by $\hat{\phi}(X)$, on $X$ or a subset of the covariates, like an indicator for gender if effect heterogeneity across gender is of interest.

As discussed in Semenova and Chernozhukov (2021), running an OLS regression of $\hat{\phi}(X)$ on a limited number of preselected covariates can under certain conditions yield asymptotically correct coefficient estimates and standard errors when testing whether effect heterogeneity is statistically significant. Two important conditions are that we estimate $\hat{\phi}(X)$ by cross-fitting (i.e., in a different data fold than the conditional mean outcome and the propensity score), and that the convergence rate of the latter plug-in parameters is faster than $n^{-1/4}$. Under these conditions, one can infer from the coefficient estimate on a gender indicator and its standard error whether the CATE statistically significantly differs across gender (and this applies similarly to other covariates). This is remarkable because unlike in a standard OLS regression with a known or directly observed outcome, $\hat{\phi}(X)$ needs to be estimated by machine learning first. But similar to ATE estimation in section 5.2, the machine learning step under certain conditions does not affect the asymptotic behavior of the OLS regression, such that we can assess effect heterogeneity with $\sqrt{n}$-consistency.

When investigating effect heterogeneity across continuously distributed (rather than binary or discrete) covariates, such as income, we might want to avoid imposing a linear association between effect heterogeneity and covariates. For this reason, we may prefer a nonparametric kernel or series regression (see section 4.2) of $\hat{\phi}(X)$ on the continuous covariates instead of a linear regression. In this case, the machine learning step does not affect the convergence rate and asymptotic behavior of the kernel regression if certain conditions are met, as demonstrated by Zimmert and Lechner (2019) and Fan, Hsu, Lieli, and Zhang (2020).

Rather than considering a limited number of preselected covariates for CATE estimation, we may also be interested in detecting the covariates that most importantly predict effect heterogeneity in a data-driven way. To this end, in principle, we can apply a machine learning algorithm to predict $\hat{\phi}(X)$ as a function of $X$ and assess which covariates have the best predictive power. In a lasso regression of $\hat{\phi}(X)$ on the covariates (and possibly interaction and higher-order terms), the so-called standardized coefficients, which have been standardized by the standard deviation of the covariates, permit ranking covariates according to their predictive power in terms of effect heterogeneity. As a word of caution, however, the most predictive variables could be strongly correlated with other covariates. The latter might obtain a lower rank due to this correlation but are not necessarily unimportant in terms of how they influence effect heterogeneity.

Conducting inference (e.g., computing p-values and confidence intervals) or hypothesis tests about the importance of covariates for effect heterogeneity is less straightforward under a data-driven selection of covariates and generally requires further sample splitting steps. This is due to the threat of overfitting when selecting important covariates by machine learning and assessing their statistical significance in the very same data. The overfitting issue implies that we should use different folds of the data for discovering and selecting the most influential covariates on the one hand, and for statistical inference, such as assessing whether the selected covariates statistically significantly drive effect heterogeneity, on the other hand. This principle is obeyed in the algorithm suggested in Lee, Bargagli-Stoffi, and Dominici (2020), as well as the sample-splitting approach in Athey and Imbens (2016).

As a final comment on the DML-based analysis of effect heterogeneity, it is worth noting that for the sake of an optimal estimation of the CATE with the smallest possible bound on the estimation error, our previous cross-fitting approach can be further refined, as discussed by Kennedy (2020). To this end, we use three folds of the data to estimate the propensity score model $p(X)$ in the first fold, the models of the conditional mean outcomes $\mu_1(X)$ and $\mu_0(X)$ in the second fold, and the efficient influence function $\phi(X)$ along with the CATE of interest in the third fold. Thus $p(X)$ as well as $\mu_1(X)$, and $\mu_0(X)$ are no longer estimated in the same fold. Again, we can swap the roles of the folds to obtain the final CATE estimates by averaging over the CATE estimates in the various folds. As further refinement, we may repeat random data-splitting into three folds and subsequent cross-fitting for CATE estimation multiple (e.g., 50) times to ultimately take the median CATE of the multiple cross-fitting steps as a final estimate. This likely entails a smaller variance in CATE estimation than a single cross-fitting approach.

Let us reconsider the *JC* (Job Corps) data of the *causalweight* package analyzed in sections 4.9 and 5.2 to estimate the CATE and the ATE in R based on the causal forest, using the *grf* package of Tibshirani, Athey, and Wager (2020). We define the pretreatment characteristics as covariates, *X=JC[,2:29]*, training in the first year of the Job Corps program as the treatment, *D=JC[,37]*, and the proportion of employment in the third year as the outcome, *Y=JC[,40]*. After setting a seed for the replicability of the results using *set.seed(1)*, we run the *causalforest* command with *X*, *Y*, and *D* and store the output in an R object named *cf*. The box here provides the code for the various steps.

```
# causal forest
library(grf)                      # load grf package
library(causalweight)             # load causalweight package
data(JC)                          # load JC data
X=JC[,2:29]                       # define covariates
D=JC[,37]                         # define treatment (training) in first year
Y=JC[,40]                         # outcome (proportion employed in third year)
set.seed(1)                       # set seed
cf=causal_forest(X=X, Y=Y, W=D)   # run causal forest
```

Next, we use the output in *cf* to estimate the ATE. This is based on the DML approach as described in section 5.2, however, using the random forest rather than lasso regression for estimating the propensity score and conditional mean outcomes. To this end, we wrap the *cf* object with the *average_treatment_effect* command and save the output in an object named *ATE*. We then compute the p-value in the same manner as outlined at the end of section 3.4: We first construct the t-statistic by dividing the ATE, the first element in *ATE*, by its standard error, the second element. We then take the negative of the t-statistic's absolute value, assess it on the standard normal distribution using the *pnorm* command and multiply it by 2 before saving it in an object named *pval*. Finally, we call the objects *ATE* (containing the ATE estimate and its p-value) and *pval*.

```
ATE=average_treatment_effect(cf)   # compute ATE
pval=2*pnorm(-abs(ATE[1]/ATE[2]))  # compute the p-value
ATE; pval                          # provide ATE, standard error, and p-value
```

Running the code gives an ATE of 4.27. As the outcome is measured in percentage from 0 to 100, this suggests that the training increases the proportion of weeks of employment in the third year after Job Corps assignment on average by 4.27 percentage points. The standard error amounts to just 0.89, such that the p-value is very close to zero. Next, we plot the distribution of the CATEs in our sample. To this end, we store the CATEs provided in *cf$predictions* for each observation in an object named *CATE* and wrap the latter by the *hist* command to produce a histogram of CATEs.

```
CATE=cf$predictions   # store CATEs in own variable
hist(CATE)            # distribution of CATEs
```

This gives the graph in figure 5.3. In line with the ATE, the vast majority of CATE estimates is positive. Yet effect heterogeneity appears nonneglibile, given the range of

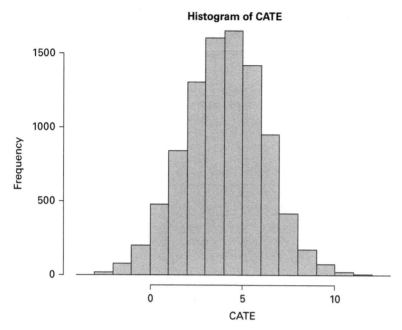

**Figure 5.3**
Distribution of CATEs.

values the CATE estimates take. In further heterogeneity analysis, we verify whether the CATEs differ importantly across a preselected covariate of interest (in our case, gender) using the approach of Semenova and Chernozhukov (2021). To this end, we apply the *best_linear_projection* command and feed in the *cf* output to linearly regress the random forest–based estimates of the efficient influence functions, $\hat{\phi}(X)$, on the variable *JC$female*.

```
best_linear_projection(forest=cf,A=JC$female) # regression of function on gender
```

Running this code yields the following output:

```
Best linear projection of the conditional average treatment effect.
Confidence intervals are cluster- and heteroscedasticity-robust (HC3):

            Estimate Std. Error t value  Pr(>|t|)
(Intercept)  5.0253    1.1889   4.2267  2.394e-05 ***
A1          -1.7110    1.7905  -0.9556     0.3393
```

The results suggest that the CATEs are on average $-1.71$ percentage points lower among females than among males (whose average CATE amounts to 5.03 percent). This difference is not statistically significant at any conventional level, however, as the p-value amounts to 0.339 (or 33.9 percent). Finally, we aim at detecting the best predictors of the CATEs, and thus the covariates that most importantly drive effect heterogeneity in data-driven way using a random forest. To this end, we load the *randomForest* package by Liaw and Wiener (2002) and use the *data.frame* command to generate a data matrix called *dat*, which contains the CATEs in the object *CATE* and the covariates X.

We then run the *randomForest* command with *CATE*~. as its first argument, which means that *CATE* is to be predicted based on all other variables in the data matrix, and *data=dat* as the second argument used to define the data matrix. We store the output in an R object named *randomf* and wrap the latter by the *importance* command to investigate the importance of the various covariates for predicting the CATEs. The importance measure is defined as the decrease in the sum of squared residuals in a tree-based out-of-sample prediction of the CATE when including versus not including the covariate for splitting, averaged over all trees in the forest. A larger number, therefore, means that the respective covariate is more relevant for assessing effect heterogeneity.

```
library(randomForest)              # load randomForest package
dat=data.frame(CATE,X)             # define data frame
randomf=randomForest(CATE~. ,data=dat)   # predict CATE as a function of X
importance(randomf)                # show predictive importance of X
```

Running the code gives the following results:

|            | IncNodePurity |
|------------|---------------|
| female     | 905.76500     |
| age        | 3551.24485    |
| white      | 1238.68232    |
| black      | 1168.65759    |
| hispanic   | 517.16448     |
| educ       | 2830.98822    |
| educmis    | 189.76240     |
| geddegree  | 324.06883     |
| hsdegree   | 523.10615     |
| english    | 339.23107     |
| cohabmarried | 739.19153   |
| haschild   | 651.00984     |
| everwkd    | 538.78978     |
| mwearn     | 1281.51420    |
| hhsize     | 9706.61434    |
| ....       |               |

We find that the covariates *hhsize*, *age*, and *educ* are by far the most relevant predictors of the CATE, as their values of the residual-based importance measure clearly exceed those of the other elements in $X$.

## 5.5  Optimal Policy Learning

A concept related to the CATE and its heterogeneity across covariates is optimal policy learning, as discussed in Manski (2004), Hirano and Porter (2009), Stoye (2009), Qian and Murphy (2011), Bhattacharya and Dupas (2012), and Kitagawa and Tetenov (2018). It aims at optimally allocating a (possibly costly) treatment in a population as a function of covariates $X$ when also taking the costs into account, which may vary across covariates, too (just as the effects do). For instance, a company might aim at adjusting its pricing policy for products or services to optimally target customers by offering discounts only to clients if the benefits (e.g., in terms of additional sales) on average outweigh the costs (e.g., in terms of a reduced profit margin) given observed characteristics $X$ (like age, education, or previous buying behavior). A further example is the optimal selection of job seekers or employees to participate in a training program to maximize their employment probability or productivity. We will henceforth focus on policy learning that aims at maximizing average outcomes under a binary treatment based on covariate dependent-treatment assignment. However, the framework can be extended to multivalued or dynamic treatments (see, for instance, Zhou, Athey, and Wager (2018) and Kallus (2017)), as well as to optimality criteria other than the average (e.g., the median or another quantile of the outcome).

To formalize the discussion, let us denote by $\pi(X)$ a specific treatment policy defined as a function of $X$. To give an example, a policy could require that a medical treatment, such as obtaining versus not obtaining a scarce vaccine, is set to $D = 1$ (vaccination) for all observations aged 65 or older and $D = 0$ (no vaccination) otherwise. This would correspond to the policy rule $\pi(X) = I\{\text{age} \geq 65\}$, such as $\pi(\text{age} = 30) = 0$ and $\pi(\text{age} = 80) = 1$. In this example, the policy depends on only one covariate (age), but in practice, it might be a function of several covariates $X$. The average effect of policy $\pi(X)$, denoted by $\Delta(\pi(X))$, corresponds to the difference in mean potential outcomes under $\pi(X)$ versus nontreatment of everyone:

$$\Delta(\pi(X)) = E[Y(\pi(X)) - Y(0)] = E[\pi(X) \cdot (Y(1) - Y(0))]$$

$$= E[\pi(X) \cdot E[Y(1) - Y(0)|X]] = E[\pi(X) \cdot \Delta_X], \tag{5.9}$$

where the third equality follows from the law of iterated expectations and highlights the close relationship of optimal policy learning based on covariates and the identification of the CATE $\Delta_X$. The optimal policy, denoted by $\pi^*(X)$, maximizes the average effect among the set of all feasible policies, denoted by $\Pi$, where we assume a countable (i.e., finite) number of policies in this set:

$$\pi^*(X) = \max_{\pi \in \Pi} \Delta(\pi(X)). \tag{5.10}$$

Based on equations (5.9) and (5.10), we can define the regret function associated with treatment policy $\pi(X)$. The regret function corresponds to the undesirable reduction in the average policy effect due to implementing the suboptimal policy $\pi(X)$, rather than the optimal policy $\pi^*(X)$, and is denoted by $R(\pi(X))$:

$$R(\pi(X)) = \Delta(\pi^*(X)) - \Delta(\pi(X)). \tag{5.11}$$

Therefore, finding the optimal policy among the set of feasible policies $\Pi$ implies that the average policy effect is maximized and regret $R$ is equal to zero. Furthermore, finding the optimal policy can be shown to amount to solving the following maximization problem:

$$\pi^*(X) = \max_{\pi \in \Pi} E[(2\pi(X) - 1) \cdot \phi(X)]. \tag{5.12}$$

Equation (5.12) demonstrates that similar to effect heterogeneity analysis, the efficient influence function $\phi(X)$ in equation (4.45) of section 4.6 is also useful for optimal policy learning, as considered in Dudík, Langford, and Li (2011); Zhang et al. (2012); and Zhou, Mayer-Hamblett, Khan, and Kosorok (2017). The term $(2\pi(X) - 1)$ implies that the CATEs of treated and nontreated subjects enter the expectation positively and negatively, respectively. Maximizing the expectation, therefore, requires optimally trading off treated and nontreated subjects in terms of their CATEs when choosing the optimal treatment policy among all feasible policies. In the presence of a large set of covariates $X$, we may base the estimation of the optimal policy on the sample analog of equation (5.12), with $\phi(X)$ being estimated by cross-fitting and machine learning–based prediction of the plug-in parameters, as outlined in section 5.2.

Similar to ATE estimation, basing policy learning on DML to obtain an estimate of the optimal policy in the data, denoted by $\hat{\pi}^*(X)$, has desirable properties under specific conditions, even if the important elements in $X$ driving confounding, effect heterogeneity, or both are a priori unknown. Namely, the upper bound on (i.e., the worst case value of) $R(\hat{\pi}^*(X)) = \Delta(\pi^*(X)) - \Delta(\hat{\pi}^*(X))$, the regret of the estimated optimal policy $\hat{\pi}^*(X)$ versus the truly optimal policy $\pi^*(X)$, can go to zero at the $\sqrt{n}$-rate, as demonstrated in Athey and Wager (2021). One condition for this property is that the plug-in parameters (i.e., the propensity score and conditional mean outcomes) in $\phi(X)$ are estimated at a convergence rate faster than $n^{-1/4}$, as already discussed in the context of ATE estimation in section 5.2. A further important condition is that the set of possible policies $\Pi$ is not too complex, meaning that it is limited to a countable number.

This is the case, for instance, if we define a limited number of predefined policies $\pi(X)$ (e.g., the three alternatives that all natives, all migrants, or everyone gets some training) to be assessed or alternatively, a limited number of subgroups of observations across which the treatment may vary (e.g., at most eight customer segments when assessing a marketing campaign). In the latter case, we may determine the subgroups in a data-driven way as a function of $X$, as well as the respective optimal treatment in each subgroup. Loosely related to (but yet different from) the causal tree of section 5.4, this approach corresponds to a decision tree using the estimate of $\phi(X)$ as the outcome for optimally splitting the covariate space and assigning (possibly distinct) treatments to the various subgroups in a way that maximizes the average policy effect in the data. Such a policy tree, therefore, yields optimal treatment assignment rules for subgroups (e.g., customer segments) that can be intuitively described by the tree structure.

We have so far abstracted from any costs related to implementing the policies. In reality, however, the costs of treatment provision need to be taken into account for a proper assessment of the optimal policy. A first reason is that any policy should be implemented only if the benefits outweigh the costs; otherwise, the optimal policy is nontreatment. Second, costs may vary across policy rules as a function of covariates $X$, thus affecting the net benefits of the various policies that could be implemented in principle. For instance, the costs of running a marketing campaign could vary across two geographic regions, which are included by means of a regional dummy variable among covariates $X$. In this case, the marketing expenditures differ depending on whether $\pi(X)$ targets one region, the other region, or both. More formally, let us denote the costs associated with policy $\pi(X)$, which might depend on the values of covariates $X$, as $C_X(\pi(X))$. Then, the following modification of equation (5.9) accounts for the costs, and therefore corresponds to the net (rather than gross) benefits, of the treatment policy $\pi(X)$:

$$\Delta(\pi(X)) = E[\pi(X) \cdot \Delta_X - C_X(\pi(X))]. \tag{5.13}$$

To implement policy learning in R, let us reconsider the NSW data previously analyzed at the end of sections 4.2, 4.3, and 4.4 in chapter 4. We load the *Matching*, *policytree*, and *DiagrammeR* packages, which contain the data of interest, the policy learning commands, and $C_X(\pi(X))$ procedures for diagrams and graphs, respectively. We apply the *data* and *attach* commands to the *lalonde* data set to load the latter and store all variables in own R objects.

As the procedure that we are going to use requires the treatment to be coded as a *factor*, we define *D=factor(treat)*. Our outcome of interest are real earnings in 1978, *Y=re78*, and also covariates $X$ are defined in the same way as in the empirical example at the end of section 4.2. In the next step, we feed $X$, $Y$, and $D$ into the *multi_arm_causal_forest* command to estimate the plug-in parameters (i.e., the propensity scores and conditional mean outcomes) by means of random forests and store the output in an object named *forest*. We wrap the latter by the *double_robust_scores* command to compute estimates of the efficient influence functions $\phi(X)$, which we save in an object named *influence*.

We then define those covariates based on which the optimal treatment policy shall be determined, which can be a different set than the variables occurring in $X$. For instance, we might want to find a policy rule that does not discriminate with regard to gender or ethnicity, such that these variables are not included. In our example, we specify the treatment policy–relevant characteristics to consist of age and education, by defining *Xpol=cbind(age,educ,nodegr)*. Finally, we feed *Xpol* and *influence* into the *policy_tree* command and set *depth=2*. This entails the estimation of a policy tree as discussed in Athey and Wager (2021), with optimal policies for four subgroups (as a depth of 2 entails $2^2 = 4$ subgroups), which are determined in a data-driven way based on the variables in *Xpol*. We save the output in an object named *tree*, which we wrap by the *plot* command to investigate the tree structure, giving the optimal treatment policy for each subgroup. The box here provides the R code for the various steps.

```
library(Matching)                          # load Matching package
library(policytree)                        # load policytree package
library(DiagrammeR)                        # load DiagrammeR package
data(lalonde)                              # load lalonde data
attach(lalonde)                            # store all variables in own objects
D=factor(treat)                            # define treatment (training)
Y=re78                                     # define outcome
X=cbind(age,educ,nodegr,married,black,hisp,re74,re75,u74,u75) # covariates
forest=multi_arm_causal_forest(X=X, Y=Y, W=D) #estimate treatment+outcome models
influence=double_robust_scores(forest)     # obtain efficient influence functions
Xpol=cbind(age,educ,nodegr)                # relevant X for optimal policy
tree=policy_tree(X=Xpol, Gamma=influence, depth=2) # policies for 4 subgroups
plot(tree)                                 # plot the tree with optimal policies
```

Running the code returns the policy tree in figure 5.4. It suggests that any individual with six years of education or less should be trained because action=2, which corresponds to training participation under the definition of treatment $D$. In contrast, we should not train individuals with seven or eight years of education, as action=1 (no treatment). Among those with more than eight years of education, only those no more than 42 years old should be assigned to training according to the optimal treatment policy in our data. However, we keep in mind that our analysis abstracts from any costs of training when computing the effects of the policies $\Delta(\pi(X))$, but in practice, such costs should also be taken into account.

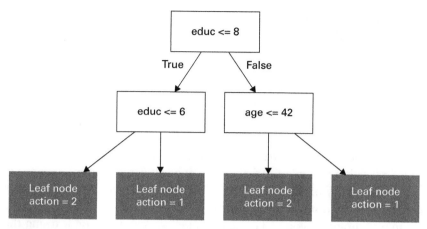

**Figure 5.4**
Policy tree.

## 5.6 Reinforcement Learning

In this section, we consider a further machine learning–based method for learning optimal treatment policies, which is yet different from the one described in the previous section in that treatment assignment is dynamic across time periods. In fact, reinforcement learning, as discussed in Sutton and Barto (1998) aims at learning the most effective treatment, such as yielding the highest ATE among a set of feasible treatments, by repeated assignment of multiple treatments and their evaluation across various periods. Such a scenario is sometimes referred to as the multiarmed bandit problem, alluding to a gambler in a casino playing a slot machine with multiple levers (or arms) that produce a payout and can be pulled in multiple periods.

To account for this dynamic framework, we denote by $D_t$ and $Y_t$ the treatment and the outcome in period $T = t$, with $t \in \{1, 2, \ldots, \mathcal{T}\}$ and $\mathcal{T}$ denoting the total number of periods. Any treatment $D_t$ may take values $d_t \in \{0, 1, \ldots, J\}$, with 0 indicating no treatment and $1, \ldots, J$ indexing different nonzero treatments, like advertisement campaigns on an online platform. Let us denote the mean potential outcome of treatment $d$ in a particular period $t$ by $\mu_t(d_t) = E[Y_t(d_t)|T = t]$, and the average of the mean potential outcomes for a fixed (i.e., the same) treatment $d = d_1 = \cdots = d_{\mathcal{T}}$ across all treatment periods by $\mu(d) = E[Y_T(d)]$.

Furthermore, we impose two assumptions, which rule out that the treatment effects interact with the time periods. First, we assume the treatments will only affect the outcomes in the same period $t$, such that, for instance, an online advertisement presumably only influences the buying behavior in the period the advertisement was placed, but not future buying behavior. This rules out dynamic treatment effects by requiring that treatments in earlier periods do not affect outcomes in later periods. Therefore, $\mu_t(d_t)$ does not depend on previous treatment assignments, which may be plausible if different individuals (i.e., repeated, nonoverlapping cross sections) are considered in different periods $t$ and no interaction of individuals across time periods (e.g., customers who see the advertisement in earlier periods informing other individuals later) takes place. However, we critically acknowledge that ruling out interactions (or interference) between individuals might not be realistic in many situations, as more thoroughly discussed in chapter 11. Second, we assume ATEs to be stationary, in the sense that the treatment effects do not change over time, such that placing an advertisement in an earlier period is as effective as in a later period. Under such a homogeneity of ATEs across time, it follows that $\mu_t(d_t) = \mu_{t'}(d_{t'})$, for two time periods $t \neq t'$, and furthermore, $\mu_t(d_t) = \mu(d)$.

In this setup with homogeneous ATEs across time, we may focus on learning the optimal treatment with regard to the overall mean potential outcome $\mu(d)$, rather than $\mu_t(d_t)$, in different outcome periods. Formally, our goal is to find the treatment that maximizes $\mu(d)$, which we denote by $d^* = \max_{d \in \{0,1,\ldots,J\}} \mu(d)$. Somewhat related to the discussion on optimal policies in section 5.4, we define a regret function $R_\mathcal{T}(d)$, which corresponds to the difference in mean potential outcomes (or ATEs) under the optimal treatment $d^*$ versus some other treatment assignment $d$, when taking into account (i.e., summing over) all periods $\mathcal{T}$:

$$R_\mathcal{T}(d) = \sum_{t=1}^{\mathcal{T}} \mu(d^*) - \mu_t(d_t). \tag{5.14}$$

If we knew the optimal treatment a priori, then we could exploit this knowledge for assigning $d^*$ in all treatment periods. In this case, the regret function $R_\mathcal{T}(d)$ would be zero, while the mean outcome (or ATE) would be maximized, by selecting the most effective advertisement throughout the evaluation window. However, the motivation for reinforcement learning is exactly that we typically do not know the optimal treatment, but need to learn it by assigning various treatments and exploring their performance across multiple periods. Ideally, we would explore (i.e., compare various treatments in terms of their performance) and find the optimal treatment based on relatively few periods to assign $d^*$ in the remaining periods in order to minimize regret. However, if too few periods are used for exploration, we may end up with a suboptimal treatment choice due to the variance related to the estimation of $\mu(D)$ for various treatment assignments in the data, which would entail a nonzero regret. For this reason, reinforcement learning faces an exploration-exploitation trade-off. Basing exploration on more periods and data increases the chance to find the optimal treatment and improve future performance, but it also reduces the number of individuals to which the currently optimal treatment can be assigned to maximize immediate performance.

Let us consider the case that treatments are randomly assigned in an experiment (as frequently conducted for advertisement campaigns on online platforms), such that the assumptions in expression (3.53) in chapter 3 are satisfied. Ideally, we would like to find an assignment rule across periods that allows optimally trading off exploration and exploitation in a way that minimizes regret, or at least permits keeping it acceptably small, through learning over time. One rule that does not appear appropriate in this context would be to randomly assign each treatment with equal probability (i.e., with proportion $1/(J+1)$ in all time periods. While this might permit learning $\mu(d)$ for any treatment rather well based on averaging the estimates of $E[Y_t|D_t = d]$ across time periods and therefore might perform well in terms of exploration, it is suboptimal in terms of exploitation. The reason is that the share of individuals assigned to any treatment is fixed and thus does not depend on the treatment performance in the data. After starting off by randomly assigning treatments with equal probability to explore their performance in the absence of any prior knowledge about their effectiveness, we would prefer to gradually assign ever more individuals to treatments that up to the current period look most promising in our data, a process known as *adaptive randomization*.

We confine our discussion to one popular adaptive randomization scheme called *Thompson sampling* (see Thompson (1933)). It is based on Bayesian updating to modify the treatment assignment as more information on the effectiveness of the various treatments from previous periods becomes available. Before getting started, we need to define a presumed prior distribution of the potential outcomes under each treatment $d$ in the nonobserved period $t = 0$, denoted by $\tilde{F}_{Y(d)_0}$; that is, prior to the periods in which we observe the treatments and outcomes. Likewise, let us denote by $\tilde{F}_{Y(\cdot)_0}$ the prior distribution in period $t = 0$ across all treatment states. A neutral or uninformative prior distribution, for instance, could consist of assuming a constant mean across the potential outcome distributions of any treatment, in line with a null hypothesis of ATEs that are zero for any of the treatments. After defining the prior, we conduct the following steps for each period $t \in \{1, 2, \ldots, \mathcal{T}\}$, in which we observe the treatments and outcomes:

1. In each period $t$, we compute the probability that a specific treatment assignment $d$ is optimal based on the performance in the previous periods $\tilde{p}_{d,t-1} = \Pr(\mu(d) = \mu(d^*)|\tilde{F}_{Y(\cdot)_{t-1}})$. Note that in the first period $t = 1$, this probability is solely based on the prior distribution, and our uninformative prior implies that every treatment is assigned with equal probability in period $t = 1$. This is generally no longer the case if $t > 1$ and learning about the performance of the various treatments sets in.

2. We randomly draw treatments $D_t$ from a multinomial distribution in which the treatment assignment probabilities correspond to their probabilities to be optimal, $\tilde{p}_{d,t-1}$. This implies that treatments that performed better in previous periods tend to be assigned to a higher proportion of individuals in the current period than previously worse-performing treatments (but the proportion also depends on the variance and thus the uncertainty in effect estimation).

3. We assess the outcomes in the various treatment groups, $Y_t = Y_t(D_t)$, to obtain a Bayesian update of the joint potential outcome distribution, $\tilde{F}_{Y(\cdot)_t}$, which is known as *posterior distribution* and will serve as the prior in the next period, $t+1$.

Thompson sampling relies on probability matching in the sense that it bases treatment assignment on the probability that a specific treatment is optimal, according to the treatment's performance observed in past data (and the prior distribution). The procedure continues to explore all treatments that might be optimal according to their respective posterior probability, but shifts sampling away from (and thus, gradually discards) those treatments that clearly underperform. Thompson sampling can attain a near-optimal regret bound. This implies that Thompson sampling may almost attain the theoretical lower bound of the regret in equation (5.14) inherent even in the best-performing methods of reinforcement learning due to the exploration-exploitation trade-off; see Lai and Robbins (1985) and Agrawal and Goyal (2013) for further discussion.

We may adapt Thompson sampling to more complicated setups than the one considered so far, to nonstationary systems where $\mu_t(d_t)$ is time-varying such that treatment effectiveness may change across periods. In this case, exploring should never fully stop in order to take account for time-induced changes in mean potential outcomes. So long as $\mu_t(d_t)$ changes moderately enough over time to still permit distinguishing effective from less effective treatments within a minimum amount of periods, one feasible approach is to simply discard observations beyond a specific number of periods in the past when applying Thompson sampling. We would then rely only on more recent observations (and the original prior distribution) for forming the posterior distribution $\tilde{F}_{Y(\cdot)_t}$. A further option would be to integrate a model with a discount factor that puts less weight on periods further in the past, as discussed in Russo et al. (2020).

For the sake of causal analysis, we might not only be interested in learning the optimal treatment, but also in inference, such as conducting hypothesis tests to verify if one treatment (particularly the best one) yields a significantly higher or different mean outcome than another one (such as the second best one). However, due to the adaptive randomization scheme and the time dependence it introduces, the probably most intuitive estimator of $\mu(d)$ based on averaging the observed outcomes of all observations $i$ with $D_i = d$ across all periods is asymptotically not normally distributed.

Therefore, conventional t-statistics do not yield valid p-values and confidence intervals. As discussed in Hadad et al. (2021), we can solve this problem by using an adaptively weighted estimator of $\mu(d)$ that satisfies asymptotic normality. An example is the following IPW-based approach with normalized weights, which relies on the optimal treatment probabilities $\tilde{p}_{d,t}$:

$$\hat{\mu}(d) = \sum_{t=1}^{T} \frac{I\{D_t = d\} \cdot Y_t}{\sqrt{\tilde{p}_{d,t}}} \bigg/ \sum_{t=1}^{T} \frac{I\{D_t = d\}}{\sqrt{\tilde{p}_{d,t}}}. \tag{5.15}$$

We note that for notational convenience, index $i$ for a specific observation (as well as summation over $i$) has been dropped from the variables in equation (5.15).

Related to the discussion in section 5.5, we may apply reinforcement learning in subsets of the data defined upon values of observed covariates $X$ (e.g., gender) to find the optimal treatment within subsets; for instance, see Caria et al. (2020). This permits taking into account potential effect heterogeneities across these subsets when learning the optimal treatment assignment. Also with regard to such heterogeneous effects, we face a type of exploration-exploitation trade-off. While considering more subsets (with in terms of $X$ more homogeneous subjects within subsets) permits us to better tailor treatments to specific subpopulations, the exploration process might take longer due to a smaller sample size within each subset, which entails a greater variance of the estimation of treatment effects.

# 6

## Instrumental Variables

### 6.1  Evaluation of the Local Average Treatment Effect

The selection-on-observables assumption discussed in chapter 4 fails if selection into treatment is driven by unobserved factors that affect potential outcomes even conditional on observed covariates $X$. As an example, let us consider an experiment in which access to a training program is randomly assigned, but some of the individuals who are offered the training do not comply and decide not to participate, a scenario known as *imperfect compliance*. If compliance behavior is influenced by unobserved characteristics, such as ability or motivation, which also affect the outcome, such as wages, a comparison of treated and nontreated outcomes does not yield the causal effect of the training even when controlling for covariates. Can we nevertheless exploit our broken experiment in a way that permits evaluating a causal effect?

Indeed, this is feasible if instrumental variable conditions hold, as first formulated in Wright (1928). More concisely, if the random treatment assignment satisfies an exclusion restriction, implying that it does not directly affect the outcome other than through actual treatment (e.g., training) participation, the assignment may serve as an instrumental variable (IV), henceforth denoted by $Z$. The latter permits identifying the treatment effect among those subjects complying with the assignment, the compliers. The intuition is that the causal effect of $Z$ on $Y$, which is identified by the randomization of $Z$, exclusively operates through the causal effect of $Z$ on $D$ among compliers due to the exclusion restriction. Therefore, scaling (or dividing) the average effect of $Z$ on $Y$ by the average effect of $Z$ on $D$ yields the average effect of $D$ on $Y$ among compliers; see the discussions by Imbens and Angrist (1994) and Angrist, Imbens, and Rubin (1996). This causal effect is commonly referred to as the *complier average causal effect* (CACE) or the *local average treatment effect* (LATE), as it refers to the local subpopulation of compliers rather than the total population.

To formally introduce the IV assumptions that permit identifying the LATE among compliers, let us for the sake of simplicity assume that both instrument $Z$ and treatment $D$ are binary (like randomization into and participation in a training), such that $z, d \in \{0, 1\}$, and introduce some further notation. Similar to the potential outcome notation used so far, we denote by $D(z)$ the potential treatment decision if instrument $Z$ is set to value $z$ (either 1 or 0). This permits defining four compliance types in terms of how the treatment reacts to the instrument.

Individuals satisfying $(D(1) = 1, D(0) = 0)$ are compliers, as they only take the treatment when receiving the instrument $(Z = 1)$, while abstaining when not receiving the instrument $(Z = 0)$, as intended in a randomized experiment. The remaining three compliance types are all noncompliers. One group are never takers, who do not take the treatment regardless of the instrument. Never takers thus satisfy $(D(1) = D(0) = 0)$ due to their low willingness to attend a training. Always takers, on the other hand, always take the treatment even when the instrument is zero, satisfying $(D(1) = D(0) = 1)$ due to their high willingness to receive the training. Finally, defiers counteract the instrument assignment by taking the treatment when the instrument is zero and abstaining from the treatment when the instrument is 1, satisfying $(D(1) = 0, D(0) = 1)$. For this reason, they defy the random assignment in an experiment by always choosing the opposite treatment than foreseen by the randomization protocol. The evaluation of the LATE typically hinges on ruling out the existence of such defiers, which can be plausible in certain, but not all, empirical applications, as discussed further later in this chapter. Table 6.1 summarizes the compliance types in terms of their treatment behavior.

As a further modification of our previous notation, let us for the moment denote the potential outcome by $Y(z, d)$: that is, as a function of both the instrument and the treatment. Then we can state the IV assumptions for the identification of the LATE formally as follows:

$$\{D(z), Y(z', d)\} \perp Z \text{ for } z, z', d \in \{0, 1\}, \quad Y(1, d) = Y(0, d) = Y(d), \tag{6.1}$$

$$\Pr(D(1) \geq D(0)) = 1, \quad E[D|Z = 1] - E[D|Z = 0] \neq 0.$$

The first line of expression (6.1), which is frequently referred to as *IV validity*, consists of two assumptions. The first states that the IV is independent of the potential

**Table 6.1**
Compliance types.

| $D(1)$ | $D(0)$ | Type |
| --- | --- | --- |
| 1 | 1 | Always takers |
| 1 | 0 | Compliers |
| 0 | 1 | Defiers |
| 0 | 0 | Never takers |

treatments as well as the potential outcomes, such that there are no variables jointly affecting $Z$ on the one hand and $D$ or $Y$ (or both) on the other hand. This assumption also implies the independence of the instrument and the compliance types, as the latter are defined upon $D(1), D(0)$. The independence assumption is satisfied by design if the instrument is successfully randomized, as in an experiment that provides access to a training program.

The second assumption in the first line of expression (6.1) states that the instrument does not affect the potential outcome conditional on the treatment, implying that $Z$ does not have a direct effect on $Y$ other than through $D$, which is the previously mentioned exclusion restriction. For this reason, we may go back to our conventional notation of the potential outcome by representing it as a function of the treatment only: $Y(d)$. The exclusion restriction holds if mere assignment to a treatment such as a training does not have a direct effect on the outcome of interest, such as through increasing motivation or frustration due to being offered the training or not, respectively. This appears particularly plausible in experiments in which study participants do not even know the assignment status, such as in medical trials where individuals in the control group are assigned to placebo treatments that they cannot distinguish from the actual treatment (like a vaccine).

But there are also examples in which the exclusion restriction can be challenged even if the instrument is randomized. For instance, Angrist (1990) considers the US draft lottery for military service during the Vietnam War as instrument ($Z$) for the treatment variable veteran status ($D$) to investigate the effect of the latter on earnings ($Y$). However, empirical evidence such as that given by Card and Lemieux (2001) suggests that the draft lottery not only affected veteran status, but also induced college enrollment because military service could be postponed or avoided through college deferments. If tertiary education affects earnings, this implies a violation of the exclusion restriction via an alternative (in this case, educational) causal mechanism through which the IV affects the outcome.

For this reason, scrutinizing the validity of IVs that appear plausible at a first glance is very important, particularly when the instrument is not randomly assigned by the analyst or researcher, such that both the independence assumption and the exclusion restriction may appear questionable. This is the case, for instance, for a well-known instrument in labor economics based on the month or quarter of birth ($Z$). The instrument arguably affects education ($D$) through regulations about school starting age and might be exploited for assessing the causal effect of education on later life earnings ($Y$) or other outcomes of interest; see Angrist and Krueger (1991). However, it is well documented (for instance in Bound, Jaeger, and Baker (1995) and Buckles and Hungerman (2013)) that seasonal birth patterns vary systematically with family background characteristics like maternal age or education, family income, and health, all of which may affect later life income, such that IV validity likely fails. As

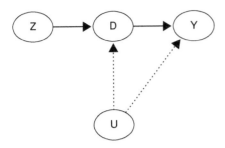

**Figure 6.1**
An instrumental variable approach.

discussed in Deaton (2010), there are many other suspicious instruments that have been applied in empirical studies. Examples range from macroeconomic evaluations of the causes of economic growth, as summarized in Bazzi and Clemens (2013), to the use of instruments assigned by "nature" like twin births; for instance, see the discussions in Rosenzweig and Wolpin (2000) and Farbmacher, Guber, and Vikström (2018).

Figure 6.1 provides a framework that satisfies IV validity, given that no unobserved characteristics omitted from the causal graph jointly affect $Z$ and $Y$, implying that the instrument is (as good as) random. Furthermore, the exclusion restriction is satisfied because $Z$ affects $Y$ only through $D$. In contrast to the instrument, however, the treatment may be affected by unobserved variables, denoted by $U$, which also influence $Y$, which is commonly referred to as *treatment endogeneity*. The causal effects of $U$ are given by dotted lines with arrows to highlight that they are due to unobserved characteristics and thus cannot be measured.

The second line of expression (6.1) imposes two assumptions on the association of the treatment and the instrument. The first says that the potential treatment state of any subject does not decrease (i.e., is weakly positive) in the instrument when switching $Z$ from 0 to 1. As demonstrated in Vytlacil (2002), this assumption is equivalent to imposing a threshold crossing model on the treatment decision, in which the treatment is an additively separable function of the instrument and an unobserved term and takes the value of 1 whenever the function on the instrument is greater than or equal to the unobserved term. The latter might therefore be interpreted as the disutility or cost of treatment, such as reluctance to participate in a training. Formally,

$$D = 1(\eta(Z) \geq V), \tag{6.2}$$

where $V$ is a scalar index of unobservables and $\eta(Z)$ is a general function of $Z$.

Alternatively to imposing such a weak positive monotonicity, we could impose weak negative monotonicity $(\Pr(D(1) \leq D(0)) = 1)$. We will, however, omit the latter case in this discussion because it is symmetric in the sense that weakly negative monotonicity can easily be turned into weakly positive simply by recoding the instrument as $1 - Z$. Weak positive monotonicity rules out the existence of defiers $(D(1) = 0, D(0) = 1)$ because for this group, it holds that $D(1) < D(0)$. By assumption, the population thus only consists of always takers, never takers, and compliers. Weak positive monotonicity is satisfied by design in randomized experiments with one-sided noncompliance (see Bloom (1984)), which guarantees that no subject randomized out of a treatment (like training) can manage to sneak in and participate anyway. This implies that neither defiers with $(D(1) = 0, D(0) = 1)$ nor always takers with $(D(1) = D(0) = 1)$ exist because $\Pr(D(0) = 1) = 0$.

Even in experiments where $\Pr(D(0) = 1) > 0$, the presence of defiers may appear implausible because it would point to counterintuitive behavior according to the randomization protocol. In other IV settings, however, the assumption might be more disputable. Reconsidering the quarter of birth instrument, positive monotonicity appears plausible in the US context at a first glance. Arguably, among students entering school in the same year, those who were born in an earlier quarter can drop out after fewer years of completed education when turning 16, which is the age when compulsory schooling ends, than those born later, particularly after the end of the academic year.

However, strategic postponement of school entry due to redshirting or unobserved school admission policies may reverse the relation of education and quarter of birth for some individuals who are thus defiers, see the discussions by Aliprantis (2012) and Barua and Lang (2009). For this reason, we need to scrutinize monotonicity of $D$ in $Z$ with similar care as IV validity. In this context, it is worth noting that the joint satisfaction of IV validity and monotonicity is partially testable in the data; for instance, see the tests proposed by Kitagawa (2015), Huber and Mellace (2015), Mourifié and Wan (2017), and Farbmacher, Guber, and Klaassen (2020).

The second assumption in the second line of expression (6.1) imposes the existence of a first-stage effect of the IV on the treatment. Together with the previously discussed assumptions, $E[D|Z=1] - E[D|Z=0] \neq 0$ implies the existence of compliers in the population such that $\Pr(D(1) = 1, D(0) = 0) > 0$, which is the subgroup for which we can assess causal effects based on the IV approach. This assumption is testable by verifying how strongly $Z$ affects $D$ in the data: that is, by investigating the size and statistical significance of the effect of the instrument on the treatment. Taken together, the assumptions in expression (6.1) permit identifying the LATE on the compliers, formally defined as

$$\Delta_{D(1)=1,D(0)=0} = E[Y(1) - Y(0)|D(1) = 1, D(0) = 0]. \tag{6.3}$$

We note that when aiming for an average effect like the LATE, we may actually replace full independence between $Z$ and $Y(z, d)$ in expression (6.1) by the weaker mean independence within compliance types: $E[Y(z, d)|D(1), D(0), Z = 1] = E[Y(z, d)|D(1), D(0), Z = 0] = E[Y(z, d)|D(1), D(0)]$ for $z, d \in \{0, 1\}$. Likewise, we can relax the exclusion restriction to an average-based version: $E[Y(1, d)| D(1), D(0)] = E[Y(0, d)|D(1), D(0)] = E[Y(d)|D(1), D(0)]$. The stronger assumptions stated in expression (6.1), however, are required for assessing distributional features like quantile treatment effects as discussed in section 4.8. Similar to the discussion in chapter 4, this distinction of assumptions frequently does not appear that relevant from a practical perspective, as setups in which mean independence holds while full independence does not might seem unnatural. For instance, if one assumes that an IV is mean independent of the potential hourly wage, it seems reasonable that it is also mean independent of the logarithm of the potential hourly wage. As the latter is a nonlinear transformation of the original potential outcome, this implies independence with respect to higher moments as well. Therefore, strengthening mean to full independence often comes with little cost in terms of credibility.

Under the assumptions in expression (6.1) and a binary treatment and instrument, the LATE is identified based on the difference in conditional mean outcomes $E[Y|Z = 1] - E[Y|Z = 0]$, which corresponds to the intention-to-treat (ITT) or reduced form effect of the instrument on the outcome, and the corresponding difference in conditional treatment means $E[D|Z = 1] - E[D|Z = 0]$. The latter gives the first-stage effect and is equal to the share of compliers (i.e., $E[D|Z = 1] - E[D|Z = 0] = E[D(1) - D(0)] = \Pr(D(1) = 1, D(0) = 0)$) because $D(1) - D(0)$ is 1 for compliers and 0 for the two remaining compliance types (not ruled out by weakly positive monotonicity) of never takers and always takers, whose treatment is unaffected by $Z$. For this reason, the effect of $Z$ on $Y$ is necessarily zero for always takers and never takers, as the instrument cannot affect the outcome other than through a change in the treatment under the exclusion restriction. It therefore follows (by an application of the law of total probability with regard to the various compliance types) that the ITT corresponds to the first-stage effect (or complier share) multiplied by the LATE among the compliers, $\Delta_{D(1)=1, D(0)=0}$, while the contribution of the never takers and always takers to the ITT is zero.

For this reason, we obtain the LATE by dividing (or scaling) the ITT by the first-stage effect. Formally,

$$E[Y|Z = 1] - E[Y|Z = 0] = \Delta_{D(1)=1, D(0)=0} \cdot [E[D|Z = 1] - E[D|Z = 0]]$$

$$\Leftrightarrow \Delta_{D(1)=1, D(0)=0} = \frac{E[Y|Z = 1] - E[Y|Z = 0]}{E[D|Z = 1] - E[D|Z = 0]}. \tag{6.4}$$

The ratio of the difference in conditional means $E[Y|Z = 1] - E[Y|Z = 0]$ and $E[D|Z = 1] - E[D|Z = 0]$ in the second line of equation (6.4) is known as the Wald estimand, which we can straightforwardly estimate based on the sample analogs of the four conditional outcome and treatment means; see Wald (1940). A numerically equivalent approach consists of (1) regressing $Y$ on a constant and $Z$ as well as $D$ on a constant and $Z$ and (2) dividing the coefficient on $Z$ in the former regression (i.e., the ITT) by the coefficient on $Z$ in the latter regression (i.e., the first stage). Yet another equivalent approach is a two-stage least squares (TSLS) regression. The latter consists of a linear (first-stage) regression of $D$ on a constant and $Z$ and a linear (second-stage) regression of $Y$ on a constant and the predicted treatment from the first stage (i.e., the estimate of $E[D|Z]$). Any of these three approaches yields the same $\sqrt{n}$-consistent and asymptotically normal estimator of $\Delta_{D(1)=1,D(0)=0}$ under specific statistical conditions (like a complier share that is not too close to zero).

However, a practical advantage of TSLS is that it directly yields the (optionally homoscedastic or heteroscedastic) standard error of the LATE estimate, which (asymptotically) appropriately accounts for estimation uncertainty in both the first- and second-stage regression. Furthermore, TSLS can also be used for the estimation of the mean potential outcomes (rather than the effect) among compliers (i.e., $E[Y(1)|D(1)=1, D(0)=0]$ and $E[Y(0)|D(1)=1, D(0)=0]$, respectively). This is obtained by regressing $Y \cdot D$ on a constant and $D$ or $Y \cdot (1-D)$ on a constant and $(1-D)$ in the second stage, while regressing $D$ or $(1-D)$ on a constant and $Z$ in the first stage. Likewise, we may estimate the cumulative distribution functions of the compliers' potential outcomes at some outcome value $y$ when replacing $Y$ by $I\{Y \leq y\}$ in such TSLS regressions; see the discussion in Imbens and Rubin (1997).

An important condition for a satisfactory performance of LATE estimation and inference (i.e., the computation of standard errors, confidence intervals, and p-values) in small or moderate samples is that the estimated first stage effect of $Z$ on $D$ is strong enough: that is, not too close to zero. It is easy to see that if $E[D|Z=1] - E[D|Z=0]$ approaches zero, the Wald estimand goes to infinity, which implies that the variance of LATE estimation explodes. This issue is known as the weak instrument problem, meaning that the impact of $Z$ on $D$ is so weak that causal inference based on conventionally computed standard errors and confidence intervals may be unreliable and misleading, particularly if the sample size is not very large, as discussed in Staiger and Stock (1997). Luckily, there is an alternative approach to inference and computing confidence intervals suggested in Anderson and Rubin (1949), which is valid even under weak instruments. Stock, Wright, and Yogo (2002) and Keane and Neal (2021), among others, provide surveys on further inference methods that are tailored to weak instruments.

It is interesting to note that even when relaxing the monotonicity assumption $\Pr(D(1) \geq D(0)) = 1$ (and thus allowing for defiers), the Wald estimand might still yield a LATE-type causal effect, given that certain conditions hold. For instance, equation (6.4) identifies the LATE on compliers if this LATE is equivalent to that on the defiers (see Angrist, Imbens, and Rubin (1996)), which, however, appears to be a disputable assumption in many applications. Further, de Chaisemartin (2017) shows that the Wald estimand corresponds to the LATE among a subpopulation of compliers, if some share of the compliers is equal to the defiers in terms of average effects and population size. The Wald estimand then yields an even more local effect among the comvivors: that is, those compliers who outnumber the defiers resembling a share of compliers in terms of size and effects. Under more general forms of defiance, the Wald estimand might at least yield the correct sign of the LATE, such as under stochastic monotonicity. The latter implies the existence of at least as many compliers as defiers conditional on any pair of potential outcomes $Y(1), Y(0)$; see the discussion by Small, Tan, Lorch, and Brookhart (2017).

To demonstrate LATE estimation in R, we reconsider the Job Corps (JC) data previously analyzed in section 3.1 in chapter 3, where we regarded random assignment to JC as the treatment. However, we will subsequently consider this variable to be the instrument, while actual participation in a training in the first year will serve as the treatment (which may deviate from random assignment). After loading the *causalweight* package and the *JC* data, we define the instrument, *Z=JC$assignment*, the treatment, *D=JC$trainy1*, and the outcome, *Y=JC$earny4* (namely, weekly earnings in the fourth year). We then compute the ITT effect of $Z$ on $Y$ based on the differences in mean outcomes of $Y$ given $Z=1$ and $Z=0$ and store the result in an R object: *ITT=mean(Y[Z==1])-mean(Y[Z==0])*. We recall that square brackets permit us to select observations that satisfy specific conditions, in our case specific values of the instrument. In an analogous manner, we calculate the first-stage effect of $Z$ on $D$: *first=mean(D[Z==1])-mean(D[Z==0])*. Following equation (6.4), we then estimate the LATE as the ratio of the *ITT* and the *first* stage effect and store it in an object named *LATE*. Finally, we call the *ITT*, *first*, and *LATE* objects to inspect the results. The box here provides the R code for these steps.

```
library(causalweight)              # load causalweight package
data(JC)                           # load JC data
Z=JC$assignment                    # define instrument (assignment to JC)
D=JC$trainy1                       # define treatment (training in 1st year)
Y=JC$earny4                        # define outcome (earnings in fourth year)
ITT=mean(Y[Z==1])-mean(Y[Z==0])    # estimate intention-to-treat effect (ITT)
first=mean(D[Z==1])-mean(D[Z==0])  # estimate first stage effect (complier share)
LATE=ITT/first                     # compute LATE
ITT; first; LATE                   # show ITT, first stage effect, and LATE
```

Running the code yields an *ITT* estimate of roughly 16 US dollar (USD), which is exactly the same as the effect obtained in the empirical example at the end of section 3.1 when regarding JC assignment as the treatment (which ignores noncompliance), rather than as the instrument, as in the current context. The *first* stage effect of $Z$ on $D$ amounts to 0.34, implying an estimated complier share of 34 percent. Therefore, the remaining 66 percent of observations are either never takers or always takers of the training program according to our estimate, given that the IV assumptions in equation (6.1) hold. Finally, the estimated *LATE* among compliers corresponds to an increase of roughly 47 USD in weekly earnings.

In the next step, we show that we obtain the very same estimate when using TSLS regression. To this end, we load the *AER* package by Kleiber and Zeileis (2008), which contains the *ivreg* command for TSLS regression. Similar to the *lm* command, the *ivreg* command requires a regression formula that specifies the outcome and the treatment (i.e., $Y \sim D$), but also an instrument for the endogenous treatment, which is separated from the regression formula by a vertical bar (i.e., $|Z$). We save the TSLS results in an object named *LATE*, which we wrap by the *summary* command, where we also set *vcov = vcovHC* for computing heteroscedasticity-robust standard errors.

```
library (AER)                          # load AER package
LATE=ivreg (Y~D|Z)                     # run two stage least squares regression
summary (LATE, vcov = vcovHC)          # results with heteroscedasticity-robust se
```

Running the code yields the following output:

```
Coefficients:
            Estimate Std. Error t value Pr(>|t|)
(Intercept)  174.039      8.665  20.086  < 2e-16 ***
D             47.194     12.027   3.924 8.77e-05 ***
```

While the LATE estimate is the same as that previously obtained from mean differences based on equation (6.4), the TSLS procedure has the practical advantage that it also yields t-statistics and p-values. We see that the LATE on compliers is highly statistically significant because the p-value is very close to zero. As an alternative to using the *ivreg* command, we can load the *ivmodel* package and run the command *ivmodel(Y=Y,D=D,Z=Z)* to obtain an estimate of the LATE based on TSLS (and two further IV approaches). Applying the *ivmodel* also yields Anderson and Rubin–type p-values and confidence intervals, which are even valid under weak instruments.

## 6.2   Instrumental Variable Methods with Covariates

In many applications, it may not appear credible that IV assumptions like random assignment hold unconditionally: that is, without controlling for observed covariates. This seems particularly relevant for observational data in which the instrument is typically not explicitly randomized like in an experiment. For instance, Card (1995) considers geographic proximity to college as IV for the likely endogenous treatment education when assessing its effect on earnings. On the one hand, proximity might induce some individuals to go to college who would otherwise not (e.g., due to housing costs associated with not living at home), implying a first-stage effect of the instrument on the treatment. On the other hand, proximity likely reflects selection into neighborhoods with a specific socioeconomic status that may have an influence on earnings, implying that the IV is not random, but associated with characteristics that have an impact on the outcome.

If all confounders that jointly affect the instrument and the outcome are plausibly observed in the data, we can implement IV-based estimation conditional on these observed covariates. For this reason, Card (1995) includes a range of control variables like parents' education, ethnicity, urbanity, and geographic region. We note that this amounts to imposing a selection-on-observables assumption, however, with regard to the instrument rather than the treatment (as in chapter 4). In fact, the treatment may be associated with unobservables affecting the outcome even conditional on the covariates, while this must not be the case for the instrument.

Let us now formally state the IV assumptions that permit identifying causal effects among compliers conditional on covariates $X$ in the binary instrument and treatment case, following the discussion in Abadie (2003):

$$\{D(z), Y(z', d)\} \perp Z | X \text{ for } z, z', d \in \{0, 1\}, \quad \Pr(Y(1, d) = Y(0, d) = Y(d)|X) = 1, \qquad (6.5)$$

$$\Pr(D(1) \geq D(0)|X) = 1, \quad E[D|Z = 1, X] - E[D|Z = 0, X] \neq 0,$$

$$X(1) = X(0) = X, \quad 0 < P(Z = 1|X) < 1.$$

The first line of expression (6.5) requires the IV validity assumptions provided in expression (6.1) to hold conditional on $X$, such that the instrument is as good as randomly assigned and satisfies the exclusion restriction among observations with the same covariate values. The second line of expression (6.5) rules out the existence of defiers, but it also requires the existence of compliers conditional on $X$, due to the nonzero conditional first stage. Note that the threshold model-based representation of monotonicity due to Vytlacil (2002) (see equation (6.2)) now may also incorporate $X$ as a factor driving the treatment decision to become

$$D = 1(\eta(Z, X) \geq V). \qquad (6.6)$$

As in expression (4.1) of section 4.1, the first assumption in the third line of expression (6.5) invokes that $X$ is not a function of $D$ and therefore must not contain (posttreatment) characteristics that are affected by the treatment (we note that $X(1), X(0)$ refer to the states of the treatment, not the instrument). The second assumption in the third line is a common support restriction on the instrument

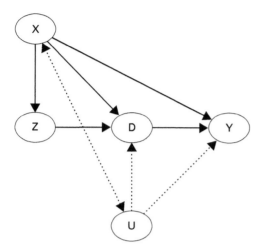

**Figure 6.2**
An instrumental variable approach with covariates.

propensity score $\Pr(Z=1|X)$, implying that $Z$ is not deterministic in $X$ such that subjects with both $Z=1$ and $Z=0$ exist for all feasible covariate values. Figure 6.2 provides a framework satisfying IV validity conditional on $X$, given that no unobserved characteristics omitted from the causal graph jointly affect $Z$ and $Y$. It is worth noting that unobservables $U$ may affect $X$ or vice versa, as indicated by the bidirectional dotted arrow, or both $U$ and $X$ might be caused by further unobservables. Conditional on $X$, however, no unobservables jointly affect $Z$ and $Y$.

As already mentioned, $\{D(z), Y(z',d)\}\perp Z|X$ in expression (6.5) is a selection-of-observables assumption similar to expression (4.1), however now with regard to the instrument rather than the treatment. Therefore, the effects of $Z$ on $Y$ and on $D$ are identified conditional on $X$, in analogy to the identification of the effect of $D$ on $Y$ given $X$ in chapter 4. For this reason, replacing $D$ by $Z$ and the treatment propensity score $p(X) = \Pr(D=1|X)$ by the instrument propensity score $\Pr(Z=1|X)$ in the identification results for the ATE in equations (4.3), (4.36), (4.40), and (4.45) yields the ITT effect of the instrument on the outcome, henceforth denoted by $\theta$. In addition, replacing $Y$ by $D$ in the identification results yields the first-stage effect of the instrument on the treatment: that is, $E[D(1) - D(0)]$, henceforth denoted by $\gamma$. In the spirit of the Wald estimand in equation (6.4), the LATE is then identified by dividing $\theta$ by $\gamma$:

$$\Delta_{D(1)=1,D(0)=0} = \frac{E[E[Y|Z=1,X] - E[Y|Z=0,X]]}{E[E[D|Z=1,X] - E[D|Z=0,X]]}$$

$$= \frac{E[Y \cdot Z / \Pr(Z=1|X) - Y \cdot (1-Z)/(1-\Pr(Z=1|X))]}{E[D \cdot Z / \Pr(Z=1|X) - D \cdot (1-Z)/(1-\Pr(Z=1|X))]}$$

$$= \frac{\theta}{\gamma}. \tag{6.7}$$

If $X$ consists of preselected covariates, the estimation of $\Delta_{D(1)=1,D(0)=0}$ under the assumptions given in expression (6.5) proceeds by (1) estimating both the ITT effect $\theta$ and the first-stage effect $\gamma$ based on one of the treatment effect estimators introduced in chapter 4 and (2) dividing the ITT estimate by the first-stage estimate. Under specific regularity conditions, such an approach can be $\sqrt{n}$-consistent and asymptotically normal. The potential methods for estimating $\frac{\theta}{\gamma}$ include matching or regression for a conditional mean based-representation of the LATE as in the first line of equation (6.7) as in Frölich (2007), inverse probability weighting (IPW) based on the second line of equation (6.7) as in Donald, Hsu, and Lieli (2014) or DR methods combining regression and IPW as in Tan (2006). We also can use such methods for estimating distributional effects among compliers like local quantile treatment effects (LQTEs); see, for instance, Frölich and Melly (2013). Finally, a fully linear approach to IV estimation based on the assumptions in expressions (6.5) is to simply add the covariates as regressors in the TSLS regression discussed in chapter 6. However, this latter estimator rules out effect heterogeneity in $X$ and is generally inconsistent if the covariates are nonlinearly associated with the treatment and the outcome.

Rather than preselecting the covariates to be controlled for, we might prefer learning the important confounders jointly affecting the instrument and the outcome from a possibly large set of variables $X$ in a data-driven way, in analogy to the discussion in chapter 5. To this end, we can apply the causal machine learning (CML) approaches outlined in section 5.2, such as double machine learning (DML) to estimate $\theta$ and $\gamma$ (and ultimately the LATE) based on equation (4.45); for instance, see Belloni, Chernozhukov, Fernández-Val, and Hansen (2017). Also, the analysis of effect heterogeneity and optimal policies discussed in sections 5.4 and 5.5 extends to the IV context by relying on doubly robust (DR) statistics (like efficient influence functions) appropriate for conditional LATE estimation given covariates $X$; for instance, see the discussion by Athey, Tibshirani, and Wager (2019).

A general criticism of the LATE is that it may be of limited practical relevance because it only refers to the subpopulation of compliers, which is likely not as interesting as the total population targeted by the ATE. This concern about external validity (i.e., the question how valid or representative the LATE is for effects in other populations) might at least partly be investigated or even alleviated with the help of observed covariates. One approach consists of verifying how similar or different compliers are in terms of covariates relative to the total population. To this end, we can apply a method suggested by Abadie (2003), which permits identifying a large range of complier-related statistics, including covariate or outcome distributions, based on weighting observations by the following function $\kappa$:

$$\kappa = 1 - \frac{D \cdot (1 - Z)}{1 - \Pr(Z = 1 | X)} - \frac{(1 - D) \cdot Z}{\Pr(Z = 1 | X)}. \tag{6.8}$$

For instance, $\frac{E(\kappa \cdot X)}{E(\kappa)} = E[X | D(1) = 1, D(0) = 0]$ yields the mean of $X$ among compliers, which permits judging how representative this subgroup is for the total population in terms of average observed characteristics. This could serve as the base for educated guesses about how strongly effects might differ across compliers and the total population, even though causal effects may generally also vary with unobserved characteristics not included in $X$.

Ruling out this latter possibility and imposing that average effects are homogeneous across various compliance types conditional on $X$ is a further way to exploit covariates in the context of external validity; see Angrist and Fernández-Val (2010) and Aronow and Carnegie (2013). Formally, such a conditional effect homogeneity assumption can be stated as follows:

$$E[Y(1) - Y(0) | D(1), D(0), X] = E[Y(1) - Y(0) | X]. \tag{6.9}$$

In contrast to our previous IV assumptions, equation (6.9) rules out that effect heterogeneity is driven by unobserved characteristics, which importantly restricts the source of treatment effect heterogeneity. This is, for instance, not consistent with the Roy model popular in economics for motivating selection into treatment based on unobserved outcome gains due to treatment participation (Roy 1951).

If we are willing to impose the assumptions in expressions (6.9) and (6.5) (where in fact monotonicity could be relaxed to stochastic monotonicity, as discussed in section 6.1, conditional on $X$), we can identify the ATE by averaging over the conditional LATE given $X$, denoted by $\Delta_{D(1)=1, D(0)=0, X} = E[Y(1) - Y(0) | D(1) = 1, D(0) = 0, X]$, in the total population:

$$\Delta = E[\Delta_{D(1)=1, D(0)=0, X}] = E\left[ \frac{E[Y | Z = 1, X] - E[Y | Z = 0, X]}{E[D | Z = 1, X] - E[D | Z = 0, X]} \right], \tag{6.10}$$

where the second equality follows from the fact if the IV assumptions hold conditional on the covariates, then a conditional version of the Wald estimand given $X$ identifies $\Delta_{D(1)=1, D(0)=0, X}$. If several instruments are available, equation (6.9) can even be tested because any instrument should entail the same conditional effect $\Delta_{D(1)=1, D(0)=0, X}$ under this homogeneity assumption.

As discussed in Angrist (2004), an alternative condition to the effect homogeneity assumption in equation (6.9), which also establishes the external validity of the LATE, is that different compliance types have the same mean potential outcomes, at least conditional on $X$. This implies a selection-on-observables assumption with regard to the treatment that is closely related to that discussed in chapter 4, such that instruments are in fact not required for identification. However, the instrument can be used for partly verifying whether potential outcomes are homogeneous across compliance types—namely, for testing the equality of mean potential outcomes across treated compliers and always takers, as well as across nontreated compliers and never takers conditional on $X$; see de Luna and Johansson (2014). In fact, a statistically significant effect of $Z$ on $Y$, conditional on $D$ and $X$, points to a violation of homogeneity in mean potential outcomes. Donald, Hsu, and Lieli (2014) suggest a related but different testing approach that verifies whether the LATE estimated based on the IV assumptions statistically significantly differs from the ATET estimated under the selection-on-observables assumption with regard to the treatment of chapter 4 under one-sided noncompliance, ruling out always takers and defiers. This test is based on the fact that under one-sided noncompliance, the LATE coincides with the ATET.

Yet another path to the external validity of IV methods is to assume that any individual's rank in the population's distribution of the potential outcome under treatment $Y(1)$ is the same as the rank under no treatment $Y(0)$, implying rank invariance, or at least not systematically different, implying rank similarity; see Chernozhukov and Hansen (2005). This approach applies to continuously distributed outcomes and means that an individual earning the median wage under treatment would also earn the respective median wage under nontreatment (rank invariance) or only randomly deviate from the median wage (rank similarity), at least when controlling for observed covariates $X$. While such outcome rank-related assumptions permit the evaluation of quantile and average treatment effects in the total population (rather than among compliers alone) without imposing monotonicity of $D$ in $Z$, the price is that they substantially restrict treatment effect heterogeneity.

As an illustration, let us consider the educational choices of a college degree versus vocational training as treatment. Rank stability or similarity assumptions rule out that a college graduate reaches a systematically higher or lower rank in the potential wage distribution under college graduation than she or he would have reached in the potential wage distribution under a vocational training (when obtaining a vocational degree instead of a college degree). However, it appears likely that given their abilities and preferences, some individuals are relatively more competitive either in the academic or vocational track compared to others, such that rank invariance or stability must be carefully scrutinized in the empirical application at hand. One approach to estimating conditional quantile treatment effects given $X$ at an outcome rank $\tau$ (e.g., $\tau = 0.5$ for the median outcome) under these assumptions is to iteratively find the value $\beta_D(\tau)$ that entails a zero coefficient on instrument $Z$ in a quantile regression of the modified outcome $Y - D\beta_D(\tau)$ on a constant, $X$, and $Z$. Estimating $\beta_D(\tau)$ across all ranks $\tau \in (0, 1)$ and averaging the estimates then yields an estimate of the conditional average treatment effect (CATE) given $X$.

To implement LATE estimation when controlling for covariates in R, we load the *causalweight* package, which contains an IPW-based estimator, and the *LARF* package by An and Wang (2016), which includes the 401(k) pension data on tax-deferred retirement plans in the US previously analyzed by Poterba, Venti, and Wise (1998). The data set named *c401k* consists of 9,275 observations with information on the eligibility for a 401(k) pension plan, which serves as instrument $Z$, as well as individual participation in such a plan, which is our treatment $D$. The outcome variable is net family financial assets in 1,000 USD. Accordingly, we define $D=c401k[,3]$ (because treatment participation is indicated in the third column of the *c401k* data), $Z=c401k[,4]$, and $Y=c401k[,2]$.

Eligibility to pension plans is decided by employers and is likely nonrandom. For this reason, we control for socioeconomic characteristics $X$ like an individual's income, age, gender, and family status when running our IV estimation, assuming that the instrument is conditionally independent of unobserved individual factors affecting the outcome (like preferences for savings). The covariates are stored in columns 5 to 11 of the data (i.e., *c401k[,5:11]*), which we wrap by the *as.matrix* command to turn the data into a numeric matrix (as required by the IV procedure) and save it in an object named $X$. We then set a seed (for the reproducability of our estimates) and feed $Y$, $D$, $Z$, and $X$ into the *lateweight* command for IPW-based LATE estimation using a probit specification for the instrument propensity score. We also set the argument *boot=299* to estimate the standard error based on 299 bootstrap replications. We store the results in an object named *LATE* and call *LATE$effect*, *LATE$se.effect*, and *LATE$pval.effect* to investigate the LATE estimate, its standard error, and the p-value. The box here provides the R code for the various steps.

```
library(LARF)                                # load LARF package
library(causalweight)                        # load causalweight package
data(c401k)                                  # load 401(k) pension data
D=c401k[,3]                                   # treatment: participation in pension plan
Z=c401k[,4]                                   # instrument: eligibility for pension plan
Y=c401k[,2]                                   # outcome: net financial assets in 1000 USD
X=as.matrix(c401k[,5:11])                     # covariates
set.seed(1)                                   # set seed
LATE=lateweight(y=Y, d=D, z=Z, x=X, boot=299)  # compute LATE (299 bootstraps)
LATE$effect; LATE$se.effect; LATE$pval.effect  # show LATE results
```

The LATE estimate amounts to 13.096, suggesting that participation in a 401(k) pension plan increases net financial assets by roughly 13,096 USD. The standard error of 2.172 is relatively small, and therefore the p-value is very close to zero, such that we can safely reject the null hypothesis of a zero treatment effect among compliers. We also take a look at the first-stage effects of the eligibility instrument $Z$ on participation *D* along with the standard error by calling *LATE$first, LATE$se.first,* and *LATE$pval.first*:

```
LATE$first; LATE$se.first; LATE$pval.first # show first stage results
```

The estimated first-stage effect corresponds to 0.684, implying a complier share of 68.4 percent. The standard error is rather small (0.008), and therefore the p-value is very close to zero. Finally, we reestimate the LATE and first-stage effect based on DML, as introduced in section 5.2 rather than IPW. To this end, we load the *npcausal* package previously considered in section 4.8, set a seed, and feed *Y, D, Z,* and *X* into the *ivlate* command for DML-based LATE estimation using an ensemble method (see section 5.3):

```
library(npcausal)           # load npcausal package
set.seed(1)                 # set seed
ivlate(y=Y, a=D, z=Z, x=X)  # estimate LATE by double machine learning
```

Running the code yields the following output:

```
  parameter         est            se        ci.ll        ci.ul pval
1      LATE 11.82973411 1.884076170 8.13694482 15.5225234    0
2 Strength  0.68281554 0.008585958 0.66598707  0.6996440   NA
3 Sharpness  0.06829792 0.020283612 0.03776416  0.1204296   NA
```

The LATE estimate of roughly 11.830 provided in the first line of the column *est* is rather similar to our previous result, and again highly statistically significant. This also applies to the first-stage effect provided in the second line, which amounts to roughly 0.683 (or 68.3 percent in terms of the complier share).

## 6.3  Nonbinary Instruments and Treatments

In contrast to the previous sections, we will henceforth consider the case that the instrument is not binary, but multivalued and possibly even continuous, while maintaining a binary treatment (which only takes the values 1 and 0). This implies that we can assess LATEs with regard to any pair of instrument values $z' > z$ (which now may be different from 1 and 0) satisfying the IV assumptions. Let us assume, for instance, that instrument *Z* is a randomized cash incentive to obtain a medical treatment like a vaccination, meaning that $z'$ and $z$ correspond to two financial incentives (e.g., 20 versus 10 USD). Such choices of such pairs of instrument values generally entail different first stages (i.e., different $E[D|Z=z'] - E[D|Z=z]$ when not controlling for any covariates), and thus different complier populations.

Particularly interesting in the context of the previous discussion about external validity in section 6.2 is choosing $z'$ and $z$ in a way that entails the largest possible complier population. In this context, this would amount to basing LATE estimation as discussed in the previous sections of this chapter on individuals with the highest and lowest cash incentives. It is worth mentioning that instead of directly conditioning on the value of instrument $Z$ in IV estimation, we may alternatively use the treatment propensity score given $Z$ and possibly $X$ as the instrument when imposing the conditional IV assumptions given $X$ in expression (6.5); see the discussion in Frölich (2007). Denoting the treatment propensity score by $p(Z, X) = \Pr(D = 1|Z, X)$, this implies that the LATE is (in analogy to equation (6.7) in section 6.2) identified by

$$\Delta_{D(1)=1,D(0)=0} = \frac{E[E[Y|p(Z,X)=p(z',X),X] - E[Y|p(Z,X)=p(z,X),X]]}{E[E[D|p(Z,X)=p(z',X),X] - E[D|p(Z,X)=p(z,X),X]]}. \tag{6.11}$$

The propensity score–based approach appears particularly useful if $Z$ consists of several instruments (e.g., cash transfers combined with geographic proximity to treatment centers), as the elements in $Z$ can then be straightforwardly collapsed into a single instrument, $p(Z, X)$. This, however, implies that the monotonicity assumption must hold with regard to this newly created combined instrument $p(Z, X)$ rather than a single instrument alone. As discussed in Mogstad, Torgovitsky, and Walters (2020), such a monotonicity condition implicitly rules out that some subjects only comply with one instrument (e.g., cash transfers) and others only with another (e.g., geographic proximity) in terms of treatment choice, which may not be plausible depending on the empirical context. For this reason, Mogstad, Torgovitsky, and Walters (2020) consider a weaker, partial monotonicity assumption, which imposes weak monotonicity of the treatment in one instrument (e.g., cash transfers) conditional on specific values of the other (e.g., proximity). We also note that the common support assumption must now hold with respect to $p(Z, X)$, implying that for any covariate value $x$ in the population, there are observations with both $p(z', x)$ and $p(z, x)$, or equivalently, $z'$ and $z$.

A continuously distributed instrument even permits us to evaluate a continuum of complier effects under appropriately adapted IV assumptions. Specifically, a marginal change in the instrument yields the marginal treatment effect (MTE); for instance, see the discussion in Heckman and Vytlacil (2001, 2005). Formally, the MTE is the average treatment effect, conditional on the covariates and the unobserved term in equation (6.6), or the unobserved term in equation (6.2) alone if there are no $X$ variables to be controlled for:

$$\Delta_{x,v} = E[Y(1) - Y(0)|X = x, V = v]. \tag{6.12}$$

Technically speaking, the MTE is the limit of the LATE when the change in the instrument (i.e., the difference in $z'$ and $z$) goes to zero. This effect can be interpreted as the average effect among individuals who are indifferent between treatment or nontreatment, given their values of instrument $Z$ and covariates $X$: that is, for whom it holds that $\eta(Z=z, X=x)=v$ in equation (6.6). In this context, it is worth noting that we can normalize the unobserved term $V$ without loss of generality so that the normalization, henceforth denoted by $\bar{V}$, is uniformly distributed and takes values between 0 and 1 (i.e., $\bar{V} \sim \text{Uniform}[0, 1]$). In fact, $\bar{V}$ corresponds to the cumulative distribution function of $V$: $\bar{V} = F_V$. As discussed in Heckman and Vytlacil (1999), we can then estimate the MTE in equation (6.12) by the local IV (LIV). The latter uses the propensity score $p(Z, X)$, as well as the normalization $\bar{V}$, and corresponds to the first derivative of the conditional mean outcome given $X$ and $p(Z, X)$ with regard to to $p(Z, X)$:

$$\Delta_{X=x, \bar{V}=p(z,x)} = \frac{\partial E[Y|X=x, p(Z,X)=p(z,x)]}{\partial p(z,x)}. \tag{6.13}$$

The LIV representation in equation (6.13) demonstrates that MTEs can be recovered based on variation in $p(Z, X)$, given that the assumptions in expression (6.5) hold. Under a sufficiently strong continuous instrument with a sufficiently large range of values, we may theoretically evaluate the MTEs for all feasible values of $X$ and $V$ to ultimately assess the ATE by averaging over the various MTEs in the population. In real-world applications, however, such strong instruments are typically unavailable. Therefore, the MTE is only identified over the common support of $p(Z, X)$ across all values of $X$: that is, those values of the propensity score that exist across all covariate values in the population. This typically limits the feasibility of flexible, nonparametric MTE estimation (e.g., by kernel regression, as discussed in section 4.2), in particular if $X$ contains many variables and $Z$ is not strong or has a limited range of values.

One approach to alleviate this problem at the cost of substantially restricting the generality of the treatment effect model is to replace the independence assumption $\{D(z), Y(z', d)\} \perp Z|X$ by a much stronger version; for instance, see Carneiro, Heckman, and Vytlacil (2011):

$$\{D(z), Y(z', d)\} \perp (Z, X) \text{ for } z, z' \text{ in the support of } Z. \tag{6.14}$$

In contrast to the previous conditional independence assumption in expression (6.5), condition (6.14) imposes independence between $X$ and unobservables affecting $D$ or $Y$. While covariates $X$ (e.g., labor market experience) may still jointly affect $Z$ on the one hand and $D$ and/or $Y$ on the other hand, they must not be associated with unobservables (e.g., ability) that affect $D$, $Y$, or both. Reconsidering the causal graph in figure 6.2 in section 6.2, this rules out the dotted causal arrows between $U$ and $X$. In this arguably special case, we can evaluate the MTE over the entire unconditional support of $p(Z, X)$ (i.e., any values the propensity score might take), such that common support across values of $X$ is not required.

Brinch, Mogstad, and Wiswall (2017) discuss further approaches for facilitating the estimation of MTEs in practice based on imposing parametric assumptions, like a linear change of the MTE across values of $p(Z, X)$. A further assumption is additive separability in treatment effect heterogeneity caused by covariates $X$ on the one hand and unobserved characteristics on the other hand. This, for instance, implies that any treatment effect heterogeneity across levels of some covariate like labor market experience must not interact with the treatment effect heterogeneity across levels of some unobserved characteristic like ability. This assumption permits easing the condition in expression (6.14) somewhat by allowing associations between $V$ (the unobservable affecting the treatment) and $X$, so long as $X$ is conditionally independent of any unobserved characteristics affecting the outcome $Y$, given $V$.

In contrast to the discussion so far, let us now assume that it is the treatment that is multivalued, while maintaining a binary instrument for the sake of simplicity. Related to section 3.5 in chapter 3, we consider an ordered treatment $D \in \{0, 1, 2, \ldots, J\}$, where $J$ denotes the number of nonzero treatment values, while receiving no treatment is coded as zero. Angrist and Imbens (1995) demonstrate that unfortunately, we cannot evaluate the effects among complier groups at specific treatment values, such as for those increasing the treatment from 1 (e.g., 1 week of training) to 2 (2 weeks of training) when increasing the instrument from 0 to 1. However, it is possible to obtain a nontrivially mixed (or weighted) average of effects of unit-level increases in the treatment for nontrivially weighted complier groups. The latter may differ in terms of how the groups change their treatment as a reaction to the instrument; for example, complier group 1 might satisfy $D(1) = 2, D(0) = 0$, while complier group 2 may satisfy $D(1) = 4, D(0) = 1$.

Given the IV assumptions in expression (6.1), Angrist and Imbens (1995) show that if $\Pr(D(1) \geq j > D(0)) > 0$ for some treatment value $j$ such that compliers exist at some treatment margin, then the Wald estimand equals a weighted average of effects of unit changes in the treatment on various complier groups defined by different margins of the potential treatments:

$$\frac{E[Y|Z=1] - E[Y|Z=0]}{E[D|Z=1] - E[D|Z=0]} = \sum_{j=1}^{J} w_j \cdot E[Y(j) - Y(j-1)|D(1) \geq j > D(0)], \tag{6.15}$$

with weights $w_j = \dfrac{\Pr(D(1) \geq j > D(0))}{\sum_{j=1}^{J} \Pr(D(1) \geq j > D(0))}$, implying that $0 \leq w_j \leq 1$ and $\sum_{j=1}^{J} w_j = 1$. However, the various treatment effects based on unit changes in the treatment, $E[Y(j) - Y(j-1)|D(1) \geq j > D(0)]$, remain unidentified. Furthermore, the complier groups contributing to the weighted effect might be overlapping. Some compliers could satisfy both $(D(1) \geq j > D(0))$ and $(D(1) \geq j + 1 > D(0))$ for treatment $j$ (e.g., 1) and therefore be accounted for multiple times, which arguably compromises the interpretability of the effect.

For this reason, it may appear tempting to binarize a multivalued treatment into just two values, such categorizing education into tertiary versus less than tertiary eduction, and to proceed with the analysis as outlined in section 6.1 or section 6.2, if controlling for covariates $X$ is required. However, such a binarization of the treatment generally violates the IV exclusion restriction; for instance, see the discussion in Andresen and Huber (2021). The reason is that the instrument also might affect the treatment at margins not captured by the redefined treatment of tertiary versus no tertiary eduction, such as at the decision of upper secondary versus lower secondary education. We note that the result in equation (6.15) and the related caveats in terms of interpretation also extend to weighted LATE estimation in setups with multiple instruments and when controlling for covariates.

LATE evaluation is also different under a treatment with unordered rather than ordered values, which is equivalent to the case of several unordered treatments that are mutually exclusive (e.g., 1 =IT course, 2 =sales training). Behaghel, Crépon, and Gurgand (2013) consider a scenario with a three-valued treatment and instrument $(D, Z \in \{0, 1, 2\})$ and modify the conventional monotonicity assumption in expression (6.1) to assess the LATEs among the two complier populations $(D(1) = 1, D(0) = 0)$ and $(D(2) = 2, D(0) = 0)$. Their assumption requires that changing $Z$ from 0 to 1 affects treatment choice 1 versus 0, but not 2, while changing $Z$ from 0 to 2 affects treatment choice 2 versus 0, but not 1. Heckman and Pinto (2018) suggest an alternative monotonicity assumption, which requires for any specific value of the unordered treatment that if some subjects move into (out of) the respective value when the instrument is switched, then no subjects can at the same time move out of (into) that value. These two examples demonstrate that monotonicity conditions under multiple unordered treatments are generally more involved compared to the binary treatment case.

Let us consider a practical illustration of MTE estimation in R, and to this end load the *localIV* package by Zhou (2020). The latter contains an artificially created data set named *toydata*, which consists of 10,000 observations and includes a binary treatment $d$, a continuous instrument $z$, an outcome $y$, and a covariate $x$. As our convention has been to use capital letters for the various variables, we define $D$=*toydata\$d*, $Z$=*toydata\$z*, $Y$=*toydata\$y*, and $X$=*toydata\$x*. We then apply the *mte* command to estimate MTEs by LIV estimation using the argument *selection=D~X+Z* for specifying the treatment propensity score to be estimated by means of a probit model, and *outcome=Y~X* for including the covariate in the kernel regression–based estimation of equation (6.13). We store the estimated model parameters in an object named *MTE*.

In the next step, we run the *mte_at* command for predicting MTEs at specific covariate values, by default at the mean of $X$. The first argument $u$ corresponds to the values of the normalized unobserved term $\bar{V}$ or the treatment propensity score $p(Z, V)$ in equation (6.13) at which the MTEs should be evaluated, which we set to *seq(0.05, 0.95, 0.01)* to predict the effects at $\bar{V} = p(Z, V) = \{0.05, 0.06, \dots, 0.94, 0.95\}$. Furthermore, we set the second argument *model=MTE* to feed in our estimated model parameters required for predicting the MTEs. We save the results in an object named *MTEs*. Finally, we use the *plot* command to plot the MTEs across values of $\bar{V}$ (or $p(Z, V)$) given the mean of $X$, with *x=MTEs$u* on the *x*-axis and *y=MTEs$value* (i.e., the values of the MTEs) on the *y*-axis. The box here provides the R code for each of the steps.

```
library(localIV)                                       # load localIV package
data(toydata)                                          # load toydata
D=toydata$d                                            # define binary treatment
Z=toydata$z                                            # define continuous instrument
Y=toydata$y                                            # define outcome
X=toydata$x                                            # define covariate
MTE=mte(selection=D~X+Z, outcome=Y~X)                  # LIV estimation of MTE
MTEs=mte_at(u=seq(0.05, 0.95, 0.01), model=MTE)        # predict MTEs at mean of X
plot(x=MTEs$u,y=MTEs$value,xlab="p(Z, mean X)",ylab="MTE at mean X") #plot
```

Running the code yields a plot of the MTEs at mean values of $X$ (i.e., estimates of $\Delta_{X=E[X], \bar{V}=p(z,E[X])}$) as a function of the probit-based propensity score estimate $\hat{p}(Z, \bar{X})$, where $\bar{X}$ denotes the average value of $X$ in the sample. The plot is provided in figure 6.3. The results suggest that the MTE is positive for any value of the propensity score (and thus, $\bar{V}$), but decreasing in its magnitude, as indicated by the negative slope of the MTE-propensity score association. We note that this association could be estimated for other values of $X$ than its mean as well.

## 6.4   Sample Selection, Dynamic and Multiple Treatments, and Causal Mechanisms

Related to the discussion in section 4.11 in chapter 4, IV-based treatment evaluation also can be complicated by nonrandom outcome attrition, such as nonresponses in follow-up surveys in which the outcome is measured, and sample selection, when wage outcomes are observed only conditional on selection into employment. One possibility to tackle this problem is to impose a missing-at-random (MAR) restriction, as considered in section 4.11, assuming conditional independence of the attrition or sample selection process and outcome $Y$ given observed variables, such as instrument $Z$, treatment $D$, and covariates $X$. An alternative to MAR that is tailored to the LATE framework is the latent ignorability (LI) assumption, suggested

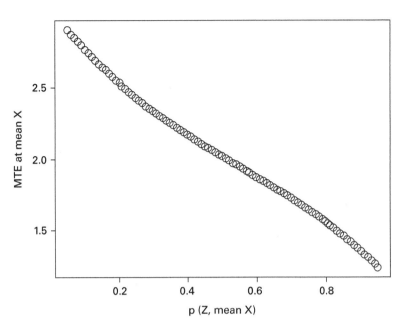

**Figure 6.3**
MTEs.

in Frangakis and Rubin (1999). LI requires outcome missingness to be as good as random, conditional on the compliance type characterizing whether a subject is a complier, an always taker, or a never taker. Furthermore, we may combine MAR and LI such that the conditional independence of the attrition or sample selection process and outcome $Y$ is assumed to hold, given both observed characteristics and the compliance type; for instance, see Mealli, Imbens, Ferro, and Biggeri (2004): $Y \perp O | Z, D(1), D(0), X$, where $O$ is a binary indicator for observing outcome $Y$.

However, even under LI (and its combination with MAR), $O$ may only in a restrictive way be related to unobservables affecting $Y$ (like ability that might have an influence on the wage outcome) conditional on observed variables—namely, solely through the compliance type. In contrast, nonignorable nonresponse or Heckman-type sample selection models allow more general associations of $O$ and unobservables affecting the outcome (see Heckman 1976), but generally require a separate instrument for $O$ (henceforth denoted by $Q$), which does not affect $Y$. Even in the latter case, we can evaluate the LATE only under further assumptions like parametric restrictions on the outcome model, specific (e.g., monotonicity) conditions concerning the effect of instrument $Q$, or both. For instance, Fricke, Frölich, Huber, and Lechner (2020) consider a continuous instrument $Q$ for influencing observability $O$, such as a randomly assigned cash incentive to participate in a follow-up survey

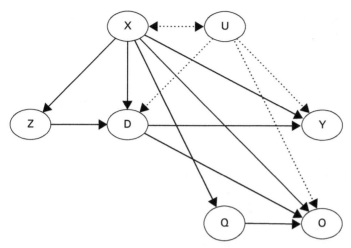

**Figure 6.4**
Causal paths with two seperate instruments for the treatment and attrition.

for measuring the outcome of interest, in addition to the binary instrument $Z$ for treatment $D$.

Figure 6.4 provides an example satisfying IV validity with two instruments in such a context, under the assumption that any unobserved variables omitted from the causal graph do not jointly affect the instruments $Z, Q$ on the one hand, and the outcome $Y$ or its observability status $O$ on the other hand. The unobservables $U$ may jointly influence $D$ and $Y$, where the dotted lines indicate the nonobservability of $U$ and its effects, thus entailing treatment endogeneity. Conditional on $X$, however, $Z$ is a valid instrument for $D$ and therefore can be used to tackle this endogeneity problem. Yet a further issue arising in the context of outcome attrition or sample selection is that we can assess the treatment effect on $Y$ only among observed outcomes: that is, conditional on $O = 1$. As both $U$ and $Z$ (via $D$) affect $O$, however, conditioning on $O = 1$ introduces a statistical association between $U$ and $Z$ even conditional on $X$. This association is problematic because $U$ also affects the outcome $Y$, such that conditioning on $O = 1$ makes the instrument $Z$ endogenous, just like the treatment.

This scenario is known as *collider bias* in statistics (in analogy to the discussion at the end of section 3.6) and sample selection bias in economics; for instance, see Pearl (2000) or Heckman (1979). The bias comes from the fact that if both $Z$ and $U$ affect $O$ even when controlling for $X$, then units with observed outcomes but different values in $Z$ (i.e., 1 or 0) must necessarily have different values in $U$. If $U$ in turn also affects the outcome, this implies that observations with $Z = 1, O = 1$ and $Z = 0, O = 1$ are not comparable in terms of their potential outcomes conditional on $X$. To tackle this second endogeneity problem, instrument $Q$ is used to exploit exogenous variation in $O$, which is not associated with $U$ conditional on $X$.

In analogy to the discussion in section 4.9 in chapter 4, we might be interested in the impact of several sequentially assigned (i.e., dynamic) treatments that take place at various points in time, rather than a single treatment. Let us, for instance, consider the effectiveness of sequences of active labor market policies like job application training followed by an information technology (IT) course. We could compare this sequence of two treatments to nonparticipation in any program or a different sequence of trainings. Such a dynamic treatment framework generally requires multiple instruments for each of the treatments and specific multiperiod, monotonicity conditions (such as, in our example, two periods). For instance, Miquel (2002) discusses various conditions under which dynamic LATEs for specific types defined in terms of first- and second-period compliance, like the compliers in the instrument of either period, can be assessed in panel data. She also demonstrates that if only a single IV is available for both treatment periods, then we can under specific assumptions assess the effects of particular sequences only for individuals that are always takers or never takers in the first treatment and compliers in the second one, or vice versa.

In a multiple treatment framework, the various treatments are not assigned sequentially, but rather at the same point in time. At first glance, the simultaneous availability of several binary treatments like alternative active labor market policies constitutes a similar evaluation problem as a treatment with multiple unordered values, as discussed in section 6.3. However, one important distinction is that multiple treatments need not be mutually exclusive, and in fact, we might be interested in the effect of assigning several treatments at the same time, like the participation in job application training and an IT course within the same time frame. In general, this requires distinct instruments for each treatment, as considered in Blackwell (2015) for the LATE evaluation of separate and joint effects of two treatments in various subpopulations defined upon compliance with either of the binary instruments.

As already discussed in section 4.10 on causal mechanisms in chapter 4, we might also be interested in disentangling the total impact of some treatment into a direct impact and an indirect effect that operates via an intermediate variable (or mediator) that also affects the outcome. As an example, let us consider the health effect of college attendance ($D_1$), which likely affects the employment state ($D_2$), which in turn may influence the health outcome. Disentangling the direct effect of $D$ and its indirect effect operating via employment sheds light on the question whether the health impact of college attendance is exclusively driven by its effect on labor market participation or also by other direct channels, such as changes in health behavior due to a person's college peers.

To formally define the effects of interest, let $D_2(d_1)$ denote the potential state of the mediator as a function of the treatment. Furthermore, we denote by $Y(d_1, d_2)$ the potential outcome under specific values of treatment $D_1$ and mediator $D_2$. We can then express the LATE of the treatment among compliers as $E[Y(1, D_2(1)) - Y(0, D_2(0))|D_1(1) = 1, D_1(0) = 0]$ and break it into direct and indirect effects, $E[Y(1, D_2(d_1)) - Y(0, D_2(d_1))|D_1(1) = 1, D_1(0) = 0]$ and $E[Y(d_1, D_2(1)) - Y(d_1, D_2(0))| D_1(1) = 1, D_1(0) = 0]$, in analogy to the decomposition in section 4.10. Assessing direct and indirect effects in general requires distinct instruments for the treatment and the mediator, along with further assumptions; for instance, see Frölich and Huber (2017). A further option not requiring a second instrument for the mediator is to invoke an LI-type assumption with respect to $D_2$ (namely, $Y(d_1, d_2) \perp D_2(d_1')| Z, D_1(1), D_1(0), X$, for $d_1, d_1' \in \{0, 1\}$), as considered in Yamamoto (2013). A more comprehensive survey on further extensions of IV-based LATE evaluation is provided in Huber and Wüthrich (2019).

# 7

---

# Difference-in-Differences

## 7.1 Difference-in-Differences without Covariates

In the presence of treatment endogeneity (even when controlling for observed covariates), an alternative strategy to using instruments, which might be hard to find in many empirical contexts, is the difference-in-differences (DiD) approach, which goes back to Snow (1855) and was more recently applied in Ashenfelter (1978). The DiD method bases treatment evaluation on the common trend assumption. The latter says that in the absence of the treatment, the average outcomes of the actually treated and nontreated subjects would experience the same change over time (i.e., a common trend) when comparing the outcomes across periods before and after the treatment. Put differently, the mean potential outcomes under nontreatment of treated and nontreated subjects follow a common trend. It is worth noting that assuming that both treatment groups would experience the same time trend in the average outcomes in the absence of the treatment nevertheless permits for differences in the average levels of potential outcomes across treatment groups rooted in treatment selection bias.

As an example, let us consider the employment effect of a minimum wage (treatment $D$), which is introduced in one geographic region, but not in another one, as discussed in Card and Krueger (1994). While the employment level (outcome $Y$) may differ in both regions due to differences in the industry structure, DiD-based evaluation requires that average changes in employment due to business cycles would be the same in both regions in the absence of a minimum wage. In this setup, a comparison of average employment in the posttreatment period across regions does not give the effect of the minimum wage due to treatment selection bias related to the industry structure. Furthermore, a before-and-after comparison of employment (i.e., before and after treatment introduction) within the treated region is biased, too, as it picks up both the treatment effect and the business cycle–related time trend.

Under the common trend assumption, however, the time trend in the average employment without treatment for either region is given by the before-and-after comparison in the nontreated region. Subtracting the before-and-after difference in employment in the nontreated region (time trend) from the before-and-after difference in the treated region (treatment effect plus time trend), therefore, gives the average treatment effect in the treated region (ATET). That is, taking the difference in the before-and-after differences across regions yields a causal effect under the common trend assumption. We may apply this estimation approach to both panel data, where the very same subjects are observed before and after treatment, or to repeated cross sections, where subjects differ across periods.

To formalize the common trend assumption, let us introduce a time index $T$, which is equal to zero in the pretreatment period, when neither group has received the treatment (yet), and one in the posttreatment period, after introducing the treatment in the treated group (but not the nontreated group). To distinguish observed outcomes in terms of pretreatment and posttreatment periods, we add the subscript $t \in \{0, 1\}$, such that $Y_0$ and $Y_1$ correspond to the pretreatment and posttreatment outcomes, respectively. Likewise, we add the time subscripts to the potential outcomes in the various periods, such that $Y_0(1), Y_0(0)$ and $Y_1(1), Y_1(0)$ correspond to the pretreatment and posttreatment potential outcomes, respectively. Using this notation, the common trend assumption corresponds to

$$E[Y_1(0) - Y_0(0)|D = 1] = E[Y_1(0) - Y_0(0)|D = 0]; \tag{7.1}$$

that is, the trend in the mean potential outcomes under nontreatment is the same across the treatment groups.

Furthermore, let us also impose a no anticipation assumption, which implies that subjects not yet treated in the pretreatment period do not anticipate their treatment in a way that already influences their pretreatment outcomes. Put differently, the treatment yet to be realized (e.g., a training program in 2021) cannot induce behavioral changes in the treated group that affect pretreatment outcomes (e.g., employment in 2020) as a reaction to the expectation of the treatment. More formally, this implies that the average treatment effect on the treated (ATET) in the pretreatment period $T = 0$ is equal to zero:

$$E[Y_0(1) - Y_0(0)|D = 1] = 0. \tag{7.2}$$

Our causal effect of interest is the ATET after treatment introduction; that is, in the posttreatment period: $\Delta_{D=1} = E[Y_1(1) - Y_1(0)|D = 1]$. To show how the latter can be evaluated based on our assumptions, we first note that $E[Y_1|D = 0] - E[Y_0|D = 0] = E[Y_1(0) - Y_0(0)|D = 0]$ (because $Y_t = Y_t(0)$ if $D = 0$), such that by the common trend assumption in equation (7.1), it follows that

$$E[Y_1|D = 0] - E[Y_0|D = 0] = E[Y_1(0) - Y_0(0)|D = 1]. \tag{7.3}$$

Therefore, we can assess the ATET based on the difference in before-and-after differences of average outcomes across treated and nontreated groups, as demonstrated here:

$$
\begin{aligned}
\Delta_{D=1} &= E[Y_1(1)|D=1] - E[Y_1(0)|D=1] \\
&= E[Y_1(1)|D=1] - E[Y_0(0)|D=1] - E[Y_1(0)|D=1] + E[Y_0(0)|D=1] \\
&= E[Y_1(1)|D=1] - E[Y_0(1)|D=1] - E[Y_1(0)|D=1] + E[Y_0(0)|D=1] \\
&= E[Y_1|D=1] - E[Y_0|D=1] - \{E[Y_1(0) - Y_0(0)|D=1]\} \\
&= \underbrace{E[Y_1|D=1] - E[Y_0|D=1]}_{\substack{\text{before-and-after change} \\ \text{among treated}}} - \underbrace{\{E[Y_1|D=0] - E[Y_0|D=0]\}}_{\substack{\text{before-and-after change} \\ \text{among nontreated}}} .
\end{aligned}
\tag{7.4}
$$

The second equality in equation (7.4) follows from subtracting and adding $E[Y_0(0)|D=1]$, and the third comes from the no anticipation assumption in equation (7.2), implying that $E[Y_0(0)|D=1] = E[Y_0(1)|D=1]$. The fourth equality follows from the fact that $Y_t = Y_t(1)$ for $t \in \{0,1\}$ conditional on $D=1$, and the fifth comes from equation (7.3). This implies that the ATET is nonparametrically identified based on the regression of the outcome on a constant, a dummy variable for the treatment group, a dummy for the posttreatment period, and an interaction of the latter two variables, as follows:

$$
E[Y_T|D] = \alpha + \beta_D D + \beta_T T + \beta_{D,T} DT,
\tag{7.5}
$$

where $\alpha = E[Y_0|D=0]$ corresponds to the mean outcome of the nontreated in the pretreatment period and $\beta_D = E[Y_0|D=1] - E[Y_0|D=0]$ to the mean difference in outcomes across treatment groups in the pretreatment period. Here, $\beta_T = E[Y_1|D=0] - E[Y_0|D=0]$ gives the presumably common time trend in mean outcomes among the nontreated, and $\beta_{D,T} = E[Y_1|D=1] - E[Y_0|D=1] - \{E[Y_1|D=0] - E[Y_0|D=0]\} = \Delta_{D=1}$ the ATET.

Figure 7.1 provides a graphical illustration of regression equation (7.5), with the conditional mean outcome $E[Y_T|D]$ depicted on the $y$-axis and time period $T$ on the $x$-axis. While $\beta_D$ provides the difference in the mean outcomes of the treated and nontreated groups in the pretreatment period $T=0$, $\beta_T$ corresponds to the change in the mean outcome over time among the nontreated. The dashed line provides the hypothetical evolution of the mean outcome for the treated group if the treatment had not been introduced (between $T=0$ and $T=1$), given that the common trend assumption holds. Under this condition, the difference between the observed change in the mean outcome among the treated over time and the hypothetical evolution under nontreatment corresponds to the ATET, given by $\beta_{D,T}$.

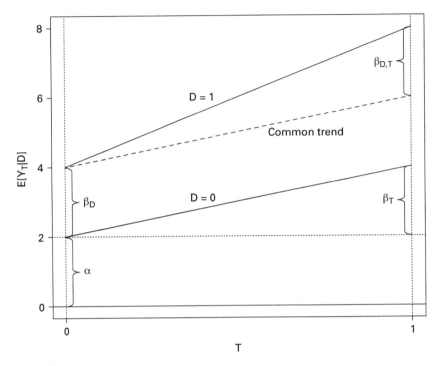

**Figure 7.1**
DiD regression.

It is worth noting that there is a second interpretation of the common trend assumption in equation (7.1), namely a statistically equivalent bias stability assumption. To see this, let us rearrange equation (7.1) in the following way:

$$E[Y_1(0)|D=1] - E[Y_1(0)|D=0] = E[Y_0(0)|D=1] - E[Y_0(0)|D=0]. \qquad (7.6)$$

This assumption implies that the difference in mean potential outcomes under nontreatment across treatment groups (i.e., the bias in the average of $Y_t(0)$ across treated and nontreated subjects due to differences in counfounders jointly affecting the treatment and the outcomes) is constant over time. For this reason, DiD can also be interpreted as taking mean differences across treated and nontreated outcomes in the posttreatment and pretreatment periods, respectively, and subtracting the difference in the pretreatment period, which gives the bias across treatment groups, from the difference in the posttreatment period. To see this formally, we note that we can rearrange equation (7.4) to

$$\Delta_{D=1} = \underbrace{E[Y_1|D=1] - E[Y_1|D=0]}_{\text{difference in posttreatment period}} - \underbrace{\{E[Y_0|D=1] - E[Y_0|D=0]\}}_{\text{difference in pretreatment period}}. \qquad (7.7)$$

Concerning statistical inference, such as the computation of p-values and confidence intervals for the ATET, a practical issue is that the subjects entering DiD-based estimation are typically not independently sampled from each other, as is conventionally assumed. In panel data, for instance, the very same units are observed prior and after the treatment, such that there is most likely a correlation of certain background characteristics (like personality traits) within subjects over time. Even in repeated cross sections with different subjects in each period, such correlations may occur, perhaps because individuals living in regions receiving or not receiving the treatment (like a change in the minimum wage) are exposed to the same institutional context of the respective region. For instance, if a treatment is introduced on the state level, such as by a federal state, then all individuals living in that state share the same regional legislation and institutions, which may entail a correlation in unobserved characteristics. For this reason, cluster-robust methods for estimating the standard error of the ATET should be considered for DiD estimation.

For panel data, cluster methods take into account that the same subjects are repeatedly sampled across time by considering multiple observations of the same subject as belonging to the same cluster. One approach is to use a cluster-corrected version of the asymptotic approximation for the standard error, such as in equation (3.40) of section 3.4; another is to use a modified version of the bootstrap introduced in expression (3.48), a cluster- or block-bootstrap. The latter consists of randomly drawing entire clusters along with all cluster-related observations from the data, rather than single observations. In a data set with repeated cross sections in which the treatment varies across regions, this amounts to randomly sampling entire regions with all its observed individuals. In panel data, the cluster bootstrap draws individuals and their observations in both the pretreatment and posttreatment periods. We, however, need to be aware that cluster-robust inference might perform satisfactorily only if sufficiently many treated and nontreated clusters (e.g., regions) are available in the data. Otherwise, we should consider more sophisticated inference methods tailored to setups with few clusters; for instance, see the discussions of DiD-related inference issues by Bertrand, Duflo, and Mullainathan (2004); Donald and Lang (2007); Cameron, Gelbach, and Miller (2008); Conley and Taber (2011); and Ferman and Pinto (2019).

DiD approaches are frequently applied to monotonically transformed outcomes, such as the logarithm of $Y$ instead of the level of $Y$, which permits interpreting the treatment effect (approximately) in terms of percentage changes of the outcome. For instance, considering the logarithm of wages permits assessing by how many percent the wages among the treated change on average due to the introduction of the treatment. As a word of caution, however, it needs to be pointed out that if the common trend assumption holds with regard to $Y$, it generally does not hold for some transformation of $Y$, and vice versa. Except in special cases discussed by Roth and Sant'Anna

(2021) (including a randomly assigned treatment $D$), the common trend assumption will depend on the functional form of the outcome, and thus for example may not hold for $Y$, but instead for the logarithm of $Y$ or some other transformation.

In practice, the plausibility of the common trend assumption might be scrutinized if several pretreatment periods are available, by running placebo tests. To this end, we apply the DiD approach in a time span prior to treatment by considering an earlier pretreatment period as $T = 0$ and a later, but still pretreatment, episode as a pseudo-treatment period $T = 1$. As the treatment has not yet been implemented in any of these periods, a statistically significant pseudotreatment effect points to a violation of the common trend (or bias stability) assumption. If the common trend assumption does not hold exactly, we may account for such violations in DiD-based causal inference by imposing restrictions on the possible differences in trends between the treated and nontreated groups, as discussed by Rambachan and Roth (2020). One restriction, for instance, could be that the observed pretreatment differences in trends across treatment groups are informative about the magnitude of the posttreatment differences in trends of mean potential outcome under nontreatment across treatment groups.

As already discussed in chapter 6 on instrumental variable (IV) methods, compliance with the treatment assignment might not be perfect, such that the actual treatment participation might differ from the assignment. For instance, if a government launches an educational program giving access to better education in some regions of a country, not necessarily all pupils in these treated regions are affected by the program. In such a fuzzy DiD case with treatment noncompliance, the conventional DiD approach only provides an intention-to-treat (ITT) effect of the treatment assignment (e.g., increasing accessibility to education) rather than the effect of treatment participation (e.g., actually receiving education).

We could therefore be tempted to apply an IV-based DiD approach, such as by running two regressions based on equation (7.4), with $D$ corresponding to the treatment assignment and $Y$ corresponding to treatment participation in the first regression and the actual outcome variable in the second regression, respectively. Inspired by the discussion in section 6.1, scaling (or dividing) the $\beta_{D,T}$ coefficient from the second regression by the $\beta_{D,T}$ coefficient from the first regression supposedly yields a local average treatment effect (LATE) among compliers: that is, those switching the treatment status from the pretreatment to the posttreatment period due to the assignment. As discussed by de Chaisemartin and D'Haultfeuille (2018), however, such an approach yields the LATE only if additional assumptions are satisfied: The share of noncomplying treated units in the group not assigned to the treatment must not change over time and the ATE among units treated in both periods (always takers) must be stable over time. For the case that the latter (and, in the context of time-varying treatment effects, often implausible) assumption is not satisfied, de Chaisemartin and D'Haultfeuille (2018) suggest a time-corrected IV approach that relies on a specific common trend assumption within subgroups of subjects sharing the same treatment in the pretreatment period.

The DiD approach can also be applied in the context of mediation analysis as discussed in section 4.10 in chapter 4. Deuchert, Huber, and Schelker (2019), for instance, assume a randomized treatment and tackle mediator endogeneity by assuming monotonicity of the binary mediator in the treatment (similar to the monotonicity assumptions in chapter 6), along with particular common trend assumptions. The latter imply that mean potential outcomes under specific treatment and mediator states share a common trend across specific subpopulations (or compliance types) that are defined in terms of how the mediator reacts to the treatment.

A further, practically relevant scenario concerns the assessment of a multivalued (rather than a binary) treatment; see the discussion in section 3.5 in chapter 3. In this case, applying the DiD approach to a pairwise comparison of a specific nonzero treatment value (e.g., $D = 3$) versus no treatment ($D = 0$) yields the ATET of that respective nonzero treatment, given that the common trend assumption holds for this specific treatment comparison. Therefore, imposing the common trend assumption with regard to all nonzero values of the treatment ($D = 1, 2, \ldots$) in principle permits assessing the ATET of each treatment value, such as various educational programs. This also applies to continuously distributed treatments; for example, see Callaway, Goodman-Bacon, and Sant'Anna (2021). As a caveat, however, such a strategy does not allow for comparing two nonzero treatments (e.g., $D = 2$ versus $D = 1$) to assess their effectiveness relative to each other. This is because the common trend assumption in equation (7.1) refers to the potential outcome under nontreatment only and makes no claims about outcome trends for nonzero treatments.

For this reason, assessing one versus another nonzero treatment requires further assumptions, such as that the average effect of at least one of the nonzero treatments (e.g., $D = 2$) versus no treatment ($D = 0$) is homogeneous across the two treatment groups considered (e.g., groups with $D = 2$ and $D = 1$), as discussed by Fricke (2017). This implies that numerous empirical DiD applications assessing multivalued, and even continuously distributed, treatments implicitly impose such a strong effect homogeneity assumption, in addition to the conventional common trend assumption. In the absence of effect homogeneity, the conventional DiD approach applied to nonzero treatments may still yield a lower bound on the absolute causal effect, such as of $D = 2$ versus $D = 1$. This is the case if an alternative (and possibly more plausible) assumption on the order of treatment effects holds—namely, that the effect of a higher treatment dose compared to no treatment (e.g., $D = 2$ versus $D = 0$) weakly dominates that of a lower treatment dose ($D = 1$ versus $D = 0$) in absolute terms.

To apply DiD estimation in R, we load the *wooldridge* package by Shea (2021), which contains a data set on house prices in North Andover, Massachusetts, prior to and after the construction of a garbage incinerator. We also load the *multiwayv-cov* package by Graham, Arai, and Hagströmer (2016) and the *lmtest* package for computing cluster-robust standard errors. We use the *data* and *attach* commands to load the *kielmc* data originally analyzed by Kiel and McClain (1995) and store its variables in own R objects. We define the outcome *Y=rprice* to be the real price of houses measured in 1978 US dollars (USD) and the treatment *D=nearinc* to be a binary indicator of whether a house is situated close to ($D = 1$) or farther away from ($D = 0$) the incinerator. Furthermore, we define the period dummy *T=y81*, which is 1 for house prices observed in 1981, after the construction of the incinerator, and 0 for those observed in 1978, prior to its construction.

In the next step, we generate an interaction term between the treatment and the time indicator, *interact=D\*T*, and run a linear regression to estimate equation (7.5) using the *lm* command, storing the output in an object named *did*. We feed the latter into the *cluster.vcov* command for computing cluster-robust standard errors and specify the argument *cluster=cbd*, thus using *cbd* (distance to the center, which is coded discretely in categories) as the cluster variable. We save the results in an object named *vcovCL* and finally run *coeftest(did, vcov=vcovCL)* to inspect the regression results with cluster-robust standard errors. The box here provides the R code for each of the steps.

```
library(wooldridge)                              # load wooldridge package
library(multiwayvcov)                            # load multiwayvcov package
library(lmtest)                                  # load lmtest package
data(kielmc)                                     # load kielmc data
attach(kielmc)                                   # attach data
Y=rprice                                         # define outcome
D=nearinc                                        # define treatment group
T=y81                                            # define period dummy
interact=D*T                                     # treatment-period interaction
did=lm(Y~D+T+interact)                           # DiD regression
vcovCL=cluster.vcov(model=did, cluster=cbd)      # cluster: distance to center (cbd)
coeftest(did, vcov=vcovCL)                       # DiD results with cluster st.error
```

Running the code yields the following output:

```
            Estimate Std. Error t value  Pr(>|t|)
(Intercept)  82517.2    3221.9  25.6112  < 2.2e-16 ***
D           -18824.4    7796.3  -2.4145  0.0163216 *
T            18790.3    5154.9   3.6452  0.0003122 ***
interact    -11863.9    6621.8  -1.7916  0.0741446 .
```

The coefficient on the interaction term in our DiD regression suggests that the construction of the garbage incinerator reduces the house price in the treatment group (i.e., among houses situated close to the incinerator) on average by 11,863.90 USD. The ATET estimate is statistically significant at the 10 percent level, but not at the 5 percent level when considering cluster-robust standard errors because the p-value amounts to 0.074 (or 7.4 percent).

## 7.2 Difference-in-Differences with Covariates

In many empirical problems, the common trend assumption, as considered in the previous section, might be debatable and appear plausible only after controlling for observed covariates $X$. When assessing the treatment effect of a policy change like access to unemployment benefits or training programs on labor market outcomes, for instance, it could be argued that the common trend assumption across treated and nontreated observations is credible only for treated and nontreated units within the same occupation or industry. This is because outcomes like employment or wages might in general develop differently across distinct occupations or industries, which poses a problem for ATET evaluation if treated and nontreated observations differ with regard to these characteristics. For this reason, we will henceforth assume that the DiD assumptions only hold conditional on $X$. This scenario is considered in Lechner (2011), who shows that the following conditions identify the ATET among the treated observations in the posttreatment period, denoted by $\Delta_{D=1,T=1} = E[Y_1(1) - Y_1(0)|D = 1, T = 1]$:

$$E[Y_1(0) - Y_0(0)|D = 1, X] = E[Y_1(0) - Y_0(0)|D = 0, X], \quad X(1) = X(0) = X, \qquad (7.8)$$

$$E[Y_0(1) - Y_0(0)|D = 1, X] = 0,$$

$$\Pr(D = 1, T = 1|X, (D, T) \in \{(d, t), (1, 1)\}) < 1 \text{ for all } (d, t) \in \{(1, 0), (0, 1), (0, 0)\}.$$

The first expression in the first line of expression (7.8) formalizes the conditional common trend assumption stating that given $X$, no unobservables jointly affect the treatment and the trend of mean potential outcomes under nontreatment. This is a selection-on-observables assumption on $D$, however, with regard to the changes in mean potential outcomes over time, rather than the levels of the outcomes as in equation (4.1) in chapter 4. The two types of selection-on-observables assumptions are not nested, in the sense that neither implies the other. Furthermore, they cannot be fruitfully combined to obtain a more flexible treatment effect model either; see the discussion by Chabé-Ferret (2017).

The second condition in the first line of expression (7.8) imposes that $X$ is not affected by $D$. In the context of DiD, this assumption deserves particular scrutiny. The reason for this is that in many empirical applications, the covariates are measured in the same period as the (pretreatment or posttreatment) outcome, implying that for $T = 1$, $X$ is measured after the introduction of the treatment. Analogous to the issues discussed in section 3.6, this jeopardizes the evaluation of the ATET if $D$ affects $X$ and $X$ affects $Y$, is associated with unobservables affecting $Y$, or both. If $X$ includes occupation, for instance, then the choice of a specific occupation in period $T = 1$ could be the result of the treatment of interest, such as a specific training program. To circumvent such issues, we might alternatively consider the covariate values in $T = 0$ as well when controlling for $X$ in the posttreatment period $T = 1$. However, even this approach might not be fully appropriate because it does not allow controlling for time trends in $X$ that differ across treatment groups between $T = 0$ and $T = 1$ (even when not affected by the treatment itself) and may thus entail a violation of the common trend assumption. As a pragmatic approach, therefore, we could consider both possibilities of covariate measurement (i.e., in the respective outcome period versus in the pretreatment period only) and check the robustness of the DiD results across both approaches. Furthermore, Caetano, Callaway, Payne, and Rodrigues (2022) discuss alternative evaluation strategies for time-varying covariates depending on whether the latter are influenced by the treatment.

A further potential problem that concerns time-varying covariates, no matter whether they are measured in pretreatment or posttreatment periods, is called *mean reversion*. The latter phenomenon implies that a time-varying covariate that takes a rather extreme value at a specific point in time (e.g., particularly bad health measured in a specific year), tends to move toward its mean value across all periods (e.g., average health) at subsequent points in time. As discussed in Daw and Hatfield (2018), for instance, controlling for such extreme values in $X$ in one period along with mean reversion in subsequent periods can generate a nonnegligible bias in DiD estimation, such that the use of time-varying covariates is to be scrutinized.

In analogy to expression (7.2), the second line in expression (7.8) rules out average anticipation effects among the treated conditional on $X$, implying that $D$ must not causally influence pretreatment outcomes in expectation of the treatment to come. The third line imposes common support. For any value of $X$ appearing in the treated group in the posttreatment period with $(D = 1, T = 1)$, subjects with such values of $X$ must also exist in the remaining three groups with $(D = 1, T = 0)$, $(D = 0, T = 1)$, and $(D = 0, T = 0)$.

Under the assumptions in expression (7.8), it holds that $E[Y_1|D = 0, X] - E[Y_0|D = 0, X] = E[Y_1(0) - Y_0(0)|D = 0, X] = E[Y_1(0) - Y_0(0)|D = 1, X]$. In analogy to the identification result without covariates in equation (7.4), the conditional ATET given $X$ corresponds to

$$E[Y_1(1) - Y_1(0)|D = 1, X] = E[Y_1(1) - Y_0(0)|D = 1, X] - E[Y_1(0) - Y_0(0)|D = 1, X]$$
$$= E[Y_1|D = 1, X] - E[Y_0|D = 1, X]$$
$$- \{E[Y_1|D = 0, X] - E[Y_0|D = 0, X]\}. \tag{7.9}$$

Therefore, averaging the conditional ATET in equation (7.9) over the distribution of $X$ among the treated in the posttreatment period yields the ATET in that period:

$$\Delta_{D=1,T=1} = E[\mu_1(1,X) - \mu_1(0,X) - (\mu_0(1,X) - \mu_0(0,X))|D=1, T=1], \tag{7.10}$$

where we use the shorthand notation $\mu_d(t,x) = E[Y_t|D=d, X=x]$ for the conditional mean outcome given the treatment, the time period, and the covariates.

An alternative and in terms of identification equivalent approach is inverse probability weighting (IPW):

$$\Delta_{D=1,T=1} = E\left[\left\{\frac{D \cdot T}{\Pi} - \frac{D \cdot (1-T) \cdot \rho_{1,1}(X)}{\rho_{1,0}(X) \cdot \Pi}\right.\right.$$
$$\left.\left. - \left(\frac{(1-D) \cdot T \cdot \rho_{1,1}(X)}{\rho_{0,1}(X) \cdot \Pi} - \frac{(1-D) \cdot (1-T) \cdot \rho_{1,1}(X)}{\rho_{0,0}(X) \cdot \Pi}\right)\right\} \cdot Y\right], \tag{7.11}$$

where $\Pi = \Pr(D=1, T=1)$ denotes the unconditional probability of being treated and observed in the posttreatment period. Here, $\rho_{d,t}(X) = \Pr(D=d, T=t|X)$ are the conditional probabilities (or propensity scores) of specific treatment group–period-combinations $D=d, T=t$, given $X$.

Combining the regression approach with IPW yields a doubly robust (DR) expression for DiD-based ATET evaluation; see Zimmert (2020):

$$\Delta_{D=1,T=1} = E\left[\left\{\frac{D \cdot T}{\Pi} - \frac{D \cdot (1-T) \cdot \rho_{1,1}(X)}{\rho_{1,0}(X) \cdot \Pi}\right.\right.$$
$$\left.\left. - \left(\frac{(1-D) \cdot T \cdot \rho_{1,1}(X)}{\rho_{0,1}(X) \cdot \Pi} - \frac{(1-D) \cdot (1-T) \cdot \rho_{1,1}(X)}{\rho_{0,0}(X) \cdot \Pi}\right)\right\}\right. \tag{7.12}$$
$$\left. \times (Y - \mu_D(T,X)) + \frac{D \cdot T}{\Pi} \cdot [\mu_1(1,X) - \mu_1(0,X) - (\mu_0(1,X) - \mu_0(0,X))]\right].$$

Depending on whether we construct estimators as sample analogs of equations (7.10), (7.11), or (7.12), $\sqrt{n}$-consistent estimation may be based on regression or matching, on IPW, or on DR estimation, respectively. Furthermore, for estimating the DR expression (7.12), we can apply double machine learning (DML) to control for covariates $X$ in a data-driven way in analogy to the methods outlined in section 5.2 in chapter 5.

It is worth noting that many DiD studies at least implicitly make an additional assumption to those in expression (7.8)—namely, that the joint distribution of treatment $D$ and covariates $X$ remains constant over time $T$, formalized by $(X,D) \perp T$; for instance, see the discussion by Hong (2013). This, for instance, rules out that the composition of $X$ changes across periods within either treatment group. Under this additional assumption, $\Delta_{D=1,T=1}$, the causal effect in the posttreatment period among the treated in the posttreatment period coincides with the ATET $\Delta_{D=1}$, the effect in the posttreatment period among all the treated in both the pretreatment and posttreatment periods (thus, including not yet treated subjects). Then, $\Delta_{D=1}$ corresponds to the following expressions:

$$\Delta_{D=1} = E[\mu_1(1,X) - \mu_1(0,X) - (\mu_0(1,X) - \mu_0(0,X))|D=1] \tag{7.13}$$

$$= E\left[\left\{\frac{D \cdot T}{P \cdot \Lambda} - \frac{D \cdot (1-T)}{P \cdot (1-\Lambda)} - \left(\frac{(1-D) \cdot T \cdot p(X)}{(1-p(X)) \cdot P \cdot \Lambda} - \frac{(1-D) \cdot (1-T) \cdot p(X)}{(1-p(X)) \cdot P \cdot (1-\Lambda)}\right)\right\} \cdot Y\right]$$

$$= E\left[\left\{\frac{D \cdot T}{P \cdot \Lambda} - \frac{D \cdot (1-T)}{P \cdot (1-\Lambda)} - \left(\frac{(1-D) \cdot T \cdot p(X)}{(1-p(X)) \cdot P \cdot \Lambda} - \frac{(1-D) \cdot (1-T) \cdot p(X)}{(1-p(X)) \cdot P \cdot (1-\Lambda)}\right)\right\}\right.$$
$$\left. \times (Y - \mu_0(T,X))\right],$$

where $p(X) = \Pr(D=1|X)$ is the treatment propensity score given $X$, $P = \Pr(D=1)$ is the unconditional treatment probability, and $\Lambda = \Pr(T=1)$ the unconditional probability of being in the posttreatment period. By exploiting the identification results after the first, second, and third equalities in equation (7.13), we can base $\sqrt{n}$-consistent ATET estimation on regression or matching, on IPW as considered in Abadie (2005), on DR estimation as in Sant'Anna and Zhao (2018), or on DML, as in Chang (2020).

A further estimation approach is based on linearly including covariates $X$ in regression equation (7.5) to control for differences in observed characteristics across treatment groups, time, or both:

$$E[Y_T|D,X] = \alpha + \beta_D D + \beta_T T + \beta_{D,T} DT + \beta_{X_1} X_1 + \cdots + \beta_{X_K} X_K, \tag{7.14}$$

where $K$ denotes the number of covariates. However, it is important to notice that estimating the ATET based on equation (7.14) imposes rather strong parametric assumptions, in particular that the treatment effect is homogeneous across different values of $X$ and the outcome is linear in $X$. In contrast, such assumptions can be avoided in the previously outlined methods.

As an illustration in R, let us reconsider the *kielmc* data set analyzed at the end of section 7.1 with the same definitions of treatment $D$, outcome $Y$, and periods $T$ as before, but now also control for covariates when running the DiD estimation. To this end, we load the *causalweight* package, which contains an IPW-based DiD procedure. We define the covariates $X$ to include the variables *area* (square footage of the house), *rooms* (number of rooms), and *baths* (number of bathrooms) using the *cbind* command. We then set a seed for the reproducability of the results and feed $Y$, $D$, $T$, and $X$ into the *didweight* command for the estimation of the average effect among the treated in the posttreatment period, $\Delta_{D=1,T=1}$, based on the IPW expression in equation (7.11). We also set the arguments *boot=399* and *cluster=cbd* to run a cluster bootstrap with 399 bootstrap samples for estimating the standard error of the ATET, where *cbd* serves as the cluster variable as in section 7.1. We store the results in an object named *out* and call *out$effect*, *out$se*, and *out$pvalue* to investigate the ATET estimate, its standard error, and the p-value. The box here provides the R code for each of the steps.

```
library(causalweight)                       # load causalweight package
X=cbind(area, rooms, baths)                 # define covariates
set.seed(1)                                 # set seed to 1
out=didweight(y=Y,d=D,t=T,x=X,boot=399,cluster=cbd)  # DiD with cluster se
out$effect; out$se; out$pvalue              # effect, se, and p-value
```

Running the code yields an ATET of −14,590.25 USD on the prices of houses close to the incinerator in the posttreatment period. However, this nonnegligible negative effect is not statistically significant at the 10 percent level due to a rather sizeable cluster-robust standard error of 11,315.72 USD, which entails a p-value of 0.197 (or 19.7 percent).

## 7.3  Multiple Periods of Treatment Introduction

In many empirical problems, there is not only one treatment group exposed to the introduction of a treatment at one specific point in time, but there are multiple groups that experience the introduction at different points in time. For instance, a smoking ban in restaurants and public places might be introduced in a subset of countries, states, or regions and then in others one or several years later. For this reason, we modify our notation to adapt the previously considered DiD framework to multiple periods and treatment groups, such as discussed in Abraham and Sun (2018), Borusyak and Jaravel (2018), Callaway and Sant'Anna (2021), Goodman-Bacon (2018), and de Chaisemartin and D'Haultfeuille (2020). Let $T$ now denote multiple periods such that $T \in \{0, 1, \ldots, \mathcal{T}\}$, with $\mathcal{T}$ corresponding to the last period. While nobody is treated in period $T = 0$, the treatment is introduced in a staggered way in later periods such that some subjects might receive the treatment in $T = 1$, others in $T = 2$, and so on. There may even be subjects who are never treated, making this group inherently nontreated in any time period. Furthermore, let $G_t$ be a dummy variable that is equal to 1 if a subject experiences treatment introduction in period $T = t$. For instance, $G_2 = 1$ implies that the treatment is introduced in period 2 for this group (e.g., the residents of a particular state), while $G_2 = 0$ implies a different period of treatment introduction.

With multiple periods, in principle, we can assess the ATET in or across various outcome periods rather than just one. For instance, we may define treatment group- and time-specific ATETs as the average effect in a specific outcome period (say $t'$), among those subjects for whom the treatment is introduced at the beginning of a specific period (say $t$):

$$\Delta_{G_t=1,T=t'} = E[Y_{t'}(1) - Y_{t'}(0)|G_t = 1], \text{ with } t' \geq t. \tag{7.15}$$

The outcome period $t'$ in which we assess the ATET may correspond to the same or a later point in time than $t$, the period at the beginning of which the treatment was introduced to a specific group. This permits investigating the evolution of the treatment effect over several follow-up periods in order to distinguish its short- and longer-term impact.

For assessing $\Delta_{G_t=1,T=t'}$, the DiD assumptions in expression (7.8) must hold when considering subjects satisfying $G_t = 1$ as the treated group and subjects not treated up to (and including) time period $t'$ as the nontreated group. The latter may consist of subjects that are never treated in any period, such that $G_0 = G_1 = \cdots = G_{\mathcal{T}} = 0$; or of subjects experiencing treatment introduction at a later point in time, such that $G_{t''} = 1$ for some $t'' > t'$; or of both types of subjects (if they exist in the data). Importantly, the common trend assumption must be plausibly satisfied in light of the selected treated and nontreated population, as well as the outcome period. When considering the never treated subjects as the nontreated group, for example, then $E[Y'_t(0) - Y_0(0)|G_t = 1, X] = E[Y'_t(0) - Y_0(0)|G_0 = G_1 = \cdots = G_{\mathcal{T}} = 0, X]$ must be satisfied. It follows that equations (7.10), (7.11), or (7.12) for time-varying covariates $X$, or equation (7.13) for time-constant covariates $X$, yields $\Delta_{G_t=1,T=t'}$ when replacing $D = 1$ with $G_t = 1$, $D = 0$ with $G_0 = G_1 = \cdots = G_{\mathcal{T}} = 0$, $D$ with $G_t$, $(1-D)$ with $(1 - G_0 - G_1 - \ldots - G_{\mathcal{T}})$, $T = 1$ with $T = t'$, $T = 0$ with $T = t - 1$, $T$ with $I\{T = t'\}$, and $1 - T$ with $I\{T = t - 1\}$ in all expressions, including the probabilities, propensity scores, and conditional mean outcomes.

If the DiD assumptions in expression (7.8) plausibly hold for several or all definitions of treatment periods $t$ and outcome periods $t'$, we may not only evaluate the ATETs $\Delta_{G_t=1,T=t'}$ in multiple treatment groups and outcome periods, but also aggregate them as discussed in Callaway and Sant'Anna (2021). For instance, we might consider the average group-specific ATET for those satisfying $G_t = 1$ across all outcome periods $t' \geq t$, formally defined as

$$\frac{1}{\mathcal{T} - t + 1} \sum_{t'=t}^{\mathcal{T}} \Delta_{G_t=1,T=t'}. \tag{7.16}$$

Computing this causal effect for various treatment groups defined in terms of $G_t$ permits investigating whether average ATETs across outcome periods importantly differ across treatment groups (and thus across the timing of the treatment introduction).

Another arguably interesting question is whether the average effect across all treated groups varies with the length of treatment exposure. We may analyze this based on a causal parameter that averages over group-specific ATETs conditional on a specific time elapsed since treatment introduction, denoted by $e$, among the respective treatment groups:

$$\sum_{t:t+e\leq\mathcal{T}} \Pr(G_t=1|T+e\leq\mathcal{T})\Delta_{G_t=1,T=t+e}, \text{ with } e\in\{0,1,\ldots,\mathcal{T}-1\}, \qquad (7.17)$$

where $\Pr(G_t=1|T+e\leq\mathcal{T})$ corresponds to the share of those facing treatment introduction in period $t$ among all treated groups whose outcome is still observed after $e$ periods in the data window—that is, prior to or in the final period $\mathcal{T}$. Setting $e=0$, for instance, yields the average of group-specific ATETs in the period of treatment introduction, while setting $e=5$ provides the corresponding longer-run effect five periods later among all treatment groups satisfying $t+5\leq\mathcal{T}$, such that their outcomes are still observed. Varying the elapsed time $e$ thus permits investigating the dynamics of treatment effects.

In the spirit of equation (7.5) or (7.14), we might be tempted to consider a linear regression approach for the staggered treatment case based on the two-way-fixed-effects (TWFE) model. The latter includes dummies for each treatment group and period, as well as a binary treatment indicator for whether the treatment has already been introduced for a specific group in the period considered. We denote this treatment indicator, which is based on treatment group–time interactions, by $\mathcal{Q}=(G_1\cdot I\{T\geq 1\}+G_2\cdot I\{T\geq 2\}+\cdots+G_\mathcal{T}\cdot I\{T=\mathcal{T}\})$, which entails the following TWFE egression model:

$$E[Y_T|G_T,X]=\alpha+\beta_{G_1=1}G_1+\cdots+\beta_{G_\mathcal{T}=1}G_\mathcal{T}+\beta_{T=1}I\{T=1\}+\cdots+\beta_{T=\mathcal{T}}I\{T=\mathcal{T}\}$$
$$+\beta_{G_T,T}\mathcal{Q}+\beta_{X_1}X_1+\cdots+\beta_{X_K}X_K. \qquad (7.18)$$

In this setup where any groups serve as nontreated observations in periods prior to their respective treatment introduction and as treated observations thereafter, $\beta_{G_T,T}$ is interpreted as the ATET. We also note that the terms $\beta_{X_1}X_1+\cdots+\beta_{X_K}X_K$ may be dropped if the common trend assumption is assumed to hold without controlling for $X$; see section 7.1.

Even in this case, however, and somewhat related to the discussion of equation (7.14), effect heterogeneity poses a threat to ATET evaluation in a framework with multiple treatment groups and periods. In particular, if the effects are heterogeneous over time (e.g., if the effectiveness of a marketing campaign varies with the business cycle), then $\beta_{G_T,T}$ can theoretically take values that do not correspond to any causal effect, not even a weighted average across any $\Delta_{G_t=1,T=t'}$. As discussed in Goodman-Bacon (2018), the reason for this is that a TWFE regression weights the group-specific ATETs, $\Delta_{G_t=1,T=t'}$, based on the group size and variance, which may entail a biased estimate of averaged group-specific ATETs if treatment effects change across time and treatment groups.

It is worth noting that the staggered treatment design considered so far in this chapter assumes the treatment to be an absorbing state, in the sense that any group entering the treatment at some point in time is assumed to remain in the treatment in its entirety until the end of the data window. However, there are empirical applications in which subjects might both switch into and out of the treatment over time, such as union membership. For instance, de Chaisemartin and D'Haultfeuille (2020) discuss the DiD-based evaluation of ATETs under nonabsorbing treatment designs and consider the fuzzy DiD framework with noncompliance in the treatment, in analogy to the issues raised at the end of section 7.1. Roth, Sant'Anna, Bilinski, and Poe (2022) provide a more comprehensive survey on further methodological advancements in DiD estimation, including estimation under potential violations of the common trend assumption, alternative approaches to statistical inference, and staggered treatment adoption. The latter topic is also reviewed in de Chaisemartin and D'Haultfoeuille (2022).

To implement DiD estimation with staggered treatment introduction in R, we consider the *did* package by Callaway and Sant'Anna (2020) and an empirical example provided in that package. Using the *data* command, we load the *mpdta* data previously analyzed by Callaway and Sant'Anna (2021). The sample contains information on the staggered introduction of a minimum wage (treatment $D$) and employment among teenagers (outcome $Y$) in 500 US counties from 2004 to 2007. We estimate the treatment group–time-specific ATETs $\Delta_{G_t=1, T=t'}$ defined in equation (7.15) using the *att_gt* command. The argument *yname* in the latter defines outcome $Y$ to be entered in quotation marks, which in our case is the log of teen employment in a county, as provided in the variable *lemp*. The argument *tname* corresponds to period $T$, which is provided in the variable *year*. The argument *gname* constructs the treatment group $G_T$ based on the period in which a county is first treated (i.e., introducing a minimum wage), as indicated in the variable *first.treat*. The argument *xformla* permits controlling for one or several covariates $X$ by including them after the $\sim$ sign, in our case the logarithm of a county's population in thousands of inhabitants provided in the variable *lpop*, which is measured only once per county, and thus is time constant.

As we consider a panel data set, we also use the argument *idname*, which is the identifier of a specific unit (in our case, the county) across various periods, as indicated in the variable *countyreal*. We also use the latter variable for clustering: that is, for computing cluster-robust standard errors by using the *clustervars* argument. We note that if the data were a repeated cross section rather than a panel, then the argument *panel* would need to be set to *FALSE* rather than the default value, *TRUE*. Finally, setting the argument *data=mpdta* feeds the data set with all the previously mentioned variables into the DiD procedure. By default, the *att_gt* uses DR estimation for computing the ATETs, but this might be changed to IPW- or regression-based approaches. We store the DR estimation results in an object named *out*, which we wrap by the *summary* command. The box here provides the R code for each of the steps.

```
library(did)                          # load did package
data(mpdta)                           # load mpdta data
out=att_gt(yname="lemp", tname="year", gname="first.treat", idname="countyreal",
  xformla=~lpop, clustervars="countyreal", data=mpdta) # doubly robust did
summary(out)                          # group-time-specific ATETs
```

Running the code yields the following output:

```
Group—Time Average Treatment Effects:
 Group Time ATT(g,t) Std. Error [95% Simult.  Conf. Band]
  2004 2004  −0.0145    0.0236     −0.0760     0.0469
  2004 2005  −0.0764    0.0328     −0.1618     0.0090
  2004 2006  −0.1404    0.0402     −0.2451    −0.0358 *
  2004 2007  −0.1069    0.0340     −0.1956    −0.0182 *
  2006 2004  −0.0005    0.0232     −0.0608     0.0598
  2006 2005  −0.0062    0.0191     −0.0560     0.0436
  2006 2006   0.0010    0.0192     −0.0489     0.0509
  2006 2007  −0.0413    0.0214     −0.0971     0.0146
  2007 2004   0.0267    0.0148     −0.0117     0.0652
  2007 2005  −0.0046    0.0161     −0.0465     0.0373
  2007 2006  −0.0284    0.0183     −0.0760     0.0191
  2007 2007  −0.0288    0.0165     −0.0716     0.0141

Signif. codes:  '*'  confidence band does not cover 0

P—value for pretest of parallel trends assumption:  0.23267
Control Group:  Never Treated,  Anticipation Periods:  0
Estimation Method:  Doubly Robust
```

The column *Group* indicates in which period the treatment is introduced, and the column *Time* in which outcome period the ATET is estimated. Therefore, the first line, *2004 2004*, provides the ATET in 2004 for counties that introduced the treatment in 2004: that is, $\Delta_{G_{2004}=1,T=2004}$ according to the definition in equation (7.15). The second line, *2004 2005*, provides the ATET in 2005 for counties that introduced the treatment in 2004, $\Delta_{G_{2004}=1,T=2005}$. The output also contains simultaneous or uniform 95 percent confidence intervals, *[95 percent Simult. Conf. Band]*. The latter appropriately account for the fact that we simultaneously estimate multiple treatment group–period-specific ATETs, and therefore test multiple hypotheses about ATETs at the same time; see also section 4.7 for a discussion of issues related to multiple hypothesis testing in a different context. As indicated by a star (*), just two ATETs are statistically significant at the 5 percent level—namely, those for group $G_{2004} = 1$ in 2006 and 2007, where the minimum wage is found to reduce teen employment by $-0.14$ and $-0.11$ log points (or roughly 14 percent and 11 percent), respectively.

It is also worth noting that whenever the outcome period (*Time*) is prior to the first treatment period (*Group*), the estimates correspond to placebo tests. For example, we see for treatment group $G_{2006} = 1$ that the placebo tests in outcome periods 2004 and 2005 are close to zero and not statistically significant, and therefore they do not point to a violation of the common trend assumption. The output also provides a joint p-value of pooled placebo tests performed across all pretreatment periods, which amounts to 0.233 (or 23.3 percent). Therefore, having a zero placebo effect in pretreatment periods is not rejected at any conventional level of significance. Finally, the output indicates that those never treated in any period serve as the nontreated group, which can be modified to consist of the not-yet-treated observations in a specific period. We may also plot the group-period-specific ATET estimates along with confidence intervals by wrapping the estimation output *out* with the *ggdid* command, which yields the graph in figure 7.2.

In the next step, we investigate particular averages over the group-time-specific ATETs by applying the *aggte* command to the *out* object and saving the results as *meanATET*. We then wrap the latter by the *summary* command.

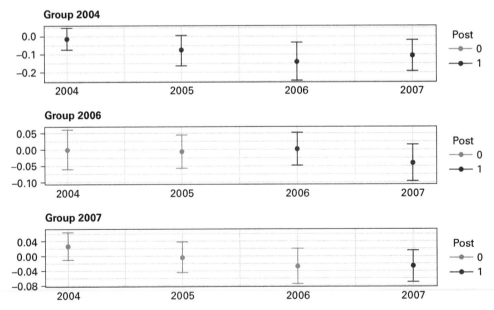

Figure 7.2
DiD regression with multiple periods.

```
meanATET=aggte(out)                # generate averages over ATETs
summary(meanATET)                  # report averaged ATETs
```

Running the code yields the following output:

```
Overall ATT:
      ATT Std. Error       [95%  Conf. Int.]
   -0.0328        0.013  -0.0583       -0.0073 *

Group Effects:
   group      ATT Std. Error [95% Simult.  Conf. Band]
    2004 -0.0846        0.0267          -0.1461       -0.0230 *
    2006 -0.0202        0.0174          -0.0604        0.0200
    2007 -0.0288        0.0173          -0.0686        0.0111
   ____

Signif. codes: '*' confidence band does not cover 0
```

The upper output, *Overall ATT*, corresponds to the simple average of all group-period-specific ATETs: that is, across all treatment groups and outcome periods. The estimate suggests that overall, the minimum wage reduces employment in treated counties by 0.03 log points (or roughly 3 percent), and this effect is statistically significant at the 5 percent level. The lower output, *Group Effects*, provides the ATETs separately for each group $G_T$, but averaged over all time periods $T$, thus corresponding to estimates of equation (7.16) for the various groups. On average (i.e., across all outcome periods), the introduction of a minimum wage reduces employment in counties starting with the treatment in 2004 ($G_{2004} = 1$) statistically significantly at the 5 percent level, but not in counties introducing the treatment in later periods.

## 7.4  Changes-in-Changes

The Changes-in-Changes (CiC) approach suggested in Athey and Imbens (2006) is related to DiD, in that it exploits differences in pretreatment and posttreatment outcomes across treated and nontreated groups, but based on different assumptions. CiC does not invoke any common trend assumption. It instead imposes that potential outcomes under nontreatment are strictly monotonic in unobserved heterogeneity (i.e., an unobserved characteristic or a function of several unobservables), and the distribution of this unobserved heterogeneity remains constant over time within treatment groups. Such a conditional independence between unobserved heterogeneity and time is satisfied if the subjects' ranks in the outcome distributions within treatment groups (e.g., the rank among wages in counties introducing a minimum wage) do not systematically change from pretreatment to posttreatment periods. This corresponds to rank similarity within treatment groups across time, rather than across treatment groups as considered in the IV context as discussed at the end of section 6.2 in chapter 6. In contrast to DiD, CiC allows for identifying both the ATET and quantile treatment effect among the treated (QTET), but (again in contrast to DiD) it generally requires a continuously distributed outcome. Similar to DiD, the CiC approach can be applied to both panel data and repeated cross sections.

For a more formal discussion, let us consider the following assumptions underlying the CiC approach:

$$Y_T(0) = \mathcal{H}(U, T), \quad U \perp T | D, \tag{7.19}$$

where $U$ is either a single (i.e., scalar) unobservable (like unobserved ability) or an index or function of unobservables, and $\mathcal{H}(u, t)$ is a general function that is assumed to be strictly monotonically increasing in the value of $u$ of unobservable $U$ for period $t$ being either 0 or 1. The model assumptions on $\mathcal{H}$ imply that the potential outcome under nontreatment is the same for all subjects with the same unobserved heterogeneity $U$ in a specific time period, independent of the actual treatment group, and a higher $U$ entails a higher potential outcome. A special case satisfying these conditions is the assumption of a linear model for the potential outcome under nontreatment, $Y_T(0) = \beta_T T + U$, with $\beta_T$ denoting the time trend under nontreatment. The conditional independence assumption $U \perp T | D$ in expression (7.19) requires that the distribution of unobserved heterogeneity is constant over time within treatment groups, while it might vary across treatment groups.

Let us introduce some further notation to discuss the identification of QTEs and ATEs. We denote by $F_{Y(d)|dt}(y) = \Pr(Y(d) \leq y | D = d, T = t)$ and $F_{dt}(y) = \Pr(Y \leq y | D = d, T = t)$ the conditional cumulative distribution functions of the potential outcome $Y(d)$ (with $d$ being either 0 or 1) and the observed outcome $Y$, given $D = d$ and $T = t$, respectively. We also note that the inverse of the conditional distribution function (namely, $F_{dt}^{-1}(y)$), corresponds to the conditional quantile function, in analogy to the discussion in section 4.8. Athey and Imbens (2006) demonstrate that under the assumptions in equation (7.19), we can identify the potential outcome distribution under nontreatment in the posttreatment period among those who actually receive the treatment, based on the observed conditional outcome distributions $F_{01}$, $F_{00}$, and $F_{10}$:

$$F_{Y(0)|11}(y) = F_{10}(F_{00}^{-1}(F_{01}(y))). \tag{7.20}$$

This in turn permits identifying the QTET at rank $\tau \in (0, 1)$, such as $\tau = 0.5$ for the effect at the median. We denote the QTET by $\Delta_{D=1}(\tau) = F_{Y(1)|11}^{-1}(\tau) - F_{Y(0)|11}^{-1}(\tau)$, which corresponds to the following equation that is based on nested quantile and distribution functions:

$$\Delta_{D=1}(\tau) = \underbrace{F_{11}^{-1}(\tau)}_{=F_{Y(1)|11}^{-1}(\tau)} - \underbrace{F_{01}^{-1}(F_{00}(F_{10}^{-1}(\tau)))}_{=F_{Y(0)|11}^{-1}(\tau)}. \tag{7.21}$$

Furthermore, we obtain the ATET by

$$\Delta_{D=1} = E[Y|D=1, T=1] - E[F_{01}^{-1}(F_{00}(Y_{10}))], \tag{7.22}$$

where $Y_{10}$ denotes the observed outcome in the group with $D=1$ and $T=0$. Intuitively, averaging the QTETs over all ranks in the treated population yields the ATET. We note that in addition to the assumptions in expression (7.19), the evaluation of the ATET relies on the common support restriction that the distribution of the unobservable $U$ among the nontreated contains all values of $U$ that exist among the treated. If the latter assumption is violated, such that some values of $U$ among the treated do not occur among the nontreated, then QTETs can be assessed only at those ranks $\tau$ that satisfy common support in $U$ across treatment groups.

Figure 7.3 provides a graphical illustration of the key property that $F_{01}^{-1}(F_{00}(F_{10}^{-1}(\tau)))$ identifies the unobserved counterfactual outcome $F_{Y(0)|11}^{-1}(\tau)$. At a quantile $\tau$ where the QTET is to be evaluated, we can directly observe $F_{Y(1)|11}^{-1}(\tau) = F_{11}^{-1}(\tau)$ in the distribution of $Y$ among the treated $(D=1)$ in the posttreatment period $(T=1)$. Next, we travel back in time to find $F_{10}^{-1}(\tau)$: that is, the corresponding quantile among the treated $(D=1)$ in the pretreatment period $(T=0)$. This is the representative outcome prior to treatment of a treated observation at rank $\tau$ in the posttreatment period due to the conditional independence $U \perp T|D$. In the next step, we assess the rank of this outcome taken from the not-yet-treated in the outcome distribution of the nontreated $(D=0)$ in the same pretreatment period $(T=0)$, yielding $\tau'$. Formally, $\tau' = F_{00}(F_{10}^{-1}(\tau))$.

We note that despite looking at pretreatment outcomes for which the treatment effect is zero in either treatment group, the rank of a specific outcome value (e.g., a monthly gross wage of 5,000 USD) among the nontreated, $\tau'$, generally differs from the rank among the treated, $\tau$, because the treated and nontreated groups may differ in terms of $U$. The CiC approach therefore allows us to match treated observations to comparable nontreated observations despite different distributions of unobservables. Finally, using $U \perp T|D$ again, we travel forward in time to find the corresponding quantile under nontreatment $(D=0)$ in the posttreatment period $(T=1)$ at rank $\tau'$—namely, $F_{10}^{-1}(\tau') = F_{01}^{-1}(F_{00}(F_{10}^{-1}(\tau)))$. The latter corresponds to the counterfactual outcome that a treated subject in the posttreatment period at rank $\tau$ in the treated outcome distribution would have expectedly obtained in the absence of treatment.

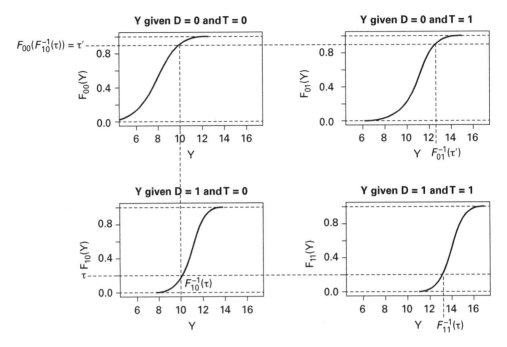

**Figure 7.3**
CiC.

While we have so far focused on the binary treatment case, the CiC framework can also be adapted to the evaluation of a multivalued, or even continuously distributed, treatment; see the discussion in D'Haultfoeuille, Hoderlein, and Sasaki (2021). In this case, the assumptions invoked in expression (7.19) have to be imposed with regard to the potential outcomes under various treatment levels rather than under nontreatment alone. Furthermore, de Chaisemartin and D'Haultfeuille (2018) extend the CiC approach to the evaluation of the LATE in scenarios with noncompliance of treatment participation with regard to treatment assignment, as discussed at the end of section 7.1 in the DiD context. Furthermore, combining random treatment assignment with CiC assumptions on intermediate variables like the actual treatment participation or some mediator permits assessing causal mechanisms as considered in section 4.10 or testing IV exclusion restrictions; for instance, see the discussions by Huber, Schelker, and Strittmatter (2020) and Sawada (2019). Finally, another important extension is QTET and ATET evaluation under the assumption that the CiC conditions hold only when controlling for observed covariates $X$, as considered by Melly and Santangelo (2015).

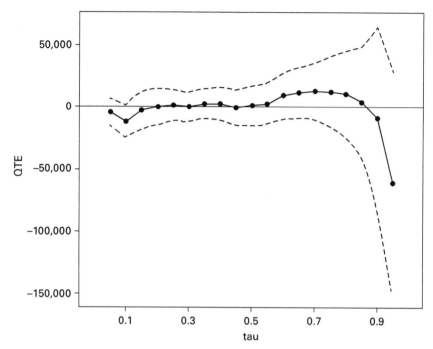

**Figure 7.4**
CiC-based QTEs across ranks.

To apply CiC estimation in R, we load the *qte* package by Callaway (2019), which contains an estimator of the QTET, as well as the *kielmc* data in the *wooldridge* package already analyzed at the end of sections 7.1 and 7.2. We apply the *CiC* command using the regression formula *rprice~nearinc*, which specifies the real house price as the outcome and the binary indicator for being close to the garbage incinerator as the treatment. Furthermore, we define the arguments *t=1981* (i.e., the posttreatment period (year 1981)), and *tmin1=1978* (i.e., the pretreatment period (year 1978)), followed by the argument *tname="year"*, which specifies the variable measuring the period to be *year*. We also specify the data to be analyzed using *data=kielmc*. By default, the *CiC* command estimates the QTETs for ranks $\tau \in \{0.05, 0.10, \ldots, 0.90, 0.95\}$, which corresponds to setting *probs=seq(0.05,0.95,0.05)* but could be changed to different ranks. Finally, we store the output in an R object named *cic*, which we wrap with the *ggqte* command to plot the results. The box here contains the R code for each step.

```
library(qte)                                            # load qte package
library(wooldridge)                                     # load wooldridge package
data(kielmc)                                            # load kielmc data
cic=CiC(rprice~nearinc,t=1981,tmin1=1978,tname="year",data=kielmc) # run CiC
ggqte(cic)
```

Running the code yields the plot provided in figure 7.4, which plots the QTEs on the $y$-axis across various outcome ranks on the $x$-axis. We find none of our QTET estimates, which correspond to the black dots, to be statistically significantly different from zero at the 5 percent level, as the 95 percent confidence intervals given by the dashed lines always include the zero.

# 8
## Synthetic Controls

### 8.1 Estimation and Inference with a Single Treated Unit

Like the difference-in-differences (DiD) and Changes-in-Changes (CiC) approaches discussed in chapter 7, the synthetic control method permits evaluating causal effects if the outcome variable is observed in both pretreatment and posttreatment periods. In contrast to the previous approaches, however, it necessarily requires panel data, implying that the same subjects or units can be followed over time. The synthetic control method was originally developed for case study setups, with the goal to assess the treatment effect on a single treated unit based on a comparison with multiple nontreated units. This appears particularly well suited for the evaluation of relatively rare (or even unique) policy interventions that affect only one specific region, country, organization, or company.

For instance, Abadie, Diamond, and Hainmueller (2015) apply the synthetic control method to evaluate the effect of the reunification of West and East Germany in 1990 on the development of the gross domestic product (GDP) per capita in West Germany. While the economic outcome under the reunification can be directly observed for West Germany in the data, the identification of the reunification's effect requires inferring the counterfactual outcome of how the West German GDP per capita would have evolved in the absence of the reunification. The idea of the synthetic control method is to impute the counterfactual outcome under nontreatment based on an appropriate combination of other countries that are sufficiently similar in economic terms to West Germany prior to the reunification, but did not experience any reunification or similar kind of policy change. More concisely, the treated unit's potential outcome under nontreatment (i.e., the West German counterfactual after the reunification) is synthetically estimated as a weighted average of the observed posttreatment outcomes coming from a donor pool of nontreated units, in this case other Organisation for Economic Co-operation and Development (OECD) countries. The weight that a nontreated country receives for computing this average called the synthetic control depends on how similar the country was to West

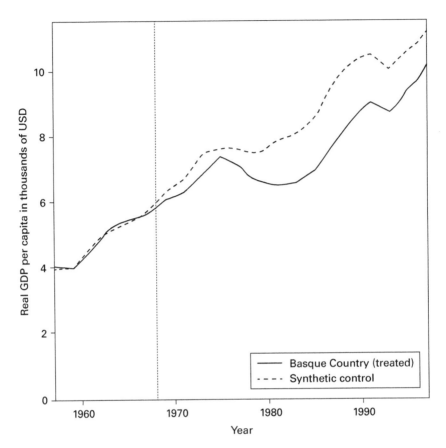

**Figure 8.1**
Synthetic control method for the terrorist conflict in Basque Country.

Germany in terms of economic conditions in the pretreatment periods: that is, prior to reunification.

A second example can be found in the seminal synthetic control study of Abadie and Gardeazabal (2003), who consider the effect of the terrorist conflict in the Basque Country, a region in Spain, on GDP per capita in that region based on a synthetic control created from other Spanish regions. Figure 8.1 provides a graphical illustration of the application of the synthetic control method in this context, which has been created using the *Synth* package by Abadie, Diamond, and Hainmueller (2011) for the statistical software R. The start of the treatment is marked by 1968, when the terrorist group ETA claimed its first victim. While the solid line provides the development of GDP per capita (in thousands of US dollars as of 1986) for the Basque Country from 1957 to 1997, the dashed line corresponds to the respective development of its synthetic control. The latter is obtained as a weighted average of other Spanish regions, such that the GDP development in these places closely matches that of the Basque Country in the pretreatment periods from 1957 to 1967. After the start of the terrorist conflict in 1968, the GDP per capita of the Basque Country and its synthetic control diverge more and more, pointing to a negative treatment effect, particularly after the mid-1970s, when ETA's terrorist activities intensified.

To discuss the synthetic control method more formally, assume that we have available a panel data set of $n$ units, such that a unit's index $i \in \{1, \ldots, n\}$, which we observe over $\mathcal{T}$ time periods such that time index $t \in \{1, \ldots, \mathcal{T}\}$. Let us denote by $Y_{it}$ the observed outcome of unit $i$ in period $t$. Furthermore, we assume (without loss of generality) that only the last unit $i = n$ in the data is treated in a period $T_0 + 1$, with $T_0$ denoting the last period prior to treatment, which satisfies $T_0 \geq 1$. Therefore, the treatment takes place in a period after the first period, implying that $1 < T_0 + 1 \leq \mathcal{T}$. For any posttreatment period $t \geq T_0 + 1$, we obtain an estimate of the treatment effect for the treated unit $i = n$, denoted by $\hat{\Delta}_{n,T=t}$, as the difference of the treated outcome and a weighted average of nontreated outcomes in that period:

$$\hat{\Delta}_{n,T=t} = Y_{nt} - \sum_{i=1}^{n-1} \hat{\omega}_i Y_{it}, \text{ for any } t \geq T_0 + 1, \tag{8.1}$$

where $\hat{\omega}_i$ is a specific weight or importance of some nontreated unit.

In its most basic version, the idea of the synthetic control method is to choose the weights such that the weighted average of pretreatment outcomes of nontreated units matches the evolution of the pretreatment outcome of the treated observation: that is, up to period $T_0$. This approach assumes that we can appropriately model the treated unit's posttreatment potential outcome under nontreatment by means of the reweighted outcomes of actually nontreated units. Put differently, controlling for pretreatment outcomes $Y_{it}$ over periods $t \in \{1, \ldots, T_0\}$ must be sufficient to control for confounders entailing diverging potential outcomes under nontreatment of the treated and nontreated units in the posttreatment periods. Formally, we choose the weights such that

$$\sum_{i=1}^{n-1} \hat{\omega}_i Y_{it} \approx Y_{nt}, \text{ for all } t = 1, \ldots, T_0. \tag{8.2}$$

This approach seems quite related to a selection-on-observables framework, as discussed in chapter 4, with pretreatment outcomes serving as covariates to be controlled for. To better see this, assume for a moment that a nontreated observation $j$ has observed pretreatment outcomes very similar to the treated observation $i$, and therefore receives the weight $\hat{\omega}_j = 1$, while all other nontreated observations have a weight of zero. This is equivalent to pair matching, as formally discussed in equation (4.20) in section 4.3 of chapter 4 when using the pretreatment outcomes as covariates and considering only one treated observation. And like the selection-on-observables framework, the synthetic control method relies on a type of common support condition concerning the existence of treated and nontreated units that are similar in terms of pretreatment outcomes, namely, the so-called convex hull condition. The latter implies that the pretreatment outcomes of the treated unit are not too extreme compared to the nontreated units—in particular, not much higher than the highest or much lower than the lowest outcome of nontreated units in any pretreatment period. Furthermore, and similar to the DiD context in chapter 7, anticipation effects must be ruled out, such that the pretreatment outcomes of the treated unit are not affected by the treatment being introduced.

In most applications, the weighted average is computed based on several, albeit usually few nontreated units to interpolate the potential outcome, implying that weights are sparse in the sense that only a few units receive nonzero weights. In the prototypical synthetic control approach, therefore, the weight of any nontreated unit $i$ is either positive or zero (i.e., $\hat{\omega}_i \geq 0$), while all weights add up to 1, as required for a properly weighted average; that is, $\sum_{i=1}^{n-1} \hat{\omega}_i = 1$. More formally, the collection of weights $\hat{\omega} = (\hat{\omega}_1, \ldots, \hat{\omega}_{n-1})$ is computed based on the following least squares approach, with $\omega^*$ denoting candidate values for the weights:

$$\hat{\omega} = \arg\min_{\omega^*} \sum_{t=1}^{T_0} (Y_{nt} - \omega_1^* Y_{1t} - \ldots - \omega_2^* Y_{n-1t})^2, \text{ subject to } \omega_i^* \geq 0, \sum_{i=1}^{n-1} \omega_i^* = 1. \tag{8.3}$$

We note that each pretreatment period is deemed equally important in the optimization problem given in equation (8.3). Put differently, all of the treated unit's pretreatment outcomes are considered equally appropriate for creating a synthetic control that permits assessing the causal effect. In practice, however, we might want to give a different importance to different periods, such as giving a higher weight to more recent pretreatment periods. This can be motivated by the assumption that having comparable outcomes across treated and nontreated units just prior to treatment introduction is more relevant to the plausibility of the selection on-observables-assumption than is similarity in rather distant pretreatment outcomes. Accounting for a differential importance of periods is easily obtained by adding a period-dependent weight, denoted by $v_t \geq 0$, that takes higher values for more important periods and a value of zero for periods that should not be considered at all:

$$\hat{\omega} = \arg\min_{\omega^*} \sum_{t=1}^{T_0} v_t (Y_{nt} - \omega_1^* Y_{1t} - \ldots - \omega_2^* Y_{n-1t})^2, \text{ subject to } \omega_i^* \geq 0, \sum_{i=1}^{n-1} \omega_i^* = 1. \tag{8.4}$$

Furthermore, it is worth noting that we may transform outcomes prior to running the synthetic control method if this appears appropriate. For instance, considering the logarithm of an outcome yields a causal effect that is to be interpreted in terms of changes in log points or percent rather than levels. As another example, we can subtract the pretreatment mean in outcomes of a unit $i$ from unit $i$'s outcome in some period $t$, in order to generate demeaned outcomes, denoted by $\tilde{Y}_{it}$:

$$\tilde{Y}_{it} = Y_{it} - \frac{1}{T_0} \sum_{t=1}^{T_0} Y_{it}. \tag{8.5}$$

Considering demeaned outcomes $\tilde{Y}_{it}$ rather than $Y_{it}$, which is equivalent to including a constant term $\alpha$ in the optimization problem in equation (8.3), implies that weighting aims to find a combination of nontreated units with similar changes or trends (rather than levels) in pretreatment outcomes like the treatment unit; see Ferman and Pinto (2021). This means that we can also apply the synthetic control method when invoking a common trend (rather than a selection-on-observables) assumption that is closely related to that of DiD estimation in chapter 7.

As a further modification to our initial setup, we may not only include pretreatment outcomes, but also (or even exclusively) observed covariates $X$ as control variables when determining the weights of nontreated units based on equation (8.3), to generate a synthetic control that is similar to the treated unit in terms of those covariates. This may be useful for making the selection-on-observables (when using outcome levels) or common trend assumption (when using outcome changes) more plausible. There are several possibilities of incorporating covariates in the synthetic control method. One obvious option is to include them in an analogous way as the pretreatment outcomes in the optimization problem in equation (8.3). An alternative approach is to regress the outcomes on a constant and $X$, and to only consider the outcome residuals (from which the influence of $X$ has been purged, similar to the discussion in section 5.2) in the optimization problem in equation (8.3), as suggested by Doudchenko and Imbens (2016).

Concerning inference, we note that determining the statistical significance of treatment effects based on synthetic controls is not straightforward due to the peculiar setup of only one treated unit for which the effect is to be assessed. One feasible approach is randomization inference based on permutation (see Abadie, Diamond, and Hainmueller (2010)), which is based on estimating placebo effects among nontreated units for whom the true effect is known to be zero. To this end, each of the nontreated units is iteratively considered to be the pseudotreated unit for the purpose of estimating the placebo effect, while all remaining nontreated units are used as the donor pool. This yields a distribution of as many placebo effects as there are nontreated units and permits assessing how extreme the treatment effect on the treated unit is relative to these placebo effects. This approach appears to be related to the computation of p-values based on bootstrap distributions; see equation (3.49) and the related discussion in section 3.4 in chapter 3. For instance, the share of placebo effects in period $t$ that is larger than the effect on the treated unit $\hat{\Delta}_{n,T=t}$ provides a p-value for a one-sided hypothesis test under the alternative hypothesis that the true effect is larger than zero (i.e., $H_1 : \Delta_{n,T=t} > 0$), and the null hypothesis that the effect is no more than zero (i.e., $H_0 : \Delta_{n,T=t} \leq 0$).

It is important to note that such tests may serve just as an approximation to actual statistical significance because randomization inference is, strictly speaking, only valid under a randomly assigned treatment (see Fisher (1935)), a condition not likely to hold in most applications of synthetic controls. A further possible issue with the computation of placebo effects is that we might lack a sufficiently similar donor pool for all of the nontreated units, which would violate the convex hull condition and entail a relatively poor estimation of the placebo effect. As a practical solution, we could drop such problematic nontreated units from the inference procedure. An alternative is to normalize the placebo effects in the posttreatment periods by the placebo effects in the pretreatment periods, namely through dividing or scaling the placebo effects in the posttreatment periods by the placebo effects in the pretreatment periods before computing the p-values. This procedure adjusts the placebo effects according to the estimation quality of the synthetic controls before assessing statistical significance; see the discussion by Abadie, Diamond, and Hainmueller (2010).

An alternative permutation method for computing p-values is conformal inference; see Chernozhukov, Wüthrich, and Zhu (2021), which focuses on the placebo effects of the treated unit rather than the nontreated donor pool. The procedure is based on the intuition that if the treatment effect in posttreatment periods equals zero, then the distribution of the treated unit's pretreatment placebo effects (i.e., the difference between the observed pretreatment outcome and the predicted synthetic control) must be comparable to the distribution of the treatment effects in the posttreatment periods. Therefore, permutation is based on reassigning the pretreatment placebo effects of the treated unit to the posttreatment periods and vice versa, in order to compute test statistics based on these permutations of posttreatment periods. We can then verify how extreme the test statistic based on the actual posttreatment effects in the original, nonpermuted data is, relative to the distribution of the permutation-based test statistics.

The tests may either be implemented by randomly permuting single effects (i.e., reassigning them to pretreatment or posttreatment periods) or by block permutation. The latter implies that the effects are sequentially rotated to enter the pretreatment or posttreatment periods, to account for the ordering of and autocorrelation in the outcomes over time. We note that such permutation-based inference with sufficiently long panel data are not exclusive to synthetic control methods, but it might also be applied to other estimators, like DiD approaches. Finally, and as an alternative to permutation, we might also consider asymptotic (i.e., large sample) variance approximations in the spirit of the methods outlined in section 3.4 for the mean effects across all posttreatment periods if they are numerous enough, as discussed in Li (2020).

## 8.2    Alternative Estimators and Multiple Treated Units

We will subsequently consider several extensions and modifications of the prototypical synthetic control method. To this end, let us first investigate how the latter is related to standard ordinary least squares (OLS) regression, which solves the following minimization problem for estimating the weights (or coefficients):

$$(\hat{\omega}, \hat{\alpha}) = \arg \min_{\omega^*, \alpha^*} \sum_{t=1}^{T_0} (Y_{nt} - \alpha^* - \omega_1^* Y_{1t} - \ldots - \omega_2^* Y_{n-1t})^2. \tag{8.6}$$

It is easy to see that imposing the conditions $\omega_i^* \geq 0, \sum_{i=1}^{n-1} \omega_i^* = 1, \alpha^* = 0$ gives the synthetic control approach in equation (8.3) of section 8.1 in chapter 8, with outcomes measured in levels. Allowing a nonzero constant $\alpha^*$, while maintaining $\omega_i^* \geq 0, \sum_{i=1}^{n-1} \omega_i^* = 1$, yields the previously discussed DiD-related synthetic control method. However, when dropping the conditions $\omega_i^* \geq 0, \sum_{i=1}^{n-1} \omega_i^* = 1$, the weights in equation (8.6) may also become negative. This implies that the method might (like any OLS estimator) extrapolate to make predictions for the treated unit beyond the convex hull of observed nontreated outcomes. Whether such a regression estimator entails greater accuracy of the effect estimate $\hat{\Delta}_{n,T=t}$ than the synthetic control approach in equation (8.3) depends on the application at hand, such as the bias of extrapolation relative to taking the weighted averages of nontreated outcomes.

Related to the discussion in section 5.2 in chapter 5, we can use machine learning to find the optimal model for computing the treated unit's potential outcome under nontreatment, and thus the treatment effect. For instance, adding a penalty $\lambda \sum_{i=1}^{n-1} |\omega_i^*|$ to the optimization problem in equation (8.6) entails a lasso regression in the spirit of equation (5.1). To find the optimal penalization $\lambda$ yielding the greatest accuracy according to the data, we may apply cross-validation in a way that is related to permutation inference based on nontreated units as outlined in the previous section 8.1. To this end, we repeatedly estimate placebo treatment effects for each nontreated unit with various choices of $\lambda$ and ultimately choose the penalization that minimizes the mean squared placebo treatment effects, as discussed by Doudchenko and Imbens (2016).

An alternative approach not relying on cross-validation is to constrain the sum of the absolute values of weights to be less than or equal to 1 (i.e., $\sum_{i=1}^{n-1} |\omega_i^*| \leq 1$), which is known as *constrained lasso*; for instance, see Raskutti, Wainwright, and Yu (2011). A further methodological twist involves combining the potential outcome prediction based on the prototypical synthetic control method using equation (8.3) with a possibly penalized regression using equation (8.6) as in Ben-Michael, Feller, and Rothstein (2021a), or other estimators like pair matching (see section 4.3) on the pretreatment outcomes and possibly covariates as in Abadie and L'Hour (2018). This may entail greater accuracy than each individual method alone, very much in the spirit of the regression-based bias correction of equation (4.35). We can use cross-validation as suggested by Doudchenko and Imbens (2016) to determine the weight or importance given to either method in a combination of the synthetic control method with another estimator, which bears some resemblance to the concept of ensemble methods discussed in section 5.3.

Interestingly, we may also apply the synthetic control method or related estimators to scenarios with multiple treated units to estimate their ATET in a specific outcome period, denoted as $\Delta_{D=1,T=t} = E[Y_t(1) - Y_t(0)|D=1]$. One obvious approach is to apply the estimator separately to each of the treated units, such that equation (8.2) holds for each treated unit, and then average over the effects to estimate the ATET:

$$\hat{\Delta}_{D=1,T=t} = \frac{1}{\sum_{i=1}^n D_i} \sum_{i:D_i=1} \hat{\Delta}_{i,T=t}, \tag{8.7}$$

with $\hat{\Delta}_{i,T=t}$ being the effect estimate given in equation (8.1) for a treated unit $i$ based on the nontreated donor pool. An alternative approach is to estimate the weights in a way that equation (8.2) holds on average for the treated units rather than for each treated unit separately.

To discuss this possibility more formally, let us denote by $n_1 = \sum_{i=1}^n D_i$ the number of treated units and assume (without loss of generality) that the observations are ordered in a way that all $n - n_1$ nontreated units appear at the top (or come first) and all treated units at the bottom (or come last) in the data. Then, equation (8.2) is modified as follows:

$$\sum_{i=1}^{n-n_1} \hat{\omega}_i Y_{it} + \hat{\alpha} \approx \frac{1}{n_1} \sum_{j=n-n_1+1}^n Y_{jt}, \text{ for all } t=0,\ldots,T_0, \tag{8.8}$$

where allowing the constant term $\hat{\alpha}$ to be nonzero corresponds to a DiD- rather than selection-on-observables-type synthetic control method. In a related manner, equation (8.3) changes to

$$(\hat{\omega},\hat{\alpha}) = \arg\min_{\omega^*,\alpha^*} \sum_{t=1}^{T_0} \left( \frac{1}{n_1} \sum_{j=n-n_1+1}^n Y_{jt} - \alpha^* - \omega_1^* Y_{1t} - \ldots - \omega_2^* Y_{n-n_1 t} - \alpha \right)^2, \tag{8.9}$$

$$\text{subject to } \omega_i^* \geq 0, \sum_{i=1}^{n-1} \omega_i^* = 1.$$

This synthetic DiD approach, as suggested in Arkhangelsky et al. (2019), relies on the assumption that a weighted average of the nontreated outcomes in the pretreatment periods permits replicating the trend of the average potential outcome under nontreatment among the treated units. It is worth noting that this condition about the average trend still permits the outcome trends to differ on the individual level of each unit. The fact that we also can apply synthetic control methods to multiple (rather than single) treated units in panel data implies that they constitute an alternative estimation strategy to those outlined in chapters 4 and 7 for the selection-on-observables and common trend assumptions, respectively. Also, for the case of multiple treated units, we may combine synthetic control with other (e.g., regression) methods, with the weights of each method being determined by cross-validation. Finally, the synthetic control method can be adapted to the case of staggered treatment introduction considered in section 7.3, as discussed in Ben-Michael, Feller, and Rothstein (2021b). For a comprehensive discussion of the synthetic control method, see the review article by Abadie (2021).

To apply the synthetic DiD and synthetic control methods in R, we use the procedures and data provided in the *synthdid* package by Arkhangelsky (2021). The package is available on the online software platform GitHub. Accessing GitHub requires first installing and loading the *devtools* package and then running *install_github("synth-inference/synthdid")* to install the *synthdid* package, before loading it using the *library* command. In the next step, we apply the *data* command to load the *california_prop99* data set on per capita cigarette consumption in 39 US states from 1970 to 2000, which was previously analyzed by Abadie, Diamond, and Hainmueller (2010).

The panel data's first variable, *State*, corresponds to unit *i*, the second variable, *Year*, to period *T*, the third one, *PacksPerCapita*, to outcome *Y*, and the fourth one, *treated*, to treatment *D*. The latter reflects Proposition 99, a tobacco control program launched in California in 1989, which increased cigarette taxes, channeled the tax revenues to health and anti-smoking education budgets, and entailed anti-smoking media campaigns and local clean indoor-air ordinances. Therefore, the treatment equals 1 for observations from California in 1989 or later and zero otherwise. In the next step, we transform the *california_prop99* data into a format suitable for running the synthetic DiD method by applying the *panel.matrices* command. The latter assumes by default that data columns 1 to 4 correspond to *i*, *T*, *Y*, and *D*, respectively, as is the case for our data set.

We store the transformed data in an object named *dat* and set a seed for the reproducability of the results to follow. We then apply the synthetic DiD method using *synthdid_estimate(Y=dat$Y, N0=dat$N0, T0=dat$T0)*, where argument *Y* defines the outcome, *N0* the number of nontreated units (38 states), and *T0* the number of pretreatment periods (19, from 1970 to 1988). We save the output in an object named *out* and estimate the variance of treatment effect estimation based on placebo treatments among control units by applying the *vcov(out, method='placebo')* command. Wrapping the latter by *sqrt* yields the square root of the estimated variance, the standard error, which we save as *se*. Finally, we call the first object in the output of the synthetic DiD method, *out[1]*, and *se* to investigate the results. The box here provides the R code for each step.

```
library(devtools)                                # load devtools package
install_github("synth-inference/synthdid")       # install synthdid package
library(synthdid)                                # load synthdid package
data(california_prop99)                          # load smoking data
dat=panel.matrices(california_prop99)            # prepare data
set.seed(1)                                      # set seed
out=synthdid_estimate(Y=dat$Y, N0=dat$N0, T0=dat$T0) # synthetic DiD
se = sqrt(vcov(out, method='placebo'))           # placebo standard error
out[1]; se                                       # show results
```

Running the code yields an average treatment effect of proposition 99 on the per capita cigarette consumption in California of −15.604 packs, with the average being across all posttreatment periods from 1989 to 2000. The standard error amounts to 10.053 and is thus quite substantial relative to the absolute magnitude of the causal effect.

Next, we implement the prototypical synthetic control method based on equation (8.3) in section 8.1 rather than synthetic DiD. To this end, we modify our previous application of the *synthdid_estimate* command. We set the argument *omega. intercept=FALSE* to rule out a nonzero intercept $\hat{\alpha}$ in equation (8.6), and thus avoid a DiD-type synthetic control method. We also set *weights=list(lambda=rep(0,setup $T0))* such that there are no period-specific weights (denominated by *lambda* in the procedure), which is in line with equation (8.3). We note that dropping the argument *weights=list(lambda=rep(0,setup$T0))* would allow weights to vary across pretreatment periods, as is the case in equation (8.4). Apart from these two modifications, we repeat the same steps as before:

```
set.seed(1)                                      # set seed
out=synthdid_estimate(Y=dat$Y, N0=dat$N0,T0=dat$T0, omega.intercept=FALSE,
    weights=list(lambda=rep(0,dat$T0)))          # synthetic control
se = sqrt(vcov(out, method='placebo'))           # placebo standard error
out[1]; se                                       # show results
```

Running this code yields an average reduction of cigarette consumption in California across all outcome periods of 21.717 packs per capita, with the standard error

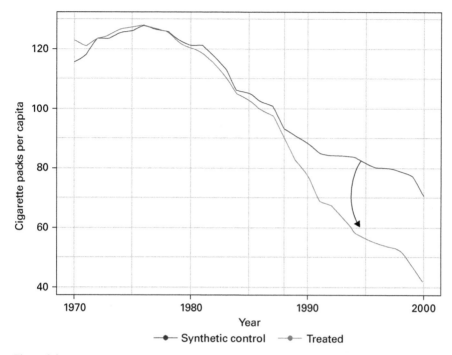

**Figure 8.2**
Effects based on the synthetic control method.

amounting to 11.506. Finally, we apply the *plot* command to the *out* object to plot the treatment effect separately for each outcome period:

```
plot(out)
```

This yields the graph in figure 8.2. The latter suggests that after the introduction of proposition 99 in 1989, the reduction of cigarette consumption in California relative to its synthetic control gradually became more important the longer the tobacco control program was in place. We also see that in the pretreatment periods, the synthetic control closely matches the development of tobacco consumption in California.

# 9

## Regression Discontinuity, Kink, and Bunching Designs

### 9.1 Sharp and Fuzzy Regression Discontinuity Designs

The previous two chapters of this book have considered methods for causal analysis that rely on observing outcomes and treatment states across several time periods. Which alternative approaches to treatment effect evaluation exist if our data does not satisfy this requirement or the previously considered assumptions (like the common trend assumption) appear implausible? In this chapter, we will take a look at the regression discontinuity design (RDD), as first suggested in Thistlethwaite and Campbell (1960). It is based on the assumption that at a particular threshold of an observed running variable, the treatment status either changes from zero to 1 for everyone (sharp design) or at least for a subpopulation (fuzzy design). As an example, let us assume that the treatment of interest is extended eligibility to unemployment benefits, to which only individuals aged 50 or older are entitled; for instance, see Lalive (2008). The idea of the RDD is to compare the outcomes (like unemployment duration) of treated and nontreated subjects close to the age threshold, such as individuals aged 50 and 49. Such individuals slightly above and below the age threshold are arguably similar in characteristics potentially affecting the outcome due to their minor difference in age, while they differ in terms of the treatment.

The RDD, therefore, aims at imitating the experimental context at the threshold to evaluate the treatment effect locally for the subpopulation at the threshold. Further examples include assigning an educational treatment (like access to a university or college) based on a threshold in the score of an admission test or the high school grade point average, providing cash transfers or welfare payments based on a threshold in a poverty index, or granting discounts in the price of a product or service to customers based on a threshold in previous sales per customer. In all these examples, it might be argued that subjects just slightly below the score, poverty, or sales threshold are very similar to subjects just slightly above in terms of background characteristics also affecting the outcome, such as ability, motivation, or other personality traits.

To formalize our discussion of the RDD, let $R$ denote the running variable and $r_0$ the threshold value. If the treatment is deterministic in $R$ such that it is one whenever the threshold is reached or exceeded (i.e., $D = I\{R \geq r_0\}$), the RDD is sharp in the sense that all individuals change their treatment status exactly at $r_0$. The evaluation of causal effects in the sharp RDD relies on the assumption that the mean potential outcomes given the running variable, $E[Y(1)|R]$ and $E[Y(0)|R]$, are continuous and sufficiently smooth around $R = r_0$; see the discussion in Hahn, Todd, and van der Klaauw (2001); Porter (2003); and Lee (2008). This requires that any (and possibly unobserved) background characteristics other than $D$ that affect the outcome are continuously distributed at the threshold.

Such a continuity implies that if treated and nontreated populations with values of $R$ exactly equal to $r_0$ existed, the treatment would be as good as randomly assigned with regard to mean potential outcomes. Therefore, treated and nontreated subjects would be comparable in terms of background characteristics affecting the outcome at $R = r_0$. This corresponds to a local selection-on-observables assumption conditional on $R = r_0$ (rather than on $X$, as in chapter 4). As a further condition, the density of the running variable $R$ must be continuous and larger than zero around the threshold, such that treated and nontreated observations are observed close to $R = r_0$. Under these assumptions, we can identify a conditional or local average treatment effect (ATE) at the threshold based on treated and nontreated outcomes in a neighborhood $\epsilon > 0$ around the threshold when letting $\epsilon$ go to zero (denoted by $\epsilon \to 0$):

$$\lim_{\epsilon \to 0} E[Y|R \in [r_0, r_0 + \epsilon)] - \lim_{\epsilon \to 0} E[Y|R \in [r_0 - \epsilon, r_0)] \tag{9.1}$$

$$= \lim_{\epsilon \to 0} E[Y(1)|R \in [r_0, r_0 + \epsilon)] - \lim_{\epsilon \to 0} E[Y(0)|R \in [r_0 - \epsilon, r_0)]$$

$$= E[Y(1) - Y(0)|R = r_0] = \Delta_{r_0}.$$

Figure 9.1 provides a graphical illustration for the sharp RDD. It plots the mean potential outcomes $E[Y(1)|R]$ and $E[Y(0)|R]$ as a function of the running variable $R$. The threshold $r_0$ at which the treatment $D$ switches from 0 to 1 is equal to 5. For this reason, the mean potential outcome under treatment $E[Y(1)|R]$ is observed only based on $E[Y|R]$ for $R \geq r_0$ such that $D = 1$, as indicated by the solid line in the graph, but not for $R < r_0$ such that $D = 0$, as indicated by the dashed line. Likewise, the mean potential outcome under nontreatment $E[Y(0)|R]$ is observed only for $R < r_0$ such that $D = 0$, but not for $R \geq r_0$ such that $D = 1$. In fact, there is no common support in $R$ across the treatment groups that would allow comparing treated and nontreated outcomes with exactly the same values in the running variable.

This contrasts with the discussion in chapter 4, where the common support condition implies that treated and nontreated units with comparable covariates $X$ to be conditioned on do exist. Even though common support fails by design in the sharp RDD, we can assess the causal effect $\Delta_{r_0}$ by the difference in $E[Y|R]$ just above the

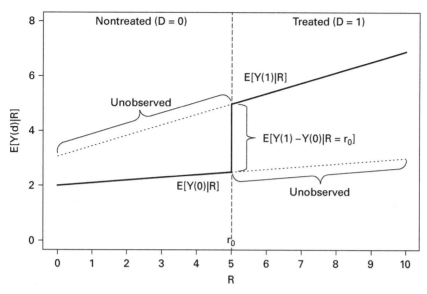

**Figure 9.1**
Sharp regression discontinuity design.

threshold $R = r_0$ such that $D = 1$, and just below the threshold such that $D = 0$. This guarantees that treated and nontreated units are similar in terms of the running variable, and thus in terms of mean potential outcomes. Figure 9.2 provides a graphical illustration for this approach, where the dots represent observations with specific values in the running variable $R$ (plotted on the $x$-axis) and the outcome $Y$ (plotted on the $y$-axis), and the solid lines correspond to the regression functions of $E[Y|R]$ above and below the threshold. The causal effect at the threshold, therefore, is given by the vertical difference (or discontinuity) of these regression functions at $R = r_0$.

In analogy to the discussion of instrumental variables in chapter 6, treatment take-up as a function of being above or below the threshold of the running variable might not be perfect. This implies that in contrast to the sharp RDD, $D$ is not deterministic in $R$ such that noncompliance in the treatment participation occurs. In this case, a fuzzy RDD approach permits assessing the causal effect on compliers at $R = r_0$ who are induced to switch their treatment state at the threshold. A precondition is that the share of treated units changes discontinuously at the threshold, which implies that compliers exist. As an illustration, let us reconsider the example that admittance to a college ($D$) depends on passing a particular threshold of the score in an admission test ($R$). While some students might decide not to attend college even if succeeding in the test, a discontinuous change in the treatment share occurs if there are compliers that are induced to go to college when passing the threshold.

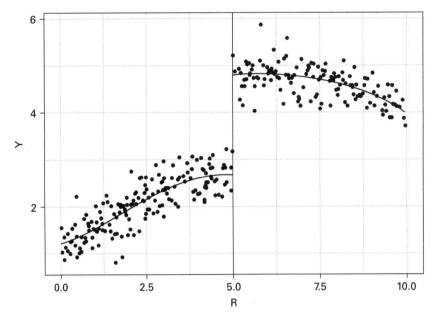

**Figure 9.2**
Observations and regression functions above and below the threshold.

To distinguish between compliance and noncompliance, we adapt our notation and, similar to chapter 6, denote by $D(z)$ the potential treatment state as a function of the binary threshold indicator $Z = I\{R \geq r_0\}$, which now serves as an instrument for actual treatment participation. Furthermore, let us assume that around the threshold, defiers do not exist and that the shares of compliers, always takers, and never takers, as well as their mean potential outcomes under treatment and nontreatment, are continuous, as discussed in Dong (2014). This implies that instrumental variable (IV)–type assumptions similar to those postulated in expression (6.5) when controlling for covariates $X$ in section 6.2 hold when conditioning on being at the threshold $R = r_0$.

Under these conditions, we can assess the first-stage effect of instrument $Z$ on treatment participation $D$ at the threshold $R = r_0$, denoted by $\gamma_{r_0}$:

$$\lim_{\epsilon \to 0} E[D|R \in [r_0, r_0 + \epsilon)] - \lim_{\epsilon \to 0} E[D|R \in [r_0 - \epsilon, r_0)] \tag{9.2}$$

$$= \lim_{\epsilon \to 0} E[D(1)|R \in [r_0, r_0 + \epsilon)] - \lim_{\epsilon \to 0} E[D(0)|R \in [r_0 - \epsilon, r_0)]$$

$$= E[D(1) - D(0)|R = r_0] = \gamma_{r_0}.$$

Furthermore, the first line of equation (9.1) yields the ITT effect of $Z$ on $Y$ at the threshold, denoted by $\theta_{r_0}$, in the fuzzy RDD (rather than $\Delta_{r_0}$ as in the sharp

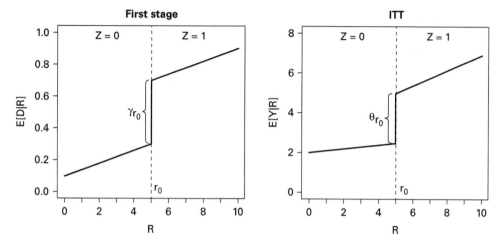

**Figure 9.3**
The fuzzy regression discontinuity design.

RDD). In analogy to equation (6.7) in section 6.2, the local average treatment effect (LATE) on compliers at the threshold, denoted by $\Delta_{D(1)=1,D(0)=0,R=r_0} = E[Y(1) - Y(0)|D(1) = 1, D(0) = 0, R = r_0]$, is identified by dividing the intention-to-treat (ITT) by the first-stage effect at the threshold:

$$\Delta_{D(1)=1,D(0)=0,R=r_0} = \frac{\theta_{r_0}}{\gamma_{r_0}}. \tag{9.3}$$

As $\Delta_{D(1)=1,D(0)=0,R=r_0}$ refers to the complier subpopulation only at the threshold, it corresponds to an even more local effect than the LATE $\Delta_{D(1)=1,D(0)=0}$ considered in chapter 6. Figure 9.3 provides a graphical illustration for the fuzzy RDD. It consists of running two sharp RDDs with $D$ and $Y$ as outcomes in order to assess $\gamma_{r_0}$ and $\theta_{r_0}$, respectively, and scaling the latter by the former to obtain $\Delta_{D(1)=1,D(0)=0,R=r_0}$.

In empirical applications of the RDD, the treatment effect is predominantly estimated by local regression around the threshold. Considering a sharp RDD, analysts or researchers frequently use a linear ordinary least squares (OLS) regression for estimating $E[Y|D = 0, R < r_0]$ and $E[Y|D = 1, R \geq r_0]$ within bandwidth $\epsilon$ around $r_0$ in order to estimate $\Delta_{r_0}$ by the difference of the regression functions at $R = r_0$. We may implement this approach by estimating equation (9.4), which characterizes the conditional mean outcome $Y$ as a function of the running variable $R$ and the treatment indicator $D = I\{R \geq r_0\}$ locally within the data window around the threshold:

$$E[Y|R \in [r_0 - \epsilon, r_0 + \epsilon)], D = I\{R \geq r_0\}]$$
$$= \alpha + \beta_R R + \underbrace{\beta_D I\{R \geq r_0\}}_{E[Y|R=r_0,D=1]-E[Y|R=r_0,D=0]} + \beta_{R,D} R \cdot I\{R \geq r_0\}. \tag{9.4}$$

It is quite common to normalize the running variable by subtracting the threshold value such that $r_0 = 0$ after the normalization. In this case, the constant term $\alpha$ corresponds to the mean outcome under nontreatment at the threshold: that is, $\alpha = E[Y|R = r_0, D = 0]$. Coefficient $\beta_D$ yields the discontinuity in the outcome functions $E[Y|R = r_0, D = 1]$ and $E[Y|R = r_0, D = 0]$, and therefore it corresponds to the causal effect $\Delta_{R=r_0}$, given that the linear model specification is correct within bandwidth $\epsilon$ around the threshold. In this context, it is worth noting that the inclusion of the interaction term $\beta_{R,D} R \cdot I\{R \geq r_0\}$ in equation (9.4) permits the linear association of $Y$ and $R$ to be different below and above the threshold (implying a nonzero coefficient $\beta_{R,D}$), as is the case in figure 9.1.

To make the model specification even more flexible, we can in principle also include higher-order terms of $R$ and interact these terms with $D$. This entails a polynomial (or series) regression, as considered in a different context in section 4.2, which can flexibly model nonlinear associations as in the example in figure 9.3, and therefore reduces the risk of misspecification relative to a linear model. However, when bandwidth $\epsilon$ is relatively small, so is the bias of the linear model from linearly approximating the nonlinear association between $Y$ and $R$. Including too many higher-order terms, on the other hand, can increase the variance due to the limited number of observations around the threshold. This may explain why local linear regression is frequently the preferred choice in empirical applications.

Such a bias-variance trade-off, as for the choice of more or fewer higher-order terms, also occurs for the selection of bandwidth $\epsilon$. A smaller bandwidth decreases estimation bias because treated and nontreated units are closer to the threshold, and thus more comparable in terms of background characteristics. Therefore, effect estimation is more robust to model misspecification of the association between $Y$ and $R$ (when assuming a linear rather than a nonlinear model); for instance, see the discussion by Gelman and Imbens (2018). On the other hand, a smaller bandwidth increases the variance due to relying on a lower number of observations. For this reason, we would like to optimally trade off the bias and the variance to select the bandwidth that minimizes the overall mean squared error (MSE). Imbens and Kalyanaraman (2012) offer such a method for optimal bandwidth selection, based on a formula of the MSE as a function of the bandwidth. The formula includes the conditional means and variances of the outcome just below and above the threshold, as well as the density of $R$ at $r_0$ as parameters.

A further approach is to use leave-one-out cross-validation, as introduced in section 4.2 in a different context. In the spirit of equation (4.18), the optimal bandwidth is determined based on checking how well a regression of $Y$ on $R$ predicts the values of $Y$ in the data when trying a range of bandwidth values. Whenever predicting the outcome of an observation $i$ based on a specific bandwidth $\epsilon$, we leave $i$ out of the sample when estimating the regression coefficients to avoid overfitting. Furthermore, and in order to mimic the fact that RDD is based on a regression at the threshold, we estimate the regression function using only observations with values of $R$ less than $R_i$ (such that $R_i - \epsilon \leq R < R_i$) if the running variable of observation $i$ is below the threshold. Accordingly, we consider only observations with values $R$ larger than $R_i$ in the estimation (such that $R_i < R \leq R_i + \epsilon$) if $R_i$ is above the threshold. Following Ludwig and Miller (2007), we then pick the bandwidth that minimizes the MSE, possibly when only considering a specific range of values in $R$ that are not too far from the threshold. As a further alternative, and related to the discussion in section 8.2 on placebo treatments for finding the optimal model specification, bandwidth $\epsilon$ (and other RDD parameters) might also be selected based on placebo zones, in which the running variable contains no treatment discontinuity. Kettlewell and Siminski (2020) suggest estimating a placebo treatment effect based on various choices concerning the bandwidth and the number of higher-order terms of $R$ as well as interaction terms to finally pick the RDD specification whose effect is closest to zero.

As a caveat against bandwidth selection, it turns out that the bandwidth which is optimal for effect estimation is generally suboptimal and too large for conducting inference, such as for computing valid confidence intervals and p-values, as discussed in Calonico, Cattaneo, and Titiunik (2014). The authors therefore suggest inference methods that are more robust to the choice of the bandwidth, such that the resulting confidence intervals are likely more accurate, meaning that they more closely match the desired nominal coverage rate of e.g., 95 percent. The findings of Calonico, Cattaneo, and Titiunik (2014) imply that when $\Delta_{r_0}$ is estimated by linear regression within a bandwidth, then quadratic regression (i.e., a regression that is one order higher) with the same bandwidth should be used for the computation of the standard error and confidence intervals.

Armstrong and Kolesár (2018) suggest an alternative approach for inference that (under certain conditions) permits computing the worst-case bias that could arise given a particular bandwidth choice. This bias can then be accounted for by an appropriate adjustment of the critical values for hypothesis testing (e.g., of whether the effect is nonzero at the threshold) or for constructing confidence intervals. This implies for the 5 percent level of statistical inference, for instance, that the asymptotic critical value is somewhat larger than the conventional value of 1.96 (as considered in section 3.4 in chapter 3). Yet another approach is randomization inference, as previously discussed in section 8.1 in chapter 8 for the synthetic control method. It consists of (1) repeatedly randomly permuting observations close to the threshold to be part of the treated or nontreated group when computing the effect and (2) verifying how extreme the RDD effect based on the actual treatment assignment is relative to the permuted effects to obtain a p-value; see Cattaneo, Frandsen, and Titiunik (2015).

It is worth noting that the identifying assumptions of the RDD are partly testable in the data. McCrary (2008) proposes a test for whether the running variable is continuous at the threshold, as a discontinuity generally points to a manipulation of $R$ and selective bunching at one side of the threshold. Considering the previously discussed example of extended eligibility to unemployment benefits in Lalive (2008), certain employees and companies might manipulate the age of entry into unemployment by agreeing on postponing layoffs such that the age requirement for extended unemployment benefits is met. In this case, we would observe a discontinuity in the density or frequency of unemployed individuals aged 49 as opposed to 50 years. As a further test, Lee (2008) suggests investigating whether observed pretreatment covariates $X$ are locally balanced at either side of the threshold, as any background characteristics that might affect the outcome must be balanced under the continuity assumptions on the potential outcomes.

However, covariates also permit weakening such continuity assumptions to only hold conditional on $X$, implying that all variables jointly affecting manipulation at the threshold and the outcome are observed. Frölich and Huber (2019) propose a kernel regression-based estimator in this context to control for differences in $X$ above and below the threshold. In contrast, Calonico, Cattaneo, Farrell, and Titiunik (2018) do not exploit covariates to tackle confounding, but rather to reduce the variance of effect estimation when linearly controlling for $X$ and provide methods for optimal bandwidth selection and robust inference for this case. In the spirit of the discussion in chapter 5, we may also apply machine learning methods to optimally control for covariates in a data-driven way, as considered by Noack, Olma, and Rothe (2021) and Arai, Otsu, and Seo (2021), or to investigate effect heterogeneity across covariates, as in Reguly (2021).

The causal effect obtained from an RDD is rather local, in the sense that it only refers to subjects close to the threshold of the running variable in the case of the sharp RDD, and to the even smaller group of compliers at the threshold in the case of the fuzzy RDD. Is it possible to extrapolate these effects to other populations farther from the threshold under certain conditions? Dong and Lewbel (2015) show that this can be achieved to some extent based on computing the derivative of the treatment effect at the threshold (in sharp or fuzzy designs), which permits evaluating the change in the effect resulting from a marginal change in the threshold.

Another approach is offered by Angrist and Rokkanen (2015), who test whether the running variable's association with the outcome vanishes on either side of the threshold conditional on covariates $X$. In the case of the sharp RDD, this implies that $X$ is sufficient to control for confounding just as under the selection-on-observables framework of chapter 4, such that we can identify treatment effects also for populations away from the threshold. In context of the fuzzy RDD, Bertanha and Imbens (2019) propose a test for the equality in mean outcomes of treated compliers and always takers, as well as of untreated compliers and never takers. This permits investigating whether the effect on compliers at the threshold may be extrapolated to all compliance types (and thus the total population) at and away from the threshold.

There are several important extensions or modifications of the conventional RDD with a continuous running variable and a single threshold. In some scenarios, there might be multiple thresholds due to varying threshold values across observations, such as in political elections under plurality rules (with vote share as running variable); for instance, see Cattaneo, Keele, Titiunik, and Vazquez-Bare (2016). In other scenarios, the running variable might be discrete (like age measured in years rather than days) rather than continuous. This generally entails challenges for effect identification and inference, as discussed in Lee and Card (2008), Dong (2015), and Kolesár and Rothe (2018). In some applications, there might even be multiple running variables (rather than just one) that determine treatment assignment, such as geographic coordinates based on longitude and latitude; for instance, see Papay, Willett, and Murnane (2011) and Keele and Titiunik (2015). In this context, Imbens and Wager (2019) propose an optimization-based inference method for computing confidence intervals that can be applied to continuous, discrete, and multiple running variables. As a further extension, Frandsen, Frölich, and Melly (2012) discuss the evaluation of quantile (rather than average) treatment effects at the threshold. Finally, Imbens and Lemieux (2008), Lee and Lemieux (2010), Melly and Lalive (2020), and Cattaneo and Titiunik (2021) provide more comprehensive surveys of the RDD literature.

Let us consider an application of the sharp RDD in R, and to this end, load the *rdrobust* package by Calonico, Cattaneo, Farrell, and Titiunik (2021). It contains the data set *rdrobust_RDsenate* with 1,390 observations on elections for the US Senate analyzed in Cattaneo, Frandsen, and Titiunik (2015), which we load using the *data* command. The effect of interest is the incumbent-party advantage: that is, the question of whether winning a Senate seat in the previous election provides an advantage for winning the same seat in the following election. We define the outcome to be the share of votes of the Democratic Party in elections for the Senate measured in percent: *Y=rdrobust_RDsenate$vote*. The running variable is the Democrats' margin of winning relative to the Republican Party in the previous elections: *R=rdrobust_RDsenate$margin*. A positive margin implies that the Democrats won the seat in the previous elections, while a negative margin means that they lost the previous elections. At the threshold around zero, previous elections in which the Democrats just won ($D = 1$) or lost ($D = 0$) by a small margin are assumably rather comparable in terms of outcome-relevant background characteristics (like regional political preferences).

To run a sharp RDD, we feed outcome $Y$ and the running variable $R$ into the *rdrobust* command, which by default considers zero as the threshold value, such that only observations with nonnegative values in the running variable are classified as treated. Furthermore, the command includes a procedure for optimal bandwidth selection in terms of minimizing the effect estimator's MSE and by default runs a local linear kernel regression (see also section 4.2) within that bandwidth using the triangular kernel. Therefore, observations within the bandwidth get a smaller weight (or less importance) the farther they are from the threshold, while observations outside the bandwidth are not considered for effect estimation at all. We save the output in an object named *results*, which we wrap by the *summary* command to inspect the results. The R code for each step is provided in the box here.

```
library(rdrobust)                   # load rdrobust library
data(rdrobust_RDsenate)             # data on elections for US Senate
Y=rdrobust_RDsenate$vote            # outcome is vote share of Democrats
R=rdrobust_RDsenate$margin          # running variable is margin of winning
results=rdrobust(y=Y, x=R)          # sharp RDD
summary(results)                    # show results
```

Running the commands gives the following output:

```
Number of Obs.            1297
BW type                   mserd
Kernel               Triangular
VCE method                  NN

Number of Obs.             595           702
Eff. Number of Obs.        360           323
Order est.   (p)             1             1
Order bias   (q)             2             2
BW est.   (h)           17.754        17.754
BW bias   (b)           28.028        28.028
rho (h/b)                0.633         0.633
Unique Obs.                595           665


===============================================================================
     Method    Coef. Std. Err.       z     P>|z|        [ 95% C.I. ]
===============================================================================
 Conventional  7.414     1.459    5.083     0.000   [4.555 , 10.273]
       Robust      -         -    4.311     0.000   [4.094 , 10.919]
===============================================================================
```

The results suggests that for margins close to the threshold, winning a seat in the Senate in previous elections increases the vote share in the following elections by 7.4 percentage points. The output also provides p-values and confidence intervals when relying on conventional or robust inference methods as suggested in Calonico, Cattaneo, and Titiunik (2014). In either case, the null hypothesis of no incumbent party effect is rejected at any conventional level of significance, as p-values are very

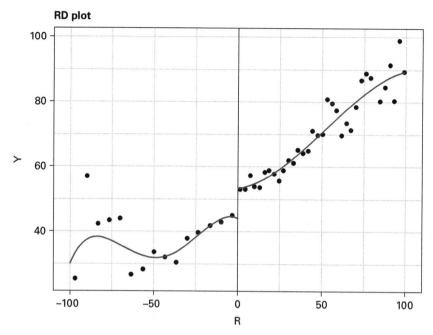

**Figure 9.4**
Plot of the discontinuity.

close to zero. The method also provides an estimate of the optimal bandwidth *BW est. (h)*, which amounts to 17.75 percentage points, and the number of nontreated and treated observations within the bandwidth (360 and 323, respectively). In the second step, we visualize the regression discontinuity by a data-driven plot of the outcome against the running variable using the *rdplot* command:

```
rdplot(y=Y, x=R)                    # plot outcome against running variable
```

Running the command yields the graph displayed in figure 9.4. The dots correspond to average outcomes within bins of values of the running variable, while the solid lines give nonlinear regression curves of the outcome as a function of the running variable above and below the threshold, respectively. The graph points to a nonnegligible discontinuity in mean outcomes at the threshold, in line with the results of our RDD estimation.

To partly check the assumptions underlying the RDD approach, in the next step we apply the McCrary (2008) test for a discontinuity in the density of the running variable at the threshold. To this end, we load the *rdd* package by Dimmery (2016) and feed the running variable *R* into the *DCdensity* command, which by default assumes the threshold to be equal to zero:

```
library(rdd)                          # load rdd library
DCdensity(runvar=R)                   # run the McCrary (2008) discontinuity test
```

Running this code yields a p-value of 0.3898 (or 38.98 percent) such that the continuity of the runnig variable around the threshold cannot be rejected at conventional levels of statistical significance. For this reason, we find no statistical evidence for a manipulation of the running variable.

Let us now consider an application of the fuzzy RDD to a data set that is part of the *RDHonest* package by Kolesár (2021), which is available on the GitHub platform. Accessing GitHub requires first loading the *devtools* package and then running *install_github("kolesarm/RDHonest")* to install the *RDHonest* package before loading it using the *library* command. In the next step, we run *data(rcp)* to load the *rcp* data, which was analyzed in Battistin, Brugiavini, Rettore, and Weber (2009) and contains information on household consumption, as well as pension eligibility and actual retirement of the household head.

In our fuzzy RDD, we consider being just above or below the age threshold for pension eligibility as an instrument for actual retirement (*D*) to estimate the effect of the latter on a measure of consumption—namely, household expenditures on nondurables (*Y*), apparently measured in euros (EUR) per year. We therefore define the outcome *Y=rcp$cn*; the running variable *R=rcp$elig_year*, which measures the age in years to/from reaching pension eligibility such that the threshold is zero; and the treatment *D=rcp$retired*. In the next step, we run the previously used *rdrobust* command of the *rdrobust* package using *Y* and *R*, but now we also specify the argument *fuzzy=D*. The latter indicates that we use a fuzzy RDD in which *D* corresponds to the actual treatment participation. We store the output in an object named *results*, which we wrap by the *summary* command:

```
library(devtools)                          # load devtools package
install_github("kolesarm/RDHonest")        # install RDHonest package
library(RDHonest)                          # load RDHonest package
data(rcp)                                  # load rcp data
Y=rcp$cn                                   # outcome is expenditures on nondurables
R=rcp$elig_year                            # running var based on eligibility to retire
D=rcp$retired                              # treatment is retirement status
results=rdrobust(y=Y, x=R, fuzzy=D)        # fuzzy RDD
summary(results)                           # show results
```

Running the code yields the following output:

```
Number of Obs.              30006
BW type                     mserd
Kernel                  Triangular
VCE method                     NN

Number of Obs.              16556        13450
Eff. Number of Obs.          1599         2078
Order est.  (p)                 1            1
Order bias  (q)                 2            2
BW est.  (h)                4.950        4.950
BW bias  (b)               15.002       15.002
rho (h/b)                   0.330        0.330
Unique Obs.                    39           49
```

| Method | Coef. | Std. Err. | z | P>\|z\| | [ 95% C.I. ] |
|---|---|---|---|---|---|
| Conventional | −5603.339 | 3072.345 | −1.824 | 0.068 | [−11625.025 , 418.346] |
| Robust | − | − | −1.837 | 0.066 | [−12222.336 , 396.082] |

We find that retirement reduces the spending on nondurable consumption goods by 5,603.34 EUR among compliers at the threshold: that is, among individuals who retire exactly when they reach the eligible pension age. Both conventional and robust inference yield p-values of roughly 7 percent, such that the reduction in consumption due to retirement is statistically significant at the 10 percent level, but not at the 5 percent level. The estimate of the optimal bandwidth around the threshold is 4.95 years, and the number of nontreated and treated observations within the bandwidth amounts to 1,599 and 2,078, respectively.

## 9.2  Sharp and Fuzzy Regression Kink Designs

A further approach to causal analysis that exploits a specific threshold in treatment assignment is the regression kink design (RKD) suggested by Card, Lee, Pei, and Weber (2015), which is technically a first derivative version of the fuzzy RDD. It can be applied to continuous (rather than binary) treatments that are a function of the running variable $R$ with a kink in that function at $r_0$, rather than a discontinuity as in the RDD. Mathematically, this implies that the first derivative of the continuous variable $D$ with regard to $R$ (rather than the level of $D$, as in the RDD) is discontinuous when crossing the threshold.

To gain some intuition, let us consider an example inspired by Landais (2015), where the amount of unemployment benefits (measured in euros, for instance) received by job seekers is treatment $D$, which is a kinked function of the previous wage, $R$. Namely, the benefit $D$ corresponds to a certain percentage (e.g., 0.8 or 80 percent) of the previous wage, $R$, up to a maximum previous wage $r_0$ (e.g., 5,000 EUR), beyond which $D$ remains constant. In this case, treatment $D$ is a piecewise linear function whose derivative with regard to $R$ (i.e., $D$'s change as a function of $R$), corresponds to the percentage (0.8) for $R < r_0$, and to zero for $R \geq r_0$. As the treatment is deterministic in the running variable in our example, this is known as *sharp RKD*.

Under specific conditions that are related to those in section 9.1, such as continuously distributed mean potential outcomes and a smooth density of the running variable $R$ (and, in particular, its first derivative) around threshold $r_0$, we can identify a causal effect at the threshold. The RKD consists of scaling the reduced form change in the first derivative of the mean outcome with regard to $R$ at the threshold by the first-stage change in the first derivative of $D$ with regard to $R$ at the threshold. This is somewhat related to an IV approach in which we scale the reduced-form effect of the instrument on the outcome by the first-stage effect of the instrument on the treatment (see section 6.1 in chapter 6). The effect that is identified by the RKD corresponds to the average derivative of the potential outcome with respect to treatment $D$ when the latter is set to its value at the threshold, denoted by $d_0$, within the local population at $R = r_0$. This is a marginal treatment effect, as previously considered for continuous treatments in sections 3.5 and 4.8, but exclusively at $R = r_0$.

Formally, the marginal treatment effect at the threshold is defined as follows:

$$\Delta_{r_0}(d_0) = \frac{\partial E[Y(d_0)|R=r_0]}{\partial d_0} = \frac{\lim\limits_{\epsilon \to 0} \frac{\partial E[Y|R\in[r_0,r_0+\epsilon)]}{\partial r_0} - \lim\limits_{\epsilon \to 0} \frac{\partial E[Y|R\in[r_0-\epsilon,r_0)]}{\partial r_0}}{\lim\limits_{\epsilon \to 0} \frac{\partial D|R\in[r_0,r_0+\epsilon)}{\partial r_0} - \lim\limits_{\epsilon \to 0} \frac{\partial D|R\in[r_0-\epsilon,r_0)}{\partial r_0}}, \tag{9.5}$$

where $\epsilon > 0$ is a neighborhood or bandwidth around the threshold that tends to zero. Figure 9.5 provides a graphical illustration for the first-stage effect of $R$ on $D$ in this sharp RKD, where the treatment (e.g., unemployment benefits) is a deterministic and (in our case) linear function of the running variable (e.g., previous earnings). The denominator in equation (9.5) corresponds to the difference in the slope of this function above and below the threshold, corresponding to 0 and 0.8, respectively, in the previously discussed example. If the association of the average outcome and $R$ is linear (at least within the bandwidth $\epsilon$), then the numerator in equation (9.5) corresponds to the difference in the slope coefficients in separate regressions of $Y$ on a constant and $R$ above and below the threshold, respectively, within $\epsilon$.

In contrast to the sharp RKD, the fuzzy RKD permits random deviations of the treatment values from the kinked function characterizing the association between

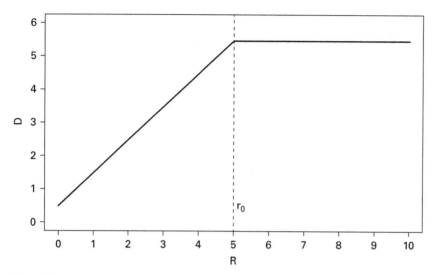

**Figure 9.5**
Sharp regression kink design.

$D$ and $R$. Therefore, $D$ is not exclusively determined by $R$, but the treatment nevertheless changes on average as a function of the running variable, which can be characterized by a regression model. As an example, consider the framework in Simonsen, Skipper, and Skipper (2016), in which the consumer price of prescription drugs in Denmark is the treatment variable $D$. The latter is a kinked function of a drug's actual costs, the running variable $R$, due to a reimbursement scheme that implies that the share of the costs borne by consumers decreases in the costs up to a specific threshold and remains constant thereafter.

Nevertheless, the actual consumer price might deviate somewhat from that kinked function due to unobserved factors, particularly nonstandard reimbursement arrangements through private (rather than public) health insurance, which motivates the application of the fuzzy RKD. Under specific continuity conditions and the monotonicity-type assumption that the kink in the association between $D$ and $R$ of any individual either goes in the same direction or is zero, we can identify a causal effect at the threshold among individuals with nonzero kinks. Requiring that the kink cannot be downward sloping for some individuals and upward sloping for others is somewhat related to ruling out the existence of defiers when applyig IV methods (see chapter 6). Furthermore, the subpopulation with nonzero kinks at the threshold is related to the notion of compliers in the IV context.

To use the fuzzy (rather than the sharp) RKD, the derivatives of the treatment in equation (9.5) (namely, $\frac{\partial D|R\in[r_0,r_0+\epsilon)}{\partial r_0}$ and $\frac{\partial D|R\in[r_0-\epsilon,r_0)}{\partial r_0}$), are to be replaced by

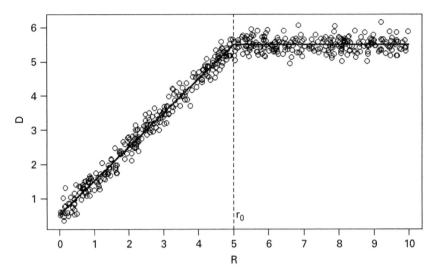

**Figure 9.6**
Fuzzy regression kink design.

the derivatives of the conditional expectations $\frac{\partial E[D|R\in[r_0,r_0+\epsilon)]}{\partial r_0}$ and $\frac{\partial E[D|R\in[r_0-\epsilon,r_0)]}{\partial r_0}$. This yields the average marginal effect of treatment $D$ at value $d_0$ among complying individuals with a nonzero kink at the threshold $r_0$:

$$\frac{\partial E[Y(d_0)|\frac{\partial D|R\in[r_0+\epsilon,r_0)}{\partial r_0} - \frac{\partial D|R\in[r_0-\epsilon,r_0)}{\partial r_0} \neq 0, R=r_0]}{\partial d_0}$$

$$= \frac{\lim_{\epsilon\to 0}\frac{\partial E[Y|R\in[r_0,r_0+\epsilon)]}{\partial r_0} - \lim_{\epsilon\to 0}\frac{\partial E[Y|R\in[r_0-\epsilon,r_0)]}{\partial r_0}}{\lim_{\epsilon\to 0}\frac{\partial E[D|R\in[r_0,r_0+\epsilon)]}{\partial r_0} - \lim_{\epsilon\to 0}\frac{\partial E[D|R\in[r_0-\epsilon,r_0)]}{\partial r_0}}. \tag{9.6}$$

Figure 9.6 provides a graphical illustration of the first-stage effect of $R$ on $D$ in the fuzzy RKD. The solid line corresponds to the conditional mean of the treatment, given the running variable, and is a piecewise linear regression function with a kink at threshold $r_0$. The dots represent actual treatment realizations of subjects in the population, which randomly deviate from the regression line.

It is worth noting that the expectation (or regression function) of a treatment may be continuous even if the treatment itself is not. For this reason, the fuzzy RKD, in contrast to the sharp RKD, may also be applied to a binary $D$, as discussed in Dong (2014). Concerning statistical inference, Calonico, Cattaneo, and Titiunik (2014) provide robust methods for computing confidence intervals and p-values for the sharp and fuzzy RKD in an analogous way as they do for the RDD, as discussed in the previous section 9.1. Furthermore, Ganong and Jäger (2018) propose a permutation method based on placebo treatments that is in the spirit of randomization inference, as discussed in sections 8.1 and 9.1.

To implement the fuzzy RKD in **R**, we apply the *rdrobust* command already considered in section 9.1 to data from Lundqvist, Dahlberg, and Mörk (2014) to analyze the effect of intergovernmental grants on local public employment. This data set, *finaldata.dta*, can be accessed via the website of the *American Economic Journal: Economic Policy* (https://www.aeaweb.org/articles?id=10.1257/pol.6.1.167). It is, however, saved in the format of another statistical software program than the one considered in this book (namely, **Stata**). To be able to read this data format, we load the *haven* library by Wickham and Miller (2021). We download the data set and save it on a hard disk, which in our case has the label *C:* (but this might differ for other computers). We then load the data into **R** and store it in an object named *data* by applying the *read_dta* command to the location (or path) of the file on our hard disk: *data=read_dta("C:/finaldata.dta")*.

We analyze the effect of intergovernmental grants (*D*) in Sweden (i.e., financial transfers from the Swedish central government to municipalities), on local public employment (*Y*): that is, the number of fulltime municipal employees per 1,000 inhabitants. The grants are partly determined by compensations aimed at supporting municipalities with diminishing population size, particularly due to out migration, which are kinked: municipalities with a decreasing population receive a positive compensation, while it is zero for growing municipalities. The population growth therefore serves as running variable *R* for the intergovernmental grants, which are, however, also determined by other components. For this reason, *D* is not fully deterministic in *R*, which motivates the application of the fuzzy RKD. To this end, we define the outcome *Y=data$pers_total*, the running variable *R=data$forcing*, and the treatment *D=data$costequalgrants* and feed them into the *rdrobust* command. Importantly, we now (and in contrast to the application at the end of section 9.1) set the argument *deriv=1* to run a fuzzy RKD rather than an RDD. We save the output in an **R** object named *results*, which we wrap by the *summary* command. The box here provides the code for each of the steps.

```
library(haven)                                    # load haven package
data=read_dta("C:/finaldata.dta")                 # load data
Y=data$pers_total                                 # define outcome (total personnel)
R=data$forcing                                    # define running variable
D=data$costequalgrants                            # define treatment (grants)
results=rdrobust(y=Y, x=R, fuzzy=D, deriv=1)      # run fuzzy RKD
summary(results)                                  # show results
```

Running the commands yields the following output:

| | | |
|---|---|---|
| Number of Obs. | 2511 | |
| BW type | mserd | |
| Kernel | Triangular | |
| VCE method | NN | |
| | | |
| Number of Obs. | 1541 | 970 |
| Eff. Number of Obs. | 476 | 425 |
| Order est. (p) | 1 | 1 |
| Order bias (q) | 2 | 2 |
| BW est. (h) | 3.501 | 3.501 |
| BW bias (b) | 6.379 | 6.379 |
| rho (h/b) | 0.549 | 0.549 |
| Unique Obs. | 1019 | 626 |

| Method | Coef. | Std. Err. | z | P>\|z\| | [ 95% C.I. ] |
|---|---|---|---|---|---|
| Conventional | 1.075 | 1.386 | 0.775 | 0.438 | [−1.642 , 3.792] |
| Robust | − | − | −0.340 | 0.734 | [−5.410 , 3.812] |

The estimate suggests that increasing intergovernmental grants by 1 unit, which corresponds to 100 krona per capita, generates on average a bit more than one additional full-time job in local public employment. This effect refers to complying municipalities whose total of intergovernmental grants is indeed kinked at the population growth–related compensation threshold as a function of these compensations. However, the effect is far from being statistically significant when relying on conventional or robust inference methods, as suggested in Calonico, Cattaneo, and Titiunik (2014). Therefore, we cannot reject the null hypothesis that the grants do not influence local public employment.

## 9.3   Bunching Designs

Somewhat related to the RDD and RKD, bunching designs exploit discontinuities or kinks in assignment variables due to a specific threshold of a running variable, as considered in Saez (2010) and Chetty, Friedman, Olsen, and Pistaferri (2011). However, while the assumptions underlying the RDD and RKD rule out self-selection around the threshold, bunching relates to exactly the opposite scenario—namely, that subjects can choose the level of the running variable and thus whether they are located above or below the threshold. Let us consider gross earnings as the running variable and a tax system in which gross earnings are not taxed up to a certain threshold while any gross earnings beyond that threshold are taxed with a positive rate (e.g., 10 percent). This will entail a kinked function of net earnings because the latter are equal to the gross earnings below the threshold, but less than the gross earnings above the threshold due to the positive tax rate.

In such a scenario, some individuals might feel that relative to the earnings just below the threshold (where no tax is imposed), working more to obtain a higher income is not worth the effort because part of the additional earnings are lost to tax. For those individuals, the tax creates a disincentive to work that is strong enough to reduce the gross earnings that they would have realized without tax to a value that is just below the threshold. In other words, those who are discouraged to work more bunch together just below the threshold, which violates the continuity of the running variable conventionally required and tested in the RDD and RKD.

Let us consider a further example involving a discontinuity or notch (rather than a kink) as discussed by Kleven and Waseem (2013). We again assume that up to a specific threshold, gross earnings are not taxed at all. Beyond the threshold, however, a tax is imposed on the total gross earnings, even those below the threshold. This implies that net earnings are discontinuously reduced at the threshold when switching from gross earnings that are not subject to taxes to slightly higher gross earnings that are taxed, thus once again inducing disincentives to work and bunching below the threshold.

The aim of bunching designs is to evaluate to which extent such bunching occurs: that is, how such a kink or notch (discontinuity) in net earnings implied by a specific tax regime affects the distribution of gross earnings. It is therefore the running variable that is the outcome of interest in bunching designs. Assessing the causal effect of the kink or notch proceeds in two steps. First, we estimate the density function (or the frequency counts) of the running variable, typically based on a polynomial function (see section 4.2 for a discussion of series or polynomial regression). However, the estimation excludes observations within a specific bandwidth around the threshold where the mentioned disincentives are likely relevant and bunching may occur. Second, we extrapolate (or predict) the estimated density function to the area within the bandwidth around the threshold. The difference in the observed and extrapolated density of gross earnings just below the threshold yields an estimate for the magnitude of bunching: that is, the excess density due to the disincentives to work. This excess density just below the threshold should be exactly matched by a reduced density over a range above the threshold, implying that the observed density is below the extrapolated density. This requirement may be exploited as a specification test for the appropriate estimation of the density function.

Figure 9.7 provides a graphical illustration of bunching based on a hypothetical example. It plots net income (on the $y$-axis) as a function of gross income (on the $x$-axis), which has a kink at a monthly gross income of 1,000 US dollars (USD) due to the imposition of an income tax on any gross income beyond the threshold $r_0 = 1,000$.

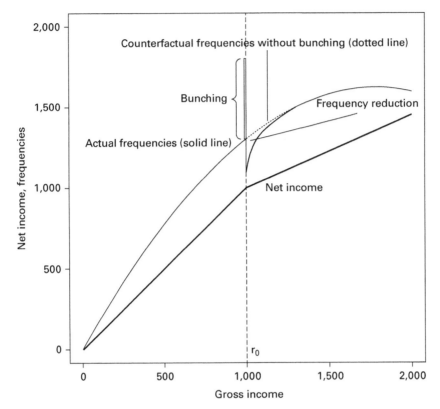

**Figure 9.7**
Bunching in the running (=outcome) variable.

The graph also plots the frequencies of workers (on the $y$-axis) with a specific gross income (on the $x$-axis), thus yielding the distribution of gross income. Due to the disincentive to work around the threshold, we observe a spike in the frequency of gross income just below the threshold: that is, bunching. Accordingly, the frequencies drop substantially over a range just above the threshold due to the lack of individuals who have been induced to bunch by the tax. The dotted line provides the counterfactual frequencies that would have been observed in the absence of the tax. The aim of bunching is to estimate this counterfactual distribution by extrapolating the actual distribution observed farther from the threshold in order to determine the amount of bunching below the threshold, which corresponds to the tax-induced reduction in frequencies above the threshold.

Yet another, conceptually different design concerns bunching in a continuous treatment as a consequence of censoring, as discussed in Caetano (2015). Censoring implies that the treatment cannot be lower or higher than a specific threshold. As an

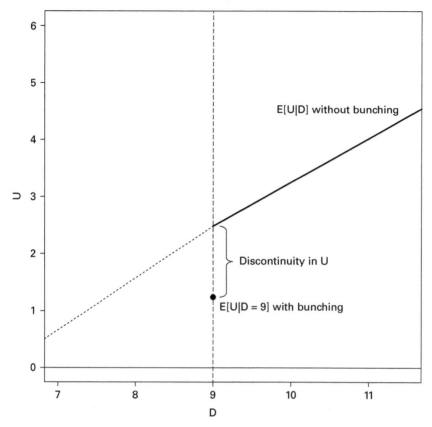

**Figure 9.8**
Bunching in the treatment variable: discontinuity in unobservables.

example, let us consider a continuous measure of education as the treatment variable and assume that by law, there are nine years of compulsory schooling. For this reason, any individuals who would have chosen a lower level of education in the absence of the compulsory schooling law bunch at exactly nine years of education. Such individuals engaging in bunching generally differ from subjects with a slightly higher level of education (where no bunching occurs) in terms of unobserved characteristics like ability and motivation, henceforth denoted by $U$. In other words, bunching induces a discontinuity in unobservables at the threshold, as individuals with rather selective (e.g., comparably low) values in $U$ are concentrated at the lower bound of the treatment (in our case, nine years of schooling).

Figure 9.8 illustrates this scenario by plotting the average of $U$, which is presumably a single unobservable to simplify the discussion, as a function of $D$. As the treatment cannot be lower than the threshold set by the compulsory schooling law,

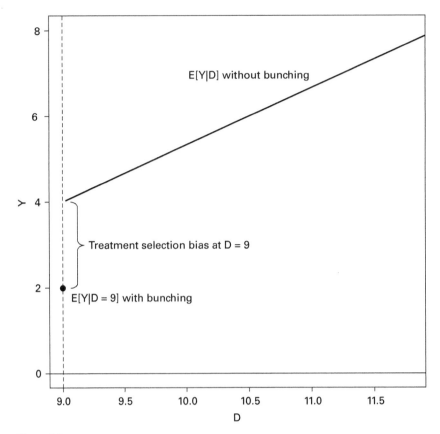

**Figure 9.9**
Bunching in the treatment variable: discontinuity in outcome.

which is indicated by the dashed line for $D < 9$, there is bunching at $D = 9$. Given that bunching individuals tend to have lower values of $U$, the mean of the latter is discontinuously lower at $D = 9$ than at slightly higher values of the treatment where bunching does not occur.

This scenario can be exploited to estimate the treatment selection bias that arises if $U$ jointly affects treatment $D$ and outcome $Y$; see the discussion in section 2.2 in chapter 2. As illustrated in figure 9.9, the approach is based on estimating the regression function $E[Y|D]$ in a data window close to but not including $D = 9$ in order to predict the regression function at $D = 9$. This is very much related to the estimation of regression functions at the threshold of the running variable in the RDD. In the next step, we subtract from the prediction (or extrapolation) of the conditional mean of $Y$ at $D = 9$ the average outcome among those previously excluded

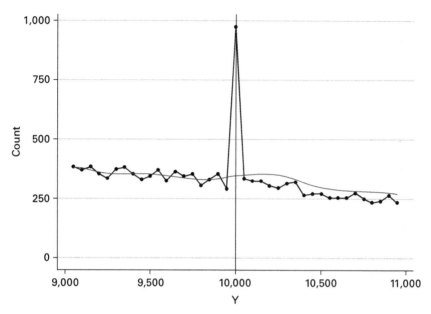

**Figure 9.10**
Bunching due to a kink in taxation.

observations whose treatment is exactly 9, which is an estimate for $E[Y|D=9]$. This difference yields an estimate of the bias due to unobserved characteristics. We may also run this approach when controlling for observed covariates $X$ to verify whether the latter permits tackling treatment selection bias. Moreover, under specific parametric assumptions, treatment selection bias can even be corrected in the estimation of treatment effects; for instance, see Caetano, Caetano, and Nielsen (2020) for a more detailed discussion of this matter.

To consider an example for the analysis of bunching related to tax brackets in R, we load the *bunching* package by Mavrokonstantis (2019), which includes an artificially created data set named *bunching_data*. The latter contains a variable named *kink_vector* consisting of 27,510 simulated earnings under a tax kink at a threshold value of 10,000, which we define as the outcome variable Y. Next, we set a seed for the reproducability of the results to follow. To estimate the magnitude of bunching, we run the *bunchit* command and set *z_vector=Y* to feed in the outcome and *zstar=10000* to define the threshold value of the tax brackets. The command also requires specifying the width of bins for outcome counts, which we set to *binwidth=50*. Further arguments are *bins_l* and *bins_r*, the number of earnings bins left and right of the threshold that are assumably affected by bunching, respectively, which we set to 20 in either case. Finally, *t0* and *t1* define the marginal tax rates below and above the threshold, respectively, which drive the financial incentives for bunching, which in our case are 0 (0 percent) and 0.2 (20 percent). We save the output in an object named *b* and call *b$B*, *b$B_sd* and *b$plot* to investigate the results. The box here provides the R code for the various steps.

```
library(bunching)                 # load bunching package
data(bunching_data)               # load bunching data
Y=bunching_data$kink_vector       # define outcome (with bunching at value 10000)
set.seed(1)                       # set seed
b=bunchit(z_vector=Y,zstar=10000,binwidth=50,bins_l=20,bins_r=20,t0=0,t1=.2)#est
b$B; b$B_sd; b$plot               # show results
```

Running the code yields an estimated excess mass of earnings right below the threshold of 630.29 observations due to bunching, with a standard error of just 82.43. The results therefore point to nonnegligible bunching due to taxing, as also illustrated in figure 9.10, which is created by calling *b$plot*. While the dots and the line connecting them correspond to the earnings counts actually observed in the outcome bins, the smooth solid line provides the estimated counterfactual counts based on observations outside the bandwidth of 20 bins around the threshold. The estimated and observed frequencies differ importantly at the threshold, suggesting bunching.

# 10
# Partial Identification and Sensitivity Analysis

## 10.1 Partial Identification

Any of the approaches to causal analysis discussed in the previous chapters are based on assumptions that permit identifying a single value for the causal effect of interest, such as an average wage effect of 30 euros (EUR), which is known as *point identification*. In contrast, *partial* or *set identification* refers to a scenario where a causal effect (or another parameter of interest) cannot be uniquely determined to take a single value but is only restricted to lie within a certain interval or set of possible values. Considering the average treatment effect (ATE) of a training program on wages, a statistical method may suggest that the effect amounts to some value between 20 and 40 EUR. The ATE is thus partially identified within the set [20, 40], while its exact value remains unknown. Partial identification arises when we impose weaker (or even no) statistical assumptions than those considered in the previous chapters, such as the selection-on-observables assumptions discussed in chapter 4. This may be preferable when stronger assumptions required for point identification do not seem plausible in a given empirical context.

We therefore face an important trade-off in data-based causal analysis: the stronger the statistical assumptions that we impose, the more concisely we can determine a causal effect, but the higher is the risk that our statistical assumptions fail to correctly describe the real-life behavior of study participants. The latter issue can entail inappropriate (i.e., biased and inconsistent) effect estimation. Partial identification makes such trade-offs explicit. The set of treatment effect values is typically large when no or few assumptions are imposed, but it becomes smaller as further assumptions are added. The set eventually collapses to point identification (i.e., a single value of the causal effect), when previously considered constraints like the selection-on-observables assumptions are imposed.

To discuss the concept of partial identification more formally, let us consider the mean of the potential outcome under treatment $Y(1)$ and recall that the latter is only observed for the treated observations, whose share in the population corresponds to the treatment probability $\text{Pr}(D=1)$. However, $Y(1)$ is not observed for the non-treated observations, whose share amounts to $\text{Pr}(D=0)=1-\text{Pr}(D=1)$. By the law of total probability, the mean potential outcome $E[Y(1)]$ in the population is a weighted average of the respective mean potential outcomes among the treated and the nontreated:

$$E[Y(1)] = E[Y(1)|D=1] \cdot \text{Pr}(D=1) + E[Y(1)|D=0] \cdot \text{Pr}(D=0), \tag{10.1}$$

$$= \underbrace{E[Y|D=1] \cdot \text{Pr}(D=1)}_{\text{observed}} + \underbrace{E[Y(1)|D=0]}_{\text{unobserved}} \cdot \underbrace{\text{Pr}(D=0)}_{\text{observed}}.$$

Analogous arguments apply to the mean of the potential outcome under nontreatment $Y(0)$, which is observed for the nontreated, but not observed for the treated observations:

$$E[Y(0)] = E[Y(0)|D=1] \cdot \text{Pr}(D=1) + E[Y(0)|D=0] \cdot \text{Pr}(D=0), \tag{10.2}$$

$$= \underbrace{E[Y(0)|D=1]}_{\text{unobserved}} \cdot \underbrace{\text{Pr}(D=1)}_{\text{observed}} + \underbrace{E[Y|D=0] \cdot \text{Pr}(D=0)}_{\text{observed}}.$$

As discussed in section 3.1 in chapter 3, we can easily identify the ATE by mean differences in observed outcomes across treated and nontreated groups if the independence assumption in expression (3.1) is satisfied as implied by the successful random assignment of the treatment. In this case, it holds that $E[Y(1)|D=0]=E[Y(1)|D=1]=E[Y|D=1]$ and $E[Y(0)|D=1]=E[Y(0)|D=0]=E[Y|D=0]$. In the subsequent discussion, however, we refrain from making such an independence assumption. We thus permit selection into treatment, which implies that $E[Y(1)|D=0] \neq E[Y(1)|D=1]$ and $E[Y(0)|D=1] \neq E[Y(0)|D=0]$; see section 2.2 in chapter 2. Without further assumptions, we cannot identify the ATE. However, by putting upper and lower bounds on the unobserved means $E[Y(1)|D=0]$ and $E[Y(0)|D=1]$ in equations (10.1) and (10.2), we can restrict the means to lie within a range or set of minimum and maximum values in order to bound the ATE within a specific set of values.

Let us, for instance, assume that there are theoretical maximum and minimum values that the outcome can take under either treatment state, which we denote by $y^{UB}$ and $y^{LB}$, respectively. Considering overnight stays in a hotel as the outcome, the latter might be bounded to lie within 0 and 100 guests (due to capacity constraints), such that $y^{LB}=0$ and $y^{UB}=100$. We may then replace the unobserved means in equations 10.1 and 10.2 by $y^{UB}$ or $y^{LB}$, respectively, to obtain upper or lower bounds on mean potential outcomes $E[Y(1)]$ and $E[Y(0)]$. We denote these bounds on the mean potential outcomes by $E[Y(1)]^{UB}$ and $E[Y(1)]^{LB}$, as well as $E[Y(0)]^{UB}$ and $E[Y(0)]^{LB}$, respectively, which are formally defined as follows:

$$E[Y(1)]^{UB} = E[Y|D=1] \cdot \text{Pr}(D=1) + y^{UB} \cdot \text{Pr}(D=0), \tag{10.3}$$

$$E[Y(1)]^{LB} = E[Y|D=1] \cdot \Pr(D=1) + y^{LB} \cdot \Pr(D=0),$$

$$E[Y(0)]^{UB} = y^{UB} \cdot \Pr(D=1) + E[Y|D=0] \cdot \Pr(D=0),$$

$$E[Y(0)]^{LB} = y^{LB} \cdot \Pr(D=1) + E[Y|D=0] \cdot \Pr(D=0).$$

As discussed in Manski (1990), taking differences between the upper bound under treatment and the lower bound under nontreatment or the lower bound under treatment and the upper bound under nontreatment yields upper and lower bounds on the ATE, denoted by $\Delta^{UB}$ and $\Delta^{LB}$, respectively:

$$\Delta^{UB} = E[Y(1)]^{UB} - E[Y(0)]^{LB}, \quad \Delta^{LB} = E[Y(1)]^{LB} - E[Y(0)]^{UB}. \tag{10.4}$$

In general, these worst-case ATE bounds provided in equations (10.4) are wide in empirical applications, such that the set of possible causal effects is quite large, and therefore not very informative. However, we may tighten the width of these bounds by imposing further assumptions. For instance, we could assume different upper and lower outcome bounds across treated and nontreated observations (rather than the same $y^{UB}$ and $y^{LB}$). As an example, it could appear plausible that under treatment (like a marketing campaign), the lower bound on an outcome (like sales) is higher than its lower bound under nontreatment. A further assumption that permits tightening the ATE bounds is monotone treatment response (MTR), as suggested in Manski (1997). For binary treatment $D$ and positive MTR, it implies that the mean potential outcome under treatment cannot be lower than under nontreatment, $E[Y(1)] \geq E[Y(0)]$, such that the ATE is assumed to be nonnegative: $E[Y(1)] - E[Y(0)] = \Delta \geq 0$. Therefore, the lower bound of the ATE in equations (10.4) is to be adjusted such that it corresponds to the maximum of $E[Y(1)]^{LB} - E[Y(0)]^{UB}$ and zero, formally: $\Delta^{LB} = \max(E[Y(1)]^{LB} - E[Y(0)]^{UB}, 0)$.

A further assumption considered by Manski and Pepper (2000) is monotone treatment selection (MTS). It postulates that subjects select themselves into treatment in a way that the mean potential outcomes of the treated and nontreated groups can be ordered. For instance, positive MTS implies that the mean potential outcomes of the treated weakly dominate those of the nontreated: that is, $E[Y(1)|D=1] \geq E[Y(1)|D=0]$ and $E[Y(0)|D=1] \geq E[Y(0)|D=0]$. In the previous example with the training program, this implies that individuals who are selected for training have weakly higher mean wages (both with and without training participation) than those not selected. This may appear reasonable if the training targets better qualified and more experienced employees with a higher wage potential. Under this assumption, the lower bound on the mean potential outcome under nontreatment in equation (10.3) simplifies to $E[Y(0)]^{LB} = E[Y|D=0]$ because $E[Y(0)|D=1]$ (which was previously bounded by $y^{LB}$) cannot be less than $E[Y|D=0]$. Likewise, the upper bound on the mean potential outcome under treatment simplifies to $E[Y(1)]^{UB} = E[Y|D=1]$

because $E[Y(1)|D=0]$ (which was previously bounded by $y^{UB}$) cannot be greater than $E[Y|D=1]$.

As discussed in Robins (1989) and Balke and Pearl (1997), the ATE bounds can also be tightened in the presence of an instrumental variable, which we henceforth denote by $Z$ in analogy to chapter 6. Let us to this end assume that the instrument and the potential outcomes satisfy a mean independence assumption, as considered by Manski (1990). For a binary instrument $Z$, this implies that $E[Y(1)|Z=1]=E[Y(1)|Z=0]$ and $E[Y(0)|Z=1]=E[Y(0)|Z=0]$. On average, the instrument must not directly affect the outcome other than through the treatment, implying a mean exclusion restriction, and not be correlated with unobserved characteristics affecting the outcome. The mean independence assumption entails the following bounds on the mean potential outcomes, which permits bounding the ATE:

$$E[Y(1)]^{UB} = \min \left( E[Y|D=1, Z=1] \cdot \Pr(D=1|Z=1) + y^{UB} \cdot \Pr(D=0|Z=1), \right.$$
$$E[Y|D=1, Z=0] \cdot \Pr(D=1|Z=0) + y^{UB} \cdot \Pr(D=0|Z=0)),$$
$$E[Y(1)]^{LB} = \max \left( E[Y|D=1, Z=1] \cdot \Pr(D=1|Z=1) + y^{LB} \cdot \Pr(D=0|Z=1), \right.$$
$$E[Y|D=1, Z=0] \cdot \Pr(D=1|Z=0) + y^{LB} \cdot \Pr(D=0|Z=0)),$$
$$E[Y(0)]^{UB} = \min \left( y^{UB} \cdot \Pr(D=1|Z=1) + E[Y|D=0, Z=1] \cdot \Pr(D=0|Z=1), \right.$$
$$y^{UB} \cdot \Pr(D=1|Z=0) + E[Y|D=0, Z=0] \cdot \Pr(D=0|Z=0)),$$
$$E[Y(0)]^{LB} = \max \left( y^{LB} \cdot \Pr(D=1|Z=1) + E[Y|D=0, Z=1] \cdot \Pr(D=0|Z=1), \right.$$
$$y^{LB} \cdot \Pr(D=1|Z=0) + E[Y|D=0, Z=0] \cdot \Pr(D=0|Z=0)). \tag{10.5}$$

The intuition of this result is that because $Z$ on average does not affect the potential outcomes, we may compute the bounds in equations (10.3) conditional on $Z$ and take the intersection of the obtained bounds across various values of the instrument to tighten them. For this reason, we take the minimum of the upper bounds and the maximum of the lower bounds of the respective mean potential outcomes across $Z=1, 0$. We also note that in contrast to the discussion in section 6.1 in chapter 6, we did not impose any assumptions on the association between the instrument and the treatment, like the existence of a first-stage effect of $Z$ on $D$ or the monotonicity of $D$ in $Z$. This implies that the local average treatment effect (LATE) on compliers is not point-identified as in section 6.1. But just like the ATE, the LATE may be bounded under specific independence assumptions concerning the instrument, as discussed in Richardson and Robins (2010).

A weaker restriction than mean independence is the assumption of a monotone instrumental variable (MIV), as considered by Manski and Pepper (2000). It implies that the instrument is monotonically (i.e., either positively or negatively) associated with the mean potential outcomes. Under a positive MIV assumption, it holds that $E[Y(1)|Z=1] \geq E[Y(1)|Z=0]$ and $E[Y(0)|Z=1] \geq E[Y(0)|Z=0]$. This generally entails wider bounds on the mean potential outcomes and the ATE than provided in equations (10.5) under stronger mean independence. On the other hand, MIV may appear more realistic in empirical applications. When considering work experience as $Z$, for instance, it appears unlikely that average potential wages do not depend on work experience, while it seems much more reasonable that additional work experience increases (or at least never decreases) wages. We may even impose several of the previously discussed assumptions jointly (e.g., both MTR and MIV), which may entail tighter bounds on the set of ATE values than either MTR or MIV alone.

A further type of assumptions in the context of instruments concerns the ordering of the potential outcomes of various compliance types introduced in figure 6.1. For instance, we might assume that the always takers $(D(1)=1, D(0)=1)$ have on average weakly higher potential outcomes under treatment than the compliers $(D(1)=1, D(0)=0)$: that is, $E[Y(1)|D(1)=1, D(0)=1] \geq E[Y(1)|D(1)=1, D(0)=0]$. This appears somewhat related to the previously discussed MTS assumption, with the difference that mean potential outcomes are now assumed to be ordered according to compliance types rather than treatment groups. Furthermore, we can impose the MTR assumption within compliance types, such that the mean potential outcome under treatment is assumed to weakly dominate that under nontreatment among always takers: $E[Y(1)|D(1)=1, D(0)=1] \geq E[Y(0)|D(1)=1, D(0)=1]$. Such dominance assumptions on the ordering of potential outcomes permit bounding the treatment effects on various compliance types, including the LATE on the compliers. This appears interesting if one is not willing to impose all of the IV assumptions in expression (6.1) required for the point identification of the LATE.

Flores and Flores-Lagunes (2013), for instance, bound the LATE when maintaining the monotonicity of $D$ in $Z$, implying $\Pr(D(1) \geq D(0)) = 1$, and the random assignment of $Z$, but they assume a violation of the exclusion restriction, such that $Z$ may directly affect $Y$. Alternatively, and depending on the application, we might invoke the random assignment of $Z$ and the exclusion restriction, but assume a violation of monotonicity, implying $\Pr(D(1) \geq D(0)) \neq 1$, as considered in Huber, Laffers, and Mellace (2017) for assessing effects among various compliance types. Finally, outcome dominance assumptions within and across types can also be used to tighten the bounds on the ATE or the average treatment effect on the treated (ATET) in the presence of a valid instrument; for instance, see Chen, Flores, and Flores-Lagunes (2018).

Partial identification approaches have also been applied to address outcome attrition and posttreatment sample selection, which imply that outcomes are observed only for a selective subpopulation; for instance, see Manski (1989). As already discussed in sections 4.11 and 6.4, sample selection creates an endogeneity/confounding problem (even under a random treatment) if both the treatment and unobserved characteristics that also affect the outcome influence the observability of the outcome. We can nevertheless derive bounds on the causal effects among specific subgroups, which are defined by how sample selection is influenced by the treatment state.

To formalize the discussion, in analogy to section 6.4, we denote by $O$ a binary selection indicator of whether outcome $Y$ is observed. Applying the potential outcome notation to the selection indicator, let $O(1)$ and $O(0)$ denote the potential selection states under treatment and nontreatment, respectively. Similar to the treatment compliance types defined in chapter 6 in terms of how the treatment reacts to some instrument, we define selection compliance types in terms of selection as a function of the treatment. Subjects satisfying $(O(1) = 1, O(0) = 1)$ are always selected because their outcome is observed independent of the treatment state. Accordingly, those satisfying $(O(1) = 0, O(0) = 0)$, $(O(1) = 1, O(0) = 0)$, and $(O(1) = 0, O(0) = 1)$ are never selected, selection compliers, and selection defiers, respectively. As an example, let us consider the effect of a training program on hourly wages, which are only observed among those in employment: that is, $O = 1$, with $O$ being a binary indicator for employment.

In this context, we might be interested in the ATE on the always selected (i.e., those employed with and without training), which is the only group whose outcomes are actually observed under both treatment and nontreatment. Related to our previous discussion in the context of instruments, we can impose specific dominance assumptions on the ordering of potential outcomes across selection compliance types, as considered by Zhang and Rubin (2003), to tighten the bounds on causal effects. One such dominance assumption states that the average potential wage of the always selected weakly dominates that of compliers under either treatment state: that is, $E[Y(1)|O(1) = 1, O(0) = 1] \geq E[Y(1)|O(1) = 1, O(0) = 0]$ and $E[Y(0)|O(1) = 1, O(0) = 1] \geq E[Y(0)|O(1) = 1, O(0) = 0]$. This might be rationalized by the fact that always selected have a higher labor market attachment than compliers, as they are employed with or without training. Another possible assumption is monotonicity of selection in the treatment (i.e., $\Pr(O(1) \geq O(0)) = 1$), as considered in Lee (2009), which rules out selection defiers.

When imposing both the dominance and monotonicity assumptions in addition to a randomized treatment assignment, the upper and lower bounds on the ATE among always selected, denoted by $\Delta^{UB}_{O(1)=1,O(0)=1}$ and $\Delta^{LB}_{O(1)=1,O(0)=1}$, correspond to the following expressions (see Zhang and Rubin (2003)):

$$\Delta^{UB}_{O(1)=1,O(0)=1} = E[Y|D = 1, O = 1, Y \geq y^*] - E[Y|D = 0, O = 1], \tag{10.6}$$

$$\Delta^{LB}_{O(1)=1,O(0)=1} = E[Y|D = 1, O = 1] - E[Y|D = 0, O = 1],$$

where outcome value $y^*$ is chosen such that the lowest outcomes in the group with $D = 1$ and $O = 1$, whose proportion matches the share of compliers in that group, are below this value. The intuition for this approach is the following: In the extreme case that all complier outcomes are concentrated at the bottom of the wage distribution under treatment among those with $D = 1$ and $O = 1$, then dropping the lower part of the wage distribution which corresponds to the complier share implies that only the treated outcomes of the always selected remain. The share of compliers conditional on $D = 1$ and $O = 1$ is identified by $\frac{\Pr(O=1|D=1)-\Pr(O=1|D=0)}{\Pr(O=1|D=1)}$, as discussed in Lee (2009), such that $y^*$ can be determined, too.

Chen and Flores (2015) extend partial identification under sample selection based on monotonicity and dominance assumptions to the instrumental variable (IV) context (e.g., noncompliance with random treatment assignment) to bound the LATE on compliers. Bounding strategies have also been applied for mediation analysis as considered in section 4.10 in chapter 4, mostly assuming a randomly assigned treatment and an endogenous mediator (such that unobservables jointly affect the mediator and the outcome), see for instance Sjölander (2009). In all those partial identification approaches, controlling for observed covariates $X$ can be useful in two dimensions. First, some assumptions (like dominance or monotonicity) might appear more plausible after making subjects comparable in observed characteristics. Second, conditioning on $X$ may further tighten the bounds. Semenova (2020), for instance, considers partial identification under sample selection and uses machine learning to control for the most important covariates (out of a potentially large number) jointly affecting sample selection $O$ and the outcome $Y$ in a data-driven way, in the spirit of the discussion in section 5.2 in chapter 5.

As a practical matter concerning the estimation of bounds and their variances, which are required for constructing confidence intervals for the estimated set of a causal effect, we need to pay attention to whether our partial identification results contain any minimum and maximum operators. Let us, for instance, consider the bounds on the mean potential outcomes in equations (10.3), which do not include such operators. For this reason, we can estimate them $\sqrt{n}$-consistently in the data with a distribution that is asymptotically normal, just like in linear regression, as discussed in section 3.3 in chapter 3. This makes the computation of confidence intervals for the estimated set straightforward. See, for instance, the discussion in Imbens and Manski (2004), who suggest the following approach for computing 95 percent confidence intervals for the estimate of the partially identified ATE, under the condition that the difference between the upper and lower bounds of the ATE is nonnegligible:

$$\left( \hat{\Delta}^{LB} - 1.645 \cdot \hat{\sigma}^{LB}, \hat{\Delta}^{UB} + 1.645 \cdot \hat{\sigma}^{UB} \right),  \tag{10.7}$$

where $\hat{\Delta}^{LB}, \hat{\Delta}^{UB}$ are the estimated lower and upper bounds on the ATE and $\hat{\sigma}^{LB}$ and $\hat{\sigma}^{UB}$ denote their respective standard errors.

As a second example, let us now consider the bounds in equations (10.5), which do contain minimum and maximum operators. It follows from results in Hirano and Porter (2012) that bounds including such nondifferentiable operators cannot be estimated without bias even when the sample size grows large. Relatedly, conventional methods for computing confidence intervals like that in expression (10.7) do generally not provide correct coverage probabilities (e.g., of 95 percent). For this reason, alternative methods of computing confidence intervals have been suggested for such partial identification problems with minimum and maximum operators, such as the half-median-unbiased confidence intervals of Chernozhukov, Lee, and Rosen (2013).

A further approach is to apply a bootstrap procedure (see section 3.4) for bias correction, as suggested by Kreider and Pepper (2007), which likely mitigates (albeit not necessarily fully eliminates) estimation bias. The procedure is based on repeatedly drawing bootstrap samples from the original data and reestimating the bound of interest in each sample. We then estimate the bias as the difference between the average of the bounds across bootstrap samples and the respective bound estimated in the original data. For instance, a bias approximation for estimating a lower bound $\Delta^{LB}$ is given by $\frac{1}{B}\sum_{b=1}^{B}\hat{\Delta}_b^{LB} - \hat{\Delta}^{LB}$, with $b$ indexing a specific bootstrap sample, $B$ denoting the number of bootstraps, $\hat{\Delta}_b^{LB}$ the estimated bound in a bootstrap sample, and $\hat{\Delta}^{LB}$ the estimated bound in the original data. We also may use such procedures of repeated sampling for directly constructing confidence intervals, as suggested by Chernozhukov, Hong, and Tamer (2007) and Romano and Shaikh (2008). More comprehensive surveys on partial identification and its subfields are provided in Tamer (2010), Molinari (2020), and Flores and Chen (2018).

To illustrate the estimation of worst-case bounds on the ATE of a randomized treatment under sample selection (or outcome attrition) in R, we load the *experiment* package by Imai, Jiang, and Li (2019). Furthermore, we reconsider the *JC* data in the *causalweight* package previously analyzed at the end of section 3.1, among others. We define random assignment to Job Corps (JC) as treatment (*D*), *treat=JC$assignment*, and weekly earnings in the fourth year after assignment as outcome (*Y*), *outcome=JC$earny4*. In contrast to our application in section 3.1, however, we are now not interested in the actual earnings outcome, which is necessarily zero for those not working in the fourth year, but in the earnings that would hypothetically be realized under employment. This information is available only for working individuals and for this reason, we define a selection indicator (*O*) for whether someone was employed for a nonzero proportion of weeks in the fourth year after assignment, *selection=JC$pworky4>0*. Accordingly, we set the outcome to "missing," which is coded as *NA* in R, whenever the selection indicator is zero: *outcome[selection==0]=NA*.

Next, we generate a data frame named *dat*, which includes the treatment, selection indicator, and outcome. We then run the *ATEbounds* command for worst-case bounds where the first argument consists of the regression formula *outcome*

*~factor(treat)*, in which the treatment must be coded as a *factor* variable, and the second argument of the data source is *data=dat*. We save the output in an object named *results* and call *results$bounds* and *results$bonf.ci* to obtain the upper and lower ATE bounds, as well as the confidence interval. The box here provides the R code for each of the steps.

```
library(experiment)                              # load experiment package
library(causalweight)                            # load causalweight package
data(JC)                                         # load JC data
treat=JC$assignment                              # random treatment (assignment to JC)
outcome=JC$earny4                                # define outcome (earnings in 4. year)
selection=JC$pworky4>0                           # sample selection: employed in 4. year
outcome[selection==0]=NA                         # recode nonselected outcomes as NA
dat=data.frame(treat,selection,outcome)          # generate data frame
results=ATEbounds(outcome~factor(treat),data=dat) # compute worst case bounds
results$bounds; results$bonf.ci                  # bounds on ATE + confidence intervals
```

Running the code yields lower and upper bounds on the ATE on weekly earnings under employment of −397.61 and 425.40 US dollars (USD), respectively, with the 95 percent convidence interval ranging from −419.62 to 453.54. Based on the worst-case bounds, the set of possible ATEs is very large and ranges from substantially negative to substantially positive values, such that we are far from rejecting the null hypothesis of a zero ATE at conventional levels of statistical signficance.

In the next step, we aim at tightening the bounds by imposing monotonicity of selection $O$ in the treatment $D$, implying that for each indiviudal, assignment to JC has either positive or zero (but not negative) effect on employment in the fourth year. To this end, we use the *leebounds* package by Semenova (2021), which is available on the GitHub platform. Accessing GitHub requires first loading the *devtools* package and then running *install_github("vsemenova/leebounds")* to install the *leebounds* package, before loading it using the *library* command. We apply the *leebounds* command to the data frame *dat* (in which the variables have the denominations *treat*, *selection*, and *outcome*, as required for the *leebounds* command) and save the output in an object named *results*. Finally, we call *results$lower_bound* and *results$upper_bound* to investigate the bounds.

```
library(devtools)                                # load devtools package
install_github("vsemenova/leebounds")            # install leebounds package
library(leebounds)                               # load leebounds package
results=leebounds(dat)                           # bounds (monotonic selection in treat)
results$lower_bound; results$upper_bound         # bounds on ATE under monotonicity
```

We now obtain lower and upper ATE bounds of $-8.33$ and $19.49$ USD, respectively, which are substantially tighter than the worst-case bounds. Yet the set includes both positive and negative values, such that the null hypothesis of a zero ATE cannot be rejected.

## 10.2  Sensitivity Analysis

The idea of sensitivity analysis follows in a sense an opposite direction from the partial identification approach outlined in the previous section 10.1. While partial identification drops point identifying assumptions altogether (or replaces them by weaker restrictions), sensitivity analysis investigates the sensitivity of the causal effect to deviations from assumptions that yield point identification. Let us, for instance, consider treatment evaluation under the selection-on-observables assumption postulated in expression (4.1) in chapter 4 based on controlling for covariates $X$: that is, $\{Y(1), Y(0)\} \perp D|X$. If there is an unobserved confounder $U$ that jointly affects treatment $D$ and outcome $Y$ even conditional on $X$, then this assumption is violated. In this case, we should also control for $U$ because the conditional independence of the potential outcomes and the treatment holds, given both $X$ and $U$:

$$\{Y(1), Y(0)\} \perp D|X, U. \tag{10.8}$$

Another way to think of this issue based on conditional treatment probabilities (i.e., propensity scores) is to acknowledge that $\Pr(D=1|X, Y(1), Y(0)) \neq \Pr(D=1|X)$. That is, the potential outcomes are associated with the treatment when controlling for $X$ alone (but not $U$), thus entailing selection bias. By expression (10.8), however, $\Pr(D=1|X, U, Y(1), Y(0)) = \Pr(D=1|X, U)$, such that the potential outcomes are not associated with the treatment when controlling for both $X$ and $U$. The problem is that we cannot control for $U$ due to its nonobservability. For this reason, sensitivity analysis makes assumptions about how strongly $U$ might be associated with $D$, $Y$, or both, in order to assess how robust an estimated causal effect is to such violations of the selection-on-observables assumption.

Several sensitivity analyses that have been suggested are based on (1) parametrically modeling the treatment propensity score $\Pr(D=1|X, U)$ or the outcome $Y$ as a function of $D$, $X$, and $U$ and (2) varying the values of $U$ over a presumably plausible range; see, for instance, Rosenbaum and Rubin (1983a), Imbens (2003), and Altonji, Elder, and Taber (2008). As an alternative, Ichino, Mealli, and Nannicini (2008) provide a nonparametric method that imposes no parametric assumptions on the treatment or outcome models, but instead restricts the distribution of $U$ to be discrete (e.g., binary, such that $U$ only takes values 1 or 0). Let us subsequently consider two approaches to sensitivity analysis that neither rely on parametric treatment/outcome models nor restrict the distribution of unobserved confounders. Furthermore, they require only a single parameter to gauge the severity of violations of the selection-on-observables assumption in expression (4.1), which appears attractive from a practical perspective.

The first approach, suggested by Rosenbaum (1995), gauges violations based on a sensitivity parameter $\Gamma \geq 1$, which characterizes the assumed worst-case degree of confounding. The latter is measured based on the odds ratios of the observed propensity score $\Pr(D=d|X)$ (with $d=1$ and $d=0$ for the conditional probabilities of treatment and nontreatment, respectively) and of the unknown propensity score $\Pr(D=d|X, Y(d))$, which may deviate from $\Pr(D=d|X)$ due to confounding by unobservables $U$. Formally, $\Gamma$ is assumed to satisfy

$$\frac{1}{\Gamma} \leq \frac{\Pr(D=d|X=x)/(1-\Pr(D=d|X=x))}{\Pr(D=d|X=x, Y(d)=y)/(1-\Pr(D=d|X=x, Y(d)=y))} \leq \Gamma \tag{10.9}$$

for any feasible covariate and outcome values $x$ and $y$. For $\Gamma = 1$, the selection-on-observables assumption holds, implying that $\Pr(D=d|X=x) = \Pr(D=d|X=x, Y(d)=y)$ and the odds $\Pr(D=d|X=x)/(1-\Pr(D=d|X=x))$ and $\Pr(D=d|X=x, Y(d)=y)/(1-\Pr(D=d|X=x, Y(d)=y))$ are the same, so their ratio is 1. By any $\Gamma > 1$, we allow a deviation from the selection-on-observables assumption. When setting $\Gamma = 3$, the treatment odds when controlling for $X$ alone are assumed to be at most 3 times higher or 2/3 lower than the true treatment odds when controlling for both $X$ and $Y(d)$.

To see how we can use expression (10.9) to assess the sensitivity of causal effects, let us first consider the mean potential outcome given $X$, which may be written as a ratio of two integrals, as discussed by Kallus, Mao, and Zhou (2019):

$$E[Y(d)|X=x] = \frac{\int y \frac{f(D=d, Y=y|X=x)}{\Pr(D=d|X=x, Y(d)=y)} dy}{\int \frac{f(D=d, Y=y|X=x)}{\Pr(D=d|X=x, Y(d)=y)} dy}. \tag{10.10}$$

Equation (10.10) consists of the observed conditional density (or probability in the case of a discrete outcome) of $D$ and $Y$ given covariates $X$, denoted by $f(D=d, Y=y|X=x)$, as well as the unobserved propensity score $\Pr(D=d|X=x, Y(d)=y)$ of expression (10.9). Therefore, we can compute upper and lower bounds on $E[Y(d)|X=x]$, denoted by $E[Y(d)|X=x]^{UB}$ and $E[Y(d)|X=x]^{LB}$, based on the maximum and minimum values of equation (10.10) when considering all values of $\Pr(D=d|X=x, Y(d)=y)$ that satisfy the constraint in expression (10.9) concerning worst-case confounding as implied by $\Gamma$.

This in turn permits computing upper an lower bounds on the conditional average treatment effect (CATE) $\Delta_x = E[Y(1) - Y(0)|X=x]$, which we denote by $\Delta_x^{UB}$ and $\Delta_x^{LB}$:

$$\Delta_x^{UB} = E[Y(1)|X=x]^{UB} - E[Y(0)|X=x]^{LB}, \quad \Delta_x^{LB} = E[Y(1)|X=x]^{LB} - E[Y(0)|X=x]^{UB}.$$

$$(10.11)$$

Averaging over these bounds also yields upper and lower bounds on the ATE, $\Delta^{UB} = E[\Delta_x^{UB}]$ and $\Delta^{LB} = E[\Delta_x^{LB}]$. Kallus, Mao, and Zhou (2019) suggest a kernel regression–based estimator for the CATE bounds in equations (10.11). However, the constraint in expression (10.9) can also be used to assess causal effect estimates coming from other estimation approaches, such as pair matching, which we use in an empirical application in R further below.

The second nonparametric approach to sensitivity analysis that we consider was suggested in Masten and Poirier (2018). It can be applied to both discrete outcomes (like the decision to buy a product) and continuous outcomes (like birth weight), even though the subsequent discussion focusses exclusively on the latter case. The method is based on characterizing the maximum violation of the selection-on-observables assumption that may occur in a population due to unobserved confounders $U$ by the maximum absolute difference in the propensity scores when controlling for $X$ alone versus both $X$ and $Y(d)$:

$$|\Pr(D=1|X=x) - \Pr(D=1|X=x, Y(d)=y)| \leq \mathcal{C}, \qquad (10.12)$$

for $d \in \{1, 0\}$ and any values $x, y$ occurring in the population. Sensitivity parameter $\mathcal{C}$ is an upper bound on the absolute difference in probabilities, and therefore necessarily between 0 and 1 (or 0 percent and 100 percent). Setting $\mathcal{C}=0$ implies the satisfaction of the selection-on-observables assumption, while larger values of $\mathcal{C}$ permit a larger degree of violation.

To see how we can use expression (10.12) for assessing the sensitivity of causal effects, we introduce some further notation to define quantile functions, similar to section 4.8. Let us denote by $F_{Y(d)|X=x}^{-1}(\tau)$ the conditional quantile of a potential outcome (the inverse of its cumulative distribution function) assessed at a particular rank $\tau \in (0, 1)$, given covariates $X=x$. Furthermore, $F_{Y|D=d,X=x}^{-1}(\tau)$ denotes the conditional quantile of the observed outcome, given treatment $D=d$ and covariates $X=x$. Masten and Poirier (2018) demonstrate that upper and lower bounds on the conditional quantiles of the potential outcomes, denoted by $F_{Y(d)|X=x}^{-1}(\tau)^{UB}$ and $F_{Y(d)|X=x}^{-1}(\tau)^{LB}$, are obtained by the following expressions:

$$F_{Y(d)|X=x}^{-1}(\tau)^{UB} = F_{Y|D=d,X=x}^{-1}(\tau'), \qquad (10.13)$$

with $\tau' = \min\left(\tau + \dfrac{\mathcal{C}}{\Pr(D=d|X=x)} \cdot \min(\tau, 1-\tau), \dfrac{\tau}{\Pr(D=d|X=x)}, 1\right),$

$$F_{Y(d)|X=x}^{-1}(\tau)^{LB} = F_{Y|D=d,X=x}^{-1}(\tau''),$$

$$\text{with } \tau'' = \max\left(\tau - \frac{\mathcal{C}}{\Pr(D=d|X=x)} \cdot \min(\tau, 1-\tau), \frac{\tau-1}{\Pr(D=d|X=x)} + 1, 0\right).$$

This permits computing upper and lower bounds on the conditional quantile treatment effect by taking the differences $F_{Y(1)|X=x}^{-1}(\tau)^{UB} - F_{Y(0)|X=x}^{-1}(\tau)^{LB}$ and $F_{Y(1)|X=x}^{-1}(\tau)^{LB} - F_{Y(0)|X=x}^{-1}(\tau)^{UB}$, respectively. Furthermore, averaging these bounds across all ranks $\tau$ from 0 to 1 yields upper and lower bounds on the CATE, $\Delta_x^{UB}$ and $\Delta_x^{LB}$, which in turn can be averaged across values of the covariates to obtain bounds on the ATE, $\Delta^{UB}$ and $\Delta^{LB}$. To see the connection between sensitivity analyses and the partial identification approaches of section 10.1 in chapter 10, we note that sufficiently large values of $\mathcal{C}$ in expression (10.12) and $\Gamma$ in expression (10.9) entail the worst-case bounds in equations (10.3) and (10.4).

Sensitivity analyses have also been suggested for the context of causal machine learning as discussed in chapter 5 (e.g., Chernozhukov et al. (2021) and Dorn, Guo, and Kallus (2021)), as well as for other evaluation strategies than the selection-on-observables framework, such as instrumental variable approaches. Let us, for instance, consider a violation of the monotonicity of treatment $D$ in instrument $Z$, such that $\Pr(D(1) \geq D(0)) = 1$ in expression (6.1) of section 6.1 does not hold and defiers (as characterized in figure 6.1) exist. We can assess the sensitivity of the LATE by making assumptions about the share of defiers and how strongly the potential outcomes differ between compliers and specific noncomplying groups (namely, always and never takers or defiers); for instance, see Huber (2014b) and Noack (2021). Also, in mediation analysis as discussed in section 4.10, sensitivity analysis can be fruitfully applied, in particular to assess the robustness of direct and indirect effects to the endogeneity of the mediator, and possibly also the treatment, if the latter is not randomized. Such approaches are among others suggested by Tchetgen Tchetgen and Shpitser (2012), Vansteelandt and VanderWeele (2012), VanderWeele and Chiba (2014), and Hong, Qin, and Yang (2018).

To implement a sensitivity analysis based on expression (10.9) in R, we load the *rbounds* package by Keele (2014). We apply the procedure in the context of direct pair matching without replacement to estimate the effect of training participation on wages among the treated. To this end, we reconsider the *lalonde* data in the *Matching* package. We use exactly the same variable definitions as in the application at the end of section 4.2 for treatment $D$ (participation in the National Supported Work (NSW) training program), outcome $Y$ (real earnings in 1978), and covariates $X$. We then set a seed for the reproducability of the steps to follow. We feed $Y$, $D$, and $X$ into the *Match* command (as previously considered in section 4.3), where we also set the argument *replace=FALSE* to run pair matching without replacement, and save the results in an object named *output*. Finally, we wrap the latter by the *hlsens* command for a Rosenbaum-type sensitivity analysis. We set the argument *Gamma=2*, which corresponds to the maximum of the sensitivity parameter $\Gamma$ in equation (10.9) that we would like to consider, as well as *GammaInc=0.25*, which defines the increment by which $\Gamma$ should be increased in our analysis.

It is important to note that the *hlsens* command does not investigate the sensitivity of the ATET (the causal effect that the *Match* command estimates by default), but rather of the median of the outcome differences between any treated and the respective matched nontreated observation. This median (rather than mean) difference in matched outcomes has also been considered by Hodges and Lehmann (1963) and motivates our application of pair matching without replacement to have unique matched pairs of treated and nontreated observations. The box here provides the R code of each step.

```
library(rbounds)                              # load rbounds package
library(Matching)                             # load Matching package
data(lalonde)                                 # load lalonde data
attach(lalonde)                               # store all variables in own objects
D=treat                                       # define treatment (training)
Y=re78                                        # define outcome
X=cbind(age,educ,nodegr,married,black,hisp,re74,re75,u74,u75) # covariates
set.seed(1)                                   # set seed
output=Match(Y=Y, Tr=D, X=X, replace=FALSE)   # pair matching (ATET),no replacement
hlsens(output, Gamma=2, GammaInc=0.25)        # sensitivity analysis
```

Running the commands gives the following output:

```
Rosenbaum Sensitivity Test for Hodges-Lehmann Point Estimate

Unconfounded estimate ....    1478.236

Gamma Lower bound Upper bound
 1.00     1478.200     1478.2
 1.25      415.740     2001.1
 1.50      -75.964     2822.8
 1.75     -608.760     3486.2
 2.00    -1071.300     3947.0
```

As discussed before, we see that specifying *Gamma=2* and *GammaInc=0.25* in the *hlsens* command computes the sensitivity of the point estimate stepwise for values of $\Gamma$ from 1 (implying no confounding due to unobservables) to 2, with an increment of 0.25. The median outcome difference under the assumption of no confounding ($\Gamma = 1$) corresponds to 1,478.2 USD. Therefore, the estimated median treatment effect on the treated is somewhat lower than the ATET of 1746.057 obtained from the *Match* package (as provided in *output$est*). For *Gamma=1.25*, the lower and upper bounds are both above zero, thus pointing to a positive median effect of training participation on participants under confounding due to unobservables. When setting $\Gamma$ to 1.5 or greater, however, the bounds include a zero effect, so the results are not robust to higher levels of confounding.

# 11

## Treatment Evaluation under Interference Effects

### 11.1   Failure of the Stable Unit Treatment Value Assumption

In all the previous chapters, we have ruled out any kind of interference or spillover effects, such that the outcome of any subject in a population must not be affected by the treatment state of any other subject. To this end, we have so far assumed the satisfaction of the stable unit treatment value assumption (SUTVA) provided in expression (2.1); see the discussion in section 2.1 in chapter 2. However, the SUTVA may appear unrealistic in many empirical problems, as discussed in Heckman, Lochner, and Taber (1998).

Let us, for instance, consider a training program as treatment, such as an information technology (IT) course. The share of individuals who receive this training in a region may have an impact on someone's employment probability even beyond their individual training status, due to an increase in the regional supply of a particular skill like IT competencies. Also, in educational interventions like the provision of free textbooks to students, spillover effects from treated to nontreated subjects may occur—namely, through sharing the books or the knowledge therein with peers who did themselves not receive the books. As a further example, the share of individuals who are vaccinated or given a medicinal drug might also influence the health status of subjects not obtaining the vaccine or drug by affecting the likelihood of disease transmission; see, for instance, Halloran and Struchiner (1991) and Miguel and Kremer (2004). In such cases that involve interference, the overall treatment effect generally differs from the average of the individual one.

In line with the discussion in section 2.1, let us express the potential outcomes of a subject $i$ as a function of the subject's own potential treatment state $D_i$ and the treatments assigned to all other subjects (but subject $i$) in the population, denoted by $\mathcal{D}_{-i}$. If the SUTVA fails such that interference occurs, the potential outcome under specific treatment assignments $D_i = d_i$ and $\mathcal{D}_{-i} = \mathbf{d}_{-i}$ is given by $Y_i(d_i, \mathbf{d}_{-i})$ rather than $Y_i(d_i)$ as under the satisfaction of the SUTVA. Denoting the potential outcomes by $Y_i(d_i, \mathbf{d}_{-i})$ bears some similarity with the causal framework of dynamic treatments in section 4.9 in chapter 4, with the first treatment being the own treatment of individual $i$, $D_i$, and the second being the treatment of other subjects, $\mathcal{D}_{-i}$. However, in the current context, the own and others' treatment need not be sequential, as in the dynamic treatment context, but may be realized at the same time. Interference also seems somewhat related to the concept of mediation analysis discussed section 4.10. In fact, we are typically interested in disentangling the direct effect of the individual treatment $D_i$ and the indirect effect of the treatment of other subjects $\mathcal{D}_{-i}$. But in contrast to mediation analysis, we do not necessarily impose a sequential association between $D_i$ and $\mathcal{D}_{-i}$.

The presence of interference generally complicates causal analysis, in particular if quite arbitrary forms of interference effects are allowed; for instance, see the discussion in Manski (2013). For this reason, evaluations aiming at separating interference from individual-specific (or direct) treatment effects typically rely on assumptions about how interference affects the outcomes of interest. One possible approach consists of imposing a partial interference assumption, which requires that interference effects are limited to occur only within but not across specific (and nonoverlapping) clusters, like geographic regions. In other words, the SUTVA may be violated on an individual level, but it must be satisfied on a cluster level such that the treatment assignments in one region must not affect the outcomes in another. Within clusters, however, the network between individuals through which interference effects materialize need not even be known, so interference can be quite general.

A second approach does not rely on the partial interference assumption, which is violated under interference across clusters, such as if training programs in one region affect labor market outcomes in other regions due to cross-regional competition for jobs. It instead uses exposure mappings, which impose assumptions on the mechanisms through which interference affects outcomes based on information about the network of peers with which some individual interacts. One might assume, for instance, that interference effects can stem only from the treatment of an individual's family members, while the treatment status of any other subject in the population is irrelevant. An alternative exposure mapping would imply that both family members and friends are relevant for interference effects. The subsequent two sections of this chapter will discuss causal analysis based on partial interference and exposure mappings, respectively. For the sake of simplicity, we will focus on a binary treatment throughout.

## 11.2   Partial Interference

Under partial interference as considered by Sobel (2006), Hong and Raudenbush (2006), and Hudgens and Halloran (2008), the SUTVA holds across clusters such that the potential outcomes of subjects depend only on the treatments of other subjects in the same cluster, but not in other clusters. Using subscript $c$ to denote a specific cluster, this implies that the potential outcome of subject $i$ in cluster $c$ can be indexed by $Y_{c,i}(d_{c,i}, \mathbf{d}_{c,-i})$. Let us now assume that we can randomize both the treatment intensity on the cluster level (i.e., the share of individuals getting a treatment in a specific cluster) and the treatment on the individual level (i.e., the decision who actually obtains the treatment within some cluster with a particular treatment share). Such a double randomization of the treatment on the cluster and the individual level permits identifying the direct, interference, and total (comprising both direct and interference) effects of a treatment.

To formalize the discussion, let us denote by $P_c = E[D|\text{cluster} = c]$ the treated proportion within cluster $c$. Under successful randomization of the proportion across clusters, the following independence assumption holds:

$$Y_{c,i}(d_{c,i}, \mathbf{d}_{c,-i}) \perp P_c \text{ for } d_{c,i} \in \{0, 1\}, \text{ any assignment } \mathbf{d}_{c,-i}, \text{ and any unit } i \text{ in any cluster } c.$$

$$(11.1)$$

Expression (11.1) states that the treated proportion in any cluster is not associated with the potential outcomes of the individuals in the respective cluster. Furthermore, under randomization of treated units within a cluster, the potential outcomes of any individual in the respective cluster and the individual treatment assignments are independent:

$$Y_{c,i}(d_{c,i}, \mathbf{d}_{c,-i}) \perp (D_{c,i}, \mathcal{D}_{c,-i})|P_c \text{ for } d_{c,i} \in \{0, 1\}, \text{ any assignment } \mathbf{d}_{c,-i},$$

$$\text{and any unit } i \text{ in any cluster } c.$$

$$(11.2)$$

Random assignment also implies that in large enough samples, observations receiving the treatment based on the cluster-varying proportions are in terms of background characteristics representative of the total population. For this reason, we subsequently consider causal effects under interference as a function of the treatment proportion $P_c = E[D|\text{cluster} = c]$ instead of any possible treatment assignment distribution $\mathcal{D}_{c,-i}$. From a practical perspective, the latter, more general interference criterion would be much harder to analyze due to the high dimensionality of $\mathcal{D}_{c,-i}$ in terms of possible treatment assignments among other units in the cluster.

We first consider the average direct effect of the individual treatment assignment $D_i$ for a given treatment share $P_c = p$, which we denote by $\theta(p)$:

$$\theta(p) = E[Y_{c,i}(1, p) - Y_{c,i}(0, p)].$$

$$(11.3)$$

This effect corresponds to the difference in mean potential outcomes when varying the individual treatment assignment, but keeping the treatment share (and thus the interference effect) fixed. We note that $\theta(0) = E[Y_{c,i}(1, 0) - Y_{c,i}(0, 0)]$ corresponds to the direct effect in the absence of any interference, which is the case if the SUTVA holds on the individual level, as in section 2.1. The average interference (or indirect) effect, denoted by $\delta(d,p,p')$, is the impact of shifting the treatment proportion in a cluster from a value $p$ (e.g., 80 percent) to $p'$ (e.g., 20 percent), while keeping the individual treatment assignment fixed at $D_{c,i} = d$:

$$\delta(d,p,p') = E[Y_{c,i}(d,p) - Y_{c,i}(d,p')]. \tag{11.4}$$

Finally, and similar to the discussion of causal mechanisms in section 4.10, the total average treatment effect, denoted by $\Delta(p,p')$, corresponds to the sum of the direct and the interference effects:

$$\begin{aligned}
\Delta(p,p') &= E[Y_{c,i}(1,p) - Y_{c,i}(0,p')] \\
&= \underbrace{E[Y_{c,i}(1,p) - Y_{c,i}(0,p)]}_{\theta(p)} + \underbrace{E[Y_{c,i}(0,p) - Y_{c,i}(0,p')]}_{\delta(0,p,p')} \\
&= \underbrace{E[Y_{c,i}(1,p') - Y_{c,i}(0,p')]}_{\theta(p')} + \underbrace{E[Y_{c,i}(1,p) - Y_{c,i}(1,p')]}_{\delta(1,p,p')}.
\end{aligned} \tag{11.5}$$

$\Delta(p,p')$ is the total treatment effect among individuals switching the treatment status from 0 to 1 when increasing the treatment proportion from $p'$ to $p$, and we bear in mind that these individuals are representative of the total population due to double randomization. Subtracting and adding $Y_{c,i}(0,p)$ after the second and $Y_{c,i}(1,p')$ after the third equality, respectively, permits disentangling $\Delta(p,p')$ into specific combinations of direct and interference effects.

The total treatment effect $\Delta(p,p')$ is to be distinguished from yet another causal parameter, which Hudgens and Halloran (2008) refer to as an overall treatment effect—namely, the average aggregate effect of assigning treatment proportions $p$ versus $p'$ to a population, denoted as $\tilde{\Delta}(p,p')$. To define this parameter, let $D_{c,i}(p)$ denote the potential treatment under treatment proportion $p$, similar in spirit to the potential treatment states defined as a function of an instrumental variable (IV), as discussed in chapter 6. The overall treatment effect is then given by

$$\tilde{\Delta}(p,p') = E[Y_{c,i}(D_{c,i}(p),p) - Y_{c,i}(D_{c,i}(p'),p')]. \tag{11.6}$$

By relying on potential individual treatment states, $\tilde{\Delta}(p,p')$ (in contrast to $\Delta(p,p')$) acknowledges that some individual treatments might not be changed when shifting the treatment proportion from $p'$ to $p$. That is, there may be always takers or never takers of the treatment (in the notation of chapter 6). The direct treatment effect of these groups is not part of the aggregate effect $\tilde{\Delta}(p,p')$ because their treatment remains unchanged, as the treatment share generally does not shift from 0 to 1 (or 100 percent).

Under double randomization, the direct, interference, total, and overall treatment effects are identified based on the following conditional mean differences:

$$\theta(p) = E[Y_{c,i}|D_{c,i}=1, P_c=p] - E[Y_{c,i}|D_{c,i}=0, P_c=p], \tag{11.7}$$

$$\delta(d,p,p') = E[Y_{c,i}|D_{c,i}=d, P_c=p] - E[Y_{c,i}|D_{c,i}=d, P_c=p'],$$

$$\Delta(p,p') = E[Y_{c,i}|D_{c,i}=1, P_c=p] - E[Y_{c,i}|D_{c,i}=0, P_c=p'],$$

$$\tilde{\Delta}(p,p') = E[Y_{c,i}|P_c=p] - E[Y_{c,i}|P_c=p'].$$

Similar to the discussion on mediation analysis in section 4.10, there may be interaction effects between the direct and the interference effects. For instance, $\delta(1,p,p')$ and $\delta(0,p,p')$ might generally differ, such that the importance of interference might differ across treatment states. In the context of the educational treatment discussed previously, for instance, the spillover effects coming from peer students' textbooks could be large among those not possessing books themselves ($D_{c,i}=0$), but zero among those possessing books themselves ($D_{c,i}=1$). Likewise, the direct effect $\theta(p)$ can depend on the proportion of treated $p$. For instance, participation in an IT training could have a larger effect on an individual's earnings if only comparably few other subjects also receive this training, such that the supply of IT skills is relatively restricted in a given region.

Concerning statistical inference, Hudgens and Halloran (2008) suggest variance estimators for the estimates of the causal effects in equations (11.7). They rely on the condition that interference effects depend only on the treatment proportion rather than who exactly receives the treatment in a cluster, which is known as the *stratified interference assumption*. In contrast, Tchetgen and VanderWeele (2012) and Liu and Hudgens (2014) provide alternative methods for assessing statistical significance and computing confidence intervals that do not rely on this assumption.

To gain some intuition about the identifiability of the different causal effects, let us consider the impact of a training program for job seekers on employment, which is presumably attended by 50 percent of job seekers in some regions, but by none (0 percent) in others. This scenario, which is inspired by Crépon et al. (2013), permits identifying the interference effect of a treatment proportion of 0.5 (50 percent are treated) versus 0 (no one is treated) on employment under individual nontreatment ($D_{c,i}=0$):

$$\delta(0,0.5,0) = E[Y_{c,i}|D_{c,i}=0, P_c=0.5] - E[Y_{c,i}|D_{c,i}=0, P_c=0]. \tag{11.8}$$

Furthermore, the direct effect of the training on employment under a treatment proportion of 0.5 is identified by

$$\theta(0.5) = E[Y_{c,i}|D_{c,i}=1, P_c=0.5] - E[Y_{c,i}|D_{c,i}=0, P_c=0.5]. \tag{11.9}$$

Here, $\theta(0.5)$ and $\delta(0,0.5,0)$ add up to the total effect of being trained and exposed to an increase in the treatment proportion from 0 to 0.5, given by

$$\Delta(0.5,0) = E[Y_{c,i}|D_{c,i}=1, P_c=0.5] - E[Y_{c,i}|D_{c,i}=0, P_c=0]. \tag{11.10}$$

In contrast to the previously discussed causal effects, however, our scenario does not permit assessing the interference effect under individual treatment, $\delta(1, 0.5, 0)$, because the lower treatment proportion of zero implies that there are no treated individuals under $P_c = 0$. For the same reason, the direct effect $\theta(0)$ (i.e., the individual treatment effect in the absence of any interference) is not identified either. Finally, the overall treatment effect in the population, which in this case corresponds to the aggregate employment effect of increasing the treatment proportion from 0 percent to 50 percent, is given by

$$\tilde{\Delta}(0.5, 0) = E[Y_{c,i}|P_c = 0.5] - E[Y_{c,i}|P_c = 0]. \tag{11.11}$$

As discussed in chapter 6, and depending on the empirical problem, not all subjects comply with their random treatment assignment. Therefore, actual treatment participation may be selective: that is, associated with unobserved background characteristics that also affect the outcome. In this case, we may apply an IV approach under specific conditions related to those discussed in section 6.1 in chapter 6. Random individual treatment assignment serves as an instrument for treatment participation $D_{c,i}$ under specific IV assumptions, particularly interference-adjusted versions of the IV exclusion restriction and monotonicity of treatment in the assignment, as discussed in Kang and Imbens (2016) and Imai, Jiang, and Malani (2021). In the spirit of expressions (11.7), we can then compute the direct, indirect, or total effects of the random assignment on treatment $D_{c,i}$ (i.e., the first-stage effects) and outcome $Y_{c,i}$ (i.e., the intention-to-treat (ITT) effects). Very much in the spirit of the discussion in section 6.1, dividing (or scaling) the respective ITT by the respective first-stage effect permits assessing the causal parameter of interest among the treatment compliers.

A further alternative to double randomization consists of imposing a selection-on-observables assumption, meaning that expressions (11.1) and (11.2) hold conditional only on cluster-specific covariates, denoted by $X_c$, which may contain both individual- and cluster-level characteristics. Let us reduce the complexity of interference associated with treatment assignment $\mathcal{D}_{c,-i}$ by assuming that only the treatment proportion drives the interference effects independent of which individuals are treated, as considered in Ferracci, Jolivet, and van den Berg (2014). This permits defining interference effects as a function of $P_c$, just as in equation (11.4). The selection-on-observables assumption implies that we can identify the conditional direct, interference, total, and overall treatment effects given $X_c$ when adding $X_c$ as control variables on the right side of equations (11.7); for instance, see the discussion in VanderWeele (2010).

The average direct effect conditional on $X_c$, for instance, is given by

$$\theta_{X_c}(p) = E[Y_{c,i}(1, p) - Y_{c,i}(0, p)|X_c] \tag{11.12}$$
$$= E[Y_{c,i}|D_{c,i} = 1, P_c = p, X_c] - E[Y_{c,i}|D_{c,i} = 0, P_c = p, X_c].$$

It follows that averaging over the covariate values of $X_c$ in the population identifies the average direct effect: $\theta(p) = E[\theta_{X_c}(p)]$. Similar to the discussion in chapter 4, we may apply regression, matching, IPW as considered by Tchetgen Tchetgen and VanderWeele (2012), or doubly robust (DR) approaches as suggested in Liu et al. (2019) for estimating the causal effects of interest.

An alternative evaluation design consists of randomly assigning treatment availability across clusters while determining individual treatment assignment based on an explicit, nonrandom eligibility criterion. This approach satisfies the notion of partial population experiments, as discussed in Moffitt (2001). As an example, let us consider a welfare program providing cash transfers for poor households as a treatment, and it is randomly assigned across geographic clusters like regions or municipalities. In treated regions where the program is available, only poor households below a specific and explicitly known poverty level are eligible for the transfers, while richer household do not have access to the treatment. In nontreated regions, the cash transfers are not available for any household, even those satisfying the poverty criterion.

As discussed in Angelucci and De Giorgi (2009), such a design permits evaluating the interference effect on ineligible subjects by comparing the outcomes of ineligibles across treated and nontreated regions. Even though they are not targeted by the cash transfers, the wealth outcomes of ineligible households in treated regions could be positively affected by the increase in consumption and spending of eligible households receiving cash transfers. Furthermore, the evaluation design also allows us to assess the total effect on eligibles by comparing the outcomes of eligibles across treated and nontreated regions. This total effect may contain the direct wealth effect of the cash transfer, as well as interference effects due to increased consumption of other eligibles. Obviously, such an evaluation approach requires the eligibility criterion to be strictly followed, such that the treatment is deterministic in the eligibility criterion, and to be observed, such that eligibles and ineligibles can be unambiguously identified in the data.

For a more formal discussion, let $P_c$ now denote a binary variable indicating whether the treatment is available in a region ($P_c = 1$) or not ($P_c = 0$). Successful randomization of $P_c$ across clusters implies that expression (11.1) is satisfied. Furthermore, we assume that individual treatment eligibility is known (e.g., as a function of an observed poverty index) and represent it as a binary indicator denoted by $\mathcal{E}$, which takes the value of 1 for eligibles and 0 for ineligibles. Furthermore, let us define the interference and total treatment effect conditional on the eligibility status as

$$\delta_{\mathcal{E}=e}(d) = E[Y_{c,i}(d,1) - Y_{c,i}(d,0)|\mathcal{E}_{c,i} = e], \tag{11.13}$$

$$\Delta_{\mathcal{E}=e} = E[Y_{c,i}(1,1) - Y_{c,i}(0,0)|\mathcal{E}_{c,i} = e],$$

respectively, where $e$ and the treatment state $d$ are either 1 or 0.

Under the evaluation design considered, we can assess the interference effect among ineligibles under nontreatment, $\delta_{\mathcal{E}=0}(0)$, based on the following mean difference:

$$\delta_{\mathcal{E}=0}(0) = E[Y_{c,i}|\mathcal{E}_{c,i}=0, P_c=1] - E[Y_{c,i}|\mathcal{E}_{c,i}=0, P_c=0]. \tag{11.14}$$

Furthermore, the total effect among eligibles, denoted by $\Delta_{\mathcal{E}=1}$, corresponds to

$$\Delta_{\mathcal{E}=1} = E[Y_{c,i}|\mathcal{E}_{c,i}=1, P_c=1] - E[Y_{c,i}|\mathcal{E}_{c,i}=1, P_c=0]. \tag{11.15}$$

Rather than assuming a randomly assigned treatment on the individual or cluster level, we may also consider alternative methods for assessing interference effects, like the difference-in-differences (DiD) approaches discussed in chapter 7. This requires that outcomes are observed in both pretreatment and posttreatment periods. To apply the DiD approach, we maintain the assumption that eligible and ineligible subjects are identifiable in the data, but we replace the independence assumption in expression (11.1) by a specific common trend assumption. The latter requires that conditional on the eligibility status, the mean potential outcome in the absence of any treatment at the cluster or individual level would change by the same magnitude from the pretreatment to the posttreatment period across treated and nontreated clusters.

To formalize this common trend assumption, we reintroduce the time index $T$ from chapter 7, which is equal to 0 in the pretreatment period, when the treatment was not available in any cluster, and 1 in the posttreatment period, when the treatment was available in treated clusters. To distinguish the outcomes and potential outcomes in terms of pretreatment and posttreatment periods, we add the subscript $t \in \{0, 1\}$ to the parameters such that they become $Y_{c,i,t}$ and $Y_{c,i,t}(d, P_c)$. This permits stating the common trend assumption as follows (for $e \in \{0, 1\}$):

$$E[Y_{c,i,1}(0,0) - Y_{c,i,0}(0,0)|\mathcal{E}_{c,i}=e, P_c=1] = E[Y_{c,i,1}(0,0) - Y_{c,i,0}(0,0)|\mathcal{E}_{c,i}=e, P_c=0]. \tag{11.16}$$

Under the satisfaction of the common trend assumption in equation (11.16) and the identifiability of eligibility types, the total effect on eligibles in treated clusters, denoted by $\Delta_{\mathcal{E}=1,P_c=1} = E[Y_{c,i}(1, 1) - Y_{c,i}(0, 0)|\mathcal{E}_{c,i}=1, P_c=1]$, is identified. The same applies to the interference effect under nontreatment among ineligibles in treated clusters, $\delta_{\mathcal{E}=0,P_c=1}(0) = E[Y_{c,i}(0, 1) - Y_{c,i}(0, 0)|\mathcal{E}_{c,i}=0, P_c=1]$. The causal parameters correspond to the following DiD expressions:

$$\Delta_{\mathcal{E}=1,P_c=1} = E[Y_{c,i,1}|\mathcal{E}_{c,i}=1, P_c=1] - E[Y_{c,i,0}|\mathcal{E}_{c,i}=1, P_c=1]$$
$$- \{E[Y_{c,i,1}|\mathcal{E}_{c,i}=1, P_c=0] - E[Y_{c,i,0}|\mathcal{E}_{c,i}=1, P_c=0]\},$$
$$\delta_{\mathcal{E}=0,P_c=1}(0) = E[Y_{c,i,1}|\mathcal{E}_{c,i}=0, P_c=1] - E[Y_{c,i,0}|\mathcal{E}_{c,i}=0, P_c=1]$$
$$- \{E[Y_{c,i,1}|\mathcal{E}_{c,i}=0, P_c=0] - E[Y_{c,i,0}|\mathcal{E}_{c,i}=0, P_c=0]\}. \tag{11.17}$$

Huber and Steinmayr (2021) apply this DiD approach to assess the effect of extended eligibility for unemployment benefits, as introduced in selected (rather than randomly chosen) regions of Austria for job seekers satisfying certain eligibility criteria in terms of age and employment history. The results suggest that the extension has an interference effect by decreasing the job-search duration of ineligible individuals in treated regions, due to reduced job competition from eligible job seekers.

Another evaluation strategy consists of applying a regression discontinuity design (RDD), as discussed in section 9.1 of chapter 9, on the cluster level, as considered in Angelucci and Maro (2016). This requires a cluster-related running variable with a threshold value, according to which the availability of a treatment in a cluster $P_c$ is assigned. As an example, let us consider a regional development fund for financing local businesses, which is provided depending on whether the regional gross domestic product (GDP) per capita is larger than a specific threshold. Regions with a GDP per capita just above and just below the threshold might arguably be comparable in terms of their background characteristics, such that a continuity assumption with regard to the potential outcomes, as discussed in section 9.1, holds for the assignment of $P_c$ around the threshold. In this case, we may evaluate the causal effects among eligibles and ineligibles based on a comparison of regions just above or below the threshold: that is, by considering equations (11.14) and (11.15) in a window around the threshold of GDP per capita. Further potential strategies include IV methods (as discussed in chapter 6), which require a valid cluster-level instrument for the assignment of $P_c$, or a selection-on-observables approach by assuming that expression (11.1) holds, conditional on covariates $X_c$; see Angelucci and Maro (2016). In the latter case, equations (11.15) and (11.14) given $X_c$ yield the respective causal parameters conditional on the covariates, because $P_c$ is conditionally independent of the potential outcomes.

Alternatively, Forastiere, Mealli, and VanderWeele (2016) assume that only a subset of potential outcomes (e.g., those under individual treatment) are independent of the eligibility status conditional on covariates and some causal effects are homogeneous across ineligible or eligible groups. In contrast to the previous methods, their approach does not rely on the identifiability of the eligibility status. We can also apply this method in scenarios where a subpopulation that is not directly observed reacts to an increase in regional treatment availability (i.e., a switch from $P_c = 0$ to $P_c = 1$) by changing the individual treatment status from zero to 1, while the treatment status of other groups remains unaffected. This is very much related to the definition of compliers, always takers, and never takers in the discussion on IVs in chapter 6. However, in the current context, $P_c$ is not a valid instrument for individual treatment take-up because it may also affect the outcomes via interference effects. For this reason, the identification of interference effects on specific compliance or eligibility types requires additional assumptions like effect homogeneity conditions.

Let us consider an empirical example for the evaluation of direct and interference effects in R under double randomization of the treatment on the regional and individual levels. To this end, we use the *interference* package by Zonszein, Samii, and Aronow (2021), which is available on the GitHub platform. This requires first loading the *devtools* package and then running *install_github("szonszein/interference")* to install the *interference* package before loading it using the *library* command. We analyze data from a randomized experiment evaluating India's National Health Insurance Scheme by Imai, Jiang, and Malani (2021), which is available in a file named india.csv on Kosuke Imai's online data repository at the Harvard Dataverse (https://doi.org/10.7910/DVN/N7D9LS, as of December 2021). We download india.csv and save it on a hard disk, which in our case has the denomination C: (but this might be different for other computers). As the data set is coded in a format called *comma-separated values (csv)*, we apply the *read.csv* command to the location (or path) of the file on the hard disk, *read.csv("C:/india.csv")*, to load it into R and save it in an object named *data*. We then run the *na.omit* command to drop observations with missing values, which reduces the number of observations from 11,089 to 10,030.

Next, we define a cluster variable named *group* to correspond to the village identifier *data$village_id* in our data set, as we consider villages as clusters. Furthermore, we generate a variable for the treatment intensity on the village level $P_c$, denominated by *group_tr*. The latter corresponds to *data$mech*, which is equal to 1 for randomly chosen villages in which 80 percent of individuals were given access to health insurance and zero for villages with a treatment proportion of just 40 percent. Furthermore, *indiv_tr=data$treat* is the individual treatment, randomly assigned access to the health insurance program, while *obs_outcome=data$EXPhosp_1* is the outcome—namely, a household's annual hospital expenditure in Indian rupees. Finally, we combine the previously defined variables in a data frame named *dat*, which we wrap using the *estimates_hierarchical* command in order to estimate the direct and interference effects of the health insurance program. The box here provides the R code for each of the steps.

```
library(devtools)                            # load devtools package
install_github("szonszein/interference")     # install interference package
library(interference)                        # load interference package
data=read.csv("C:/india.csv")                # load data
data=na.omit(data)                           # drop observations with missings
group=data$village_id                        # cluster id
group_tr=data$mech                           # indicator high treatment proportion
indiv_tr=data$treat                          # individual treatment (insurance)
obs_outcome=data$EXPhosp_1                    # outcome (hospital expenditure)
dat=data.frame(group,group_tr,indiv_tr,obs_outcome) # generate data frame
estimates_hierarchical(dat)                  # run estimation
```

Running the code provides the following results:

```
$direct_psi_hat        -137.9754        $direct_phi_hat        1668.476
$indirect_hat           125.4711        $total_hat             -12.50431
$overall_hat            -657.91
$var_direct_psi_hat 311757              $var_direct_phi_hat 515698.7
$var_indirect_hat       552775.3        $var_total_hat         499843
$var_overall_hat        496786.5
```

Here, *direct_psi_hat* corresponds to the estimate of $\theta(p)$: that is, the average direct effect for the higher treatment proportion $p = 0.8$ (or 80 percent); *direct_phi_hat* is the estimate of $\theta(p')$, the direct effect under the lower treatment proportion $p' = 0.4$ (or 40 percent); *indirect_hat* provides the estimated $\delta(0, p, p')$, the interference effect under no individual treatment; and *total_hat* and *overall_hat* are the estimates of the total effect, $\Delta(p, p')$, and the overall effect, $\tilde{\Delta}(p, p')$, respectively. The output also provides the estimated variances of the effects under the previously mentioned stratified interference assumption of Hudgens and Halloran (2008) (imposing that interference effects are fully determined by the treatment proportion); for instance, *var_direct_psi_hat*. For this reason, taking the square root of any variance yields the respective standard error.

Among our effect estimates, only that of $\theta(p')$ is statistically significantly different from zero at the 5 percent level, as its t-statistic $(1668.476/\sqrt{515698.7} = 2.32)$ exceeds the critical value of 1.96. Imai, Jiang, and Malani (2021) argue that the fact that the estimated direct effect is positive under the lower treatment proportion $p'$, but statistically insignificant (and even negative, namely $-137.9754$) under the higher treatment proportion $p$, might point to a congestion effect: the more individuals in a village are assigned to the health insurance program, the smaller is the direct effect on health spending, possibly due to increased competition for a limited amount of health services available at hospitals. However, we keep in mind that in our empirical application, we assess the ITT effects of access to the health insurance program, which in the case of noncompliance differs from actual treatment take-up: that is, enrollment in the insurance program. We note that the *experiment* package contains a command called *CADErand*, which permits estimating the direct and interference effects of actual treatment participation among treatment compliers based on an IV approach suggested by Imai, Jiang, and Malani (2021).

## 11.3   Interference Based on Exposure Mappings

In contrast to the partial interference assumption invoked in the previous section, exposure mappings do not confine interference to take place only within clusters, but instead impose restrictions on the relevant interference network. This requires information about the contacts that a specific subject interacts with to determine the strength of interference, while the partial interference approach of the previous section is agnostic about networks within clusters. Let us, for instance, consider the effects of an information or marketing campaign on opinions about a product or political issue, and assume that we can observe the parts of a subject's network that are relevant for the interference effects, such as family, friends, and other social contacts. Such knowledge about the social network appears important, as the campaign may not only directly affect someone's opinion, but also the opinions of her or his contacts in the network, which in turn may also exert an interference effect on a subject's own opinion through social interactions.

Coppock, Guess, and Ternovski (2016), for instance, investigate the impact of a mobilization social media campaign on signing an online environmental petition and find evidence for interference effects through the treatment of contacts in the virtual social network. Such interference networks, however, can be quite different in terms of size and structure for various subjects in a population of interest. Therefore, it is generally cumbersome, if not impossible, to assess the interaction effects for all possible forms of networks that might exist. For this reason, exposure mappings restrict the complexity of how interference effects materialize through a social network. One assumption reducing complexity, for instance, is that the interference effect depends only on the number of contacts who receive the treatment, but not their type (e.g., family or friends). An even more restrictive assumption is that all that matters for interference is whether at least one contact is treated, while the exact number of treated contacts is irrelevant.

Exposure mappings ultimately permit defining multiple (but not an excessive number of) kinds of interference. This allows us to assess a treatment's direct and interference effects if we appropriately control for differences in the probabilities of specific exposure mappings across subjects, particularly due to different network structures. Network features like the number of contacts generally affect both the exposure mappings and the outcome and are thus confounders, even if the treatment is randomized on the individual level. For instance, subjects with larger networks are more likely to have at least one treated contact (which might be the exposure mapping of interest) and may also systematically differ in terms of their outcomes (like opinions) from subjects with smaller networks. For this reason, adjusting for the joint probability of (1) a subject's own treatment state and (2) the exposure mapping related to the treatment assignment in a subject's network is key to a sound causal analysis with network interference.

To discuss the idea of exposure mappings more formally, let us denote by $\mathcal{G}_i$ the exposure mapping for subject $i$. $\mathcal{G}_i$ defines interference effects as a function of $i$'s interference network, denoted by $\mathcal{N}_i$ and the treatment assignment among the remaining subjects in the population, denoted by $\mathcal{D}_{-i}$. That is, the mapping is defined as

$$\mathcal{G}_i = \mathcal{F}(\mathcal{N}_i, \mathcal{D}_{-i}), \tag{11.18}$$

with $\mathcal{F}$ being a known function. Reconsidering the previous examples for exposure mappings, for instance, $\mathcal{F}$ could correspond to the number of subjects who are at the same time treated according to the assignment variable $\mathcal{D}_{-i}$ and part of individual $i$'s network according to $\mathcal{N}_i$. Alternatively, $\mathcal{F}$ could be a binary indicator for whether at least one subject who is treated according to $\mathcal{D}_{-i}$ is also part of individual $i$'s network according to $\mathcal{N}_i$. Depending on the assumed complexity of interference, $\mathcal{G}_i$ may take more or fewer different values $g$, somewhat related to the multiple treatment framework of section 3.5 in chapter 3. In the simplest case, the exposure mapping is binary, with $g = 1$ if any social contact is treated and $g = 0$ if no social contact is treated.

Under a correctly assumed exposure mapping in equation (11.18), the potential outcome $Y_i(d_i, \mathbf{d}_{-i})$ simplifies to $Y_i(d, g)$. This permits defining average direct, interference, and total effects, denoted by $\theta(g)$, $\delta(d, g, g')$, and $\Delta(g, g')$, respectively, as functions of a subject's own treatment and the exposure mapping:

$$\theta(g) = E[Y_i(1, g) - Y_i(0, g)], \tag{11.19}$$

$$\delta(d, g, g') = E[Y_i(d, g) - Y_i(d, g')],$$

$$\Delta(g, g') = E[Y_i(1, g) - Y_i(0, g')],$$

where $g$ and $g'$ are two distinct mappings (e.g., 1 and 0).

Let us assume that the treatment is randomly assigned among individuals in the population after the social networks have been formed. This is implied by the following independence assumption between the potential outcomes and the treatment assignment of subject $i$ and all remaining subjects in the population:

$$Y_i(d_i, \mathbf{d}_{-i}) \perp (D_i, \mathcal{D}_{-i}) \text{ for } d_i \in \{0, 1\} \text{ and any assignment } \mathbf{d}_{-i}. \tag{11.20}$$

Despite random treatment assignment, however, exposure mappings are not random. Subjects with a larger social network $\mathcal{N}_i$, for instance, are more likely to have at least one treated contact than subjects with smaller networks. This selection issue can be tackled by controlling for the joint probability (or propensity score) of the own potential treatment state $D_i$ and exposure mapping $\mathcal{G}_i$ as a function of network structure $\mathcal{N}_i$, which we henceforth denote by $p_i(d, g) = \Pr(D_i = d, \mathcal{G}_i = g | \mathcal{N}_i)$.

Under specific research designs and exposure mappings, we can compute this propensity score via dynamic programming (see Ugander, Karrer, Backstrom, and Kleinberg (2013)) or estimate it in the data by randomization inference, as discussed in section 8.1 in chapter 8. The latter approach is based on randomly reassigning placebo treatments $D_i, \mathcal{D}_{-i}$ many times in the data with the same share of treated subjects as observed in the original data, and then computing the share of exposure $g$ for a specific network structure $\mathcal{N}_i$ across the generated treatment assignments. For instance, we might reassign the treatment in the data 1,000 times with a treatment probability of 50 percent and verify the frequency of having a placebo treatment of both individual $i$ (such that $d = 1$) and at least one subject in the network of individual $i$ (such that $g = 1$). This permits computing the probability $p_i(1, 1) = \Pr(D_i = 1, \mathcal{G}_i = 1 | \mathcal{N}_i)$.

Similar to the IPW approach outlined in the selection-on-observables framework of section 4.5, reweighting by the inverse of the propensity score permits identifying the effects in equations (11.19). The following IPW expressions yield the direct and interference effects of interest:

$$\theta(g) = E\left[\frac{Y_i \cdot D_i \cdot I\{\mathcal{G}_i = g\}}{p_i(1, g)} - \frac{Y_i \cdot (1 - D_i) \cdot I\{\mathcal{G}_i = g\}}{p_i(0, g)}\right], \tag{11.21}$$

$$\delta(d, g, g') = E\left[\frac{Y_i \cdot I\{D_i = d\} \cdot I\{\mathcal{G}_i = g\}}{p_i(d, g)} - \frac{Y_i \cdot I\{D_i = d\} \cdot I\{\mathcal{G}_i = g'\}}{p_i(d, g')}\right].$$

Aronow and Samii (2017) provide a detailed discussion on identification, effect estimation, and a conservative method of estimating the variance of the effect estimates in this context.

Causal analysis based on exposure mappings can also be applied in a selection-on-observables framework. The latter implies that the assignment of the individual treatment $D_i$ and exposure mapping $\mathcal{G}_i$ is as good as randomly assigned when controlling for observed covariates $X$ in addition to network structure $\mathcal{N}_i$. These covariates may contain both individual- and network-specific characteristics and are to be included as control variables when computing the propensity score $p_i(d, g)$. Alternatively to IPW, we could apply matching- or regression-based approaches adjusting for the propensity scores when assessing direct and interference effects; for instance, see Forastiere, Airoldi, and Mealli (2021). Furthermore, van der Laan (2014) provides a doubly robust (DR) estimation approach in this context based on targeted maximum likelihood estimation (TMLE), as discussed in section 4.6. Qu, Xiong, Liu, and Imbens (2021) suggest a DR estimation approach that combines the frameworks of exposure mappings and partial interference. A more detailed survey on causal analysis under interference is given by Aronow, Eckles, Samii, and Zonszein (2020), who also provide an empirical example in R for IPW-based estimation (see equations (11.21)), using the *interference* package.

# 12
## Conclusion

We have come a long way by discovering a diverse toolkit of methods for causal analysis, which can easily be implemented in the statistical software R. We started with an introduction to causality and the experimental evaluation of a randomized treatment. We then moved to identification and flexible estimation under selection on observables, instrumental variables (IVs), difference-in-differences (DiD), Changes-in-Changes (CiC), synthetic controls, regression discontinuities and kinks, and bunching designs. We devoted particular attention to approaches that combine causal analysis with machine learning to provide data-driven procedures for tackling confounding related to observed covariates, investigating effect heterogeneities across subgroups, and learning optimal treatment policies. In a world with ever-increasing data availability, such causal machine learning (CML) methods aimed at optimally exploiting large amounts of information will likely be on the rise in the years to come. Finally, we also discussed the evaluation of causal effects with partial identification and interference effects.

However, our journey does not necessarily stop here, because the fields of impact evaluation and CML are ever evolving. Methodological innovations and refinements are happening at a breathtaking pace, also fueled by the increased data availability in the age of digitization. One promising area that has been gaining momentum is causal discovery. In contrast to assessing the causal effect of a predefined treatment variable (like a marketing intervention or training sessions) as is the focus in this book, causal discovery aims at learning the causal relations between two, several, or even many variables in a data-driven way.

Learning which variables affect which other variables from statistical associations alone rather than from a presupposed causal structure, in which a treatment affects the outcome but not vice versa, is a challenging task. However, a growing number of studies demonstrates under which assumptions and circumstances this is at least theoretically feasible; for instance, see Kalisch and Bühlmann (2014); Peters, Janzing, and Schölkopf (2017); Glymour, Zhang, and Spirtes (2019); and Breunig and Burauel

(2021). Should causal discovery succeed in correctly revealing causal associations on a larger scale in practically relevant cases, then it would arguably be the first artificial intelligence method that comes closer to actually doing something intelligent: finding complex causal associations, which are a fundamental part of human reasoning. Therefore, the best of causal analysis might be yet to come, so let's stay tuned for exciting developments in the future.

# References

Abadie, A. (2003). Semiparametric instrumental variable estimation of treatment response models. *Journal of Econometrics 113*, 231–263.

Abadie, A. (2005). Semiparametric difference-in-differences estimators. *Review of Economic Studies 72*, 1–19.

Abadie, A. (2021). Using synthetic controls: Feasibility, data requirements, and methodological aspects. *Journal of Economic Literature 59*, 391–425.

Abadie, A., A. Diamond, and J. Hainmueller (2010). Synthetic control methods for comparative case studies: Estimating the effect of California's tobacco control program. *Journal of the American Statistical Association 105*, 493–505.

Abadie, A., A. Diamond, and J. Hainmueller (2011). Synth: An R package for synthetic control methods in comparative case studies. *Journal of Statistical Software 42*(13), 1–17.

Abadie, A., A. Diamond, and J. Hainmueller (2015). Comparative politics and the synthetic control method. *American Journal of Political Science 59*, 495–510.

Abadie, A., and J. Gardeazabal (2003). The economic costs of conflict: A case study of the Basque country. *American Economic Review 93*, 113–132.

Abadie, A., and G. W. Imbens (2006). Large sample properties of matching estimators for average treatment effects. *Econometrica 74*, 235–267.

Abadie, A., and G. W. Imbens (2008). On the failure of the bootstrap for matching estimators. *Econometrica 76*, 1537–1557.

Abadie, A., and G. W. Imbens (2011). Bias-corrected matching estimators for average treatment effects. *Journal of Business & Economic Statistics 29*, 1–11.

Abadie, A., and G. W. Imbens (2016). Matching on the estimated propensity score. *Econometrica 84*, 781–807.

Abadie, A., and J. L'Hour (2018). A penalized synthetic control estimator for disaggregated data. Working paper, Massachusetts Institute of Technology (MIT), Cambridge, MA.

Abraham, S., and L. Sun (2018). Estimating dynamic treatment effects in event studies with heterogeneous treatment effects. Working paper, Massachusetts Institute of Technology, Cambridge, MA.

Agrawal, S., and N. Goyal (2013). Further optimal regret bounds for Thompson sampling. In C. M. Carvalho and P. Ravikumar (Eds.), *Proceedings of the Sixteenth International*

*Conference on Artificial Intelligence and Statistics*, Volume 31 of *Proceedings of Machine Learning Research*, Scottsdale, AZ, pp. 99–107. Proceedings of Machine Learning Research.

Albert, J. M., and S. Nelson (2011). Generalized causal mediation analysis. *Biometrics 67*, 1028–1038.

Aliprantis, D. (2012). Redshirting, compulsory schooling laws, and educational attainment. *Journal of Educational and Behavioral Statistics 37*, 316–338.

Altonji, J. G., T. E. Elder, and C. R. Taber (2008). Using selection on observed variables to assess bias from unobservables when evaluating Swan-Ganz catheterization. *American Economic Review 98*, 345–50.

An, W., and X. Wang (2016). LARF: Instrumental variable estimation of causal effects through local average response functions. *Journal of Statistical Software 71*, 1–13.

Anderson, T. W., and H. Rubin (1949). Estimation of the parameters of a single equation in a complete system of stochastic equations. *Annals of Mathematical Statistics 20*, 46–63.

Andresen, M. E., and M. Huber (2021). Instrument-based estimation with binarised treatments: Issues and tests for the exclusion restriction. *Econometrics Journal 24*, 536–558.

Angelucci, M., and G. De Giorgi (2009). Indirect effects of an aid program: How do cash transfers affect ineligibles' consumption? *American Economic Review 99*, 486–508.

Angelucci, M., and V. D. Maro (2016). Programme evaluation and spillover effects. *Journal of Development Effectiveness 8*, 22–43.

Angrist, J. (1990). Lifetime earnings and the Vietnam era draft lottery: Evidence from Social Security administrative records. *American Economic Review 80*, 313–336.

Angrist, J., and I. Fernández-Val (2010). Extrapolate-ing: External validity and overidentification in the late framework. NBER Working Paper 16566.

Angrist, J., and G. W. Imbens (1995). Two-stage least squares estimation of average causal effects in models with variable treatment intensity. *Journal of American Statistical Association 90*, 431–442.

Angrist, J., G. Imbens, and D. Rubin (1996). Identification of causal effects using instrumental variables. *Journal of American Statistical Association 91*, 444–472 (with discussion).

Angrist, J., and A. Krueger (1991). Does compulsory school attendance affect schooling and earnings? *Quarterly Journal of Economics 106*, 979–1014.

Angrist, J. D. (2004). Treatment effect heterogeneity in theory and practice. *Economic Journal 114*, C52–C83.

Angrist, J. D., and J.-S. Pischke (2008). *Mostly Harmless Econometrics: An Empiricist's Companion*. Princeton University Press.

Angrist, J. D., and M. Rokkanen (2015). Wanna get away? Regression discontinuity estimation of exam school effects away from the cutoff. *Journal of the American Statistical Association 110*, 1331–1344.

Arai, Y., T. Otsu, and M. H. Seo (2021). Regression discontinuity design with potentially many covariates. *arXiv preprint 2109.08351*.

Arkhangelsky, D. (2021). synthdid: Synthetic difference-in-difference estimation. *R package*.

Arkhangelsky, D., S. Athey, D. A. Hirshberg, G. W. Imbens, and S. Wager (2019). Synthetic difference in differences. Working paper, Stanford University, Stanford, CA.

Armstrong, T. B., and M. Kolesár (2018). Optimal inference in a class of regression models. *Econometrica 86*(2), 655–683.

Aronow, P. M., and A. Carnegie (2013). Beyond late: Estimation of the average treatment effect with an instrumental variable. *Political Analysis 21*, 492–506.

Aronow, P. M., D. Eckles, C. Samii, and S. Zonszein (2020). Spillover effects in experimental data. *arXiv preprint 2001.05444*.

Aronow, P. M., and C. Samii (2017). Estimating average causal effects under general interference, with application to a social network experiment. *Annals of Applied Statistics 11*, 1912–1947.

Ashenfelter, O. (1978). Estimating the effect of training programs on earnings. *Review of Economics and Statistics 6*, 47–57.

Athey, S., and G. Imbens (2006). Identification and inference in nonlinear difference-in-differences models. *Econometrica 74*, 431–497.

Athey, S., and G. Imbens (2016). Recursive partitioning for heterogeneous causal effects. *Proceedings of the National Academy of Sciences 113*, 7353–7360.

Athey, S., and G. W. Imbens (2019). Machine learning methods that economists should know about. *Annual Review of Economics 11*, 685–725.

Athey, S., G. W. Imbens, and S. Wager (2018). Approximate residual balancing: Debiased inference of average treatment effects in high dimensions. *Journal of the Royal Statistical Society Series B 80*, 597–623.

Athey, S., J. Tibshirani, and S. Wager (2019). Generalized random forests. *Annals of Statistics 47*, 1148–1178.

Athey, S., and S. Wager (2021). Policy learning with observational data. *Econometrica 89*, 133–161.

Avin, C., I. Shpitser, and J. Pearl (2005). Identifiability of path-specific effects. In *IJCAI-05, Proceedings of the Nineteenth International Joint Conference on Artificial Intelligence*, Edinburgh, UK, pp. 357–363.

Balke, A., and J. Pearl (1997). Bounds on treatment effects from studies with imperfect compliance. *Journal of the American Statistical Association 92*, 1171–1176.

Barua, R., and K. Lang (2009). School entry, educational attainment, and quarter of birth: A cautionary tale of late. NBER Working Paper 15236.

Battistin, E., A. Brugiavini, E. Rettore, and G. Weber (2009). The retirement consumption puzzle: Evidence from a regression discontinuity approach. *American Economic Review 99*, 2209–2226.

Bazzi, S., and M. A. Clemens (2013). Blunt instruments: Avoiding common pitfalls in identifying the causes of economic growth. *American Economic Journal: Macroeconomics 5*, 152–186.

Behaghel, L., B. Crépon, and M. Gurgand (2013). Robustness of the encouragement design in a two-treatment randomized control trial. IZA Discussion Paper No 7447.

Belloni, A., V. Chernozhukov, I. Fernández-Val, and C. Hansen (2017). Program evaluation and causal inference with high-dimensional data. *Econometrica 85*, 233–298.

Belloni, A., V. Chernozhukov, and C. Hansen (2014). Inference on treatment effects after selection among high-dimensional controls. *Review of Economic Studies 81*, 608–650.

Ben-Michael, E., A. Feller, and J. Rothstein (2021a). The augmented synthetic control method. *Journal of the American Statistical Association* 116, 1789–1803.

Ben-Michael, E., A. Feller, and J. Rothstein (2021b). Synthetic controls with staggered adoption. *arXiv preprint 1912.03290.*

Bertanha, M., and G. W. Imbens (2019). External validity in fuzzy regression discontinuity designs. *Journal of Business & Economic Statistics* 38, 593–612.

Bertrand, M., E. Duflo, and S. Mullainathan (2004). How much should we trust differences-in-differences estimates? *Quarterly Journal of Economics* 119, 249–275.

Bhattacharya, D., and P. Dupas (2012). Inferring welfare maximizing treatment assignment under budget constraints. *Journal of Econometrics 167*, 168–196.

Bia, M., M. Huber, and L. Lafférs (2021). Double machine learning for sample selection models. *arXiv preprint 2012.00745.*

Blackwell, M. (2015). Identification and estimation of joint treatment effects with instrumental variables. Working paper, Department of Government, Harvard University, Cambridge, MA.

Blackwell, M., and A. Strezhnev (2020). Telescope matching for reducing model dependence in the estimation of the effects of time-varying treatments: An application to negative advertising. Working paper, Harvard University, Cambridge, MA.

Bloom, H. S. (1984). Accounting for no-shows in experimental evaluation designs. *Evaluation Review 8*, 225–246.

Bodory, H., and M. Huber (2018). The causalweight package for causal inference in R. SES Working Paper 493, University of Fribourg, Fribourg, Switzerland.

Borusyak, K., and X. Jaravel (2018). Revisiting event study designs. Working paper, Harvard University, Cambridge, MA.

Boser, B. E., I. M. Guyon, and V. N. Vapnik (1992). A training algorithm for optimal margin classifiers. In *Proceedings of the Fifth Annual Workshop on Computational Learning Theory*, COLT '92, New York, pp. 144–152. Association for Computing Machinery.

Bound, J., D. A. Jaeger, and R. M. Baker (1995). Problems with instrumental variables estimation when the correlation between the instruments and the endogeneous explanatory variable is weak. *Journal of the American Statistical Association 90*, 443–450.

Breiman, L. (1996). Bagging predictors. *Machine Learning* 24(2), 123–140.

Breiman, L. (2001). Random forests. *Machine Learning 45*, 5–32.

Breiman, L., J. Friedman, R. Olshen, and C. Stone (1984). *Classification and Regression Trees.* Belmont, CA: Wadsworth.

Breunig, C., and P. Burauel (2021). Testability of reverse causality without exogeneous variation. *arXiv preprint 2107.05936.*

Brinch, C. N., M. Mogstad, and M. Wiswall (2017). Beyond late with a discrete instrument. *Journal of Political Economy 125*, 985–1039.

Brooks, J. M., and R. L. Ohsfeldt (2013). Squeezing the balloon: Propensity scores and unmeasured covariate balance. *Health Services Research 48*, 1487–1507.

Buckles, K. S., and D. M. Hungerman (2013). Season of birth and later outcomes: Old questions, new answers. *Review of Economics and Statistics 95*, 711–724.

Busso, M., J. DiNardo, and J. McCrary (2014). New evidence on the finite sample properties of propensity score matching and reweighting estimators. *Review of Economics and Statistics 96*, 885–897.

Caetano, C. (2015). A test of exogeneity without instrumental variables in models with bunching. *Econometrica 83*, 1581–1600.

Caetano, C., G. Caetano, and E. Nielsen (2020). Correcting for endogeneity in models with bunching. Finance and Economics Discussion Series 2020-080, Washington, DC: Board of Governors of the Federal Reserve System.

Caetano, C., B. Callaway, S. Payne, and H. S. Rodrigues (2022). Difference in differences with time-varying covariates. *arXiv preprint 2202.02903*.

Callaway, B. (2019). qte: Quantile treatment effects. *R package.*

Callaway, B., A. Goodman-Bacon, and P. H. Sant'Anna (2021). Difference-in-differences with a continuous treatment. *arXiv preprint 2107.02637*.

Callaway, B., and P. H. Sant'Anna (2020). did: Difference in differences. *R package.*

Callaway, B., and P. H. Sant'Anna (2021). Difference-in-differences with multiple time periods. *Journal of Econometrics 225*, 200–230.

Calonico, S., M. D. Cattaneo, M. H. Farrell, and R. Titiunik (2018). Regression discontinuity designs using covariates. *Review of Economics and Statistics 101*, 442–451.

Calonico, S., M. D. Cattaneo, M. H. Farrell, and R. Titiunik (2021). rdrobust: Robust data-driven statistical inference in regression-discontinuity designs. *R package.*

Calonico, S., M. D. Cattaneo, and R. Titiunik (2014). Robust nonparametric confidence intervals for regression-discontinuity designs. *Econometrica 82*, 2295–2326.

Cameron, A. C., J. B. Gelbach, and D. L. Miller (2008). Bootstrap-based improvements for inference with clustered errors. *Review of Economics and Statistics 90*, 414–427.

Canty, A., and B. D. Ripley (2021). boot: Bootstrap r (s-plus) functions. *R package.*

Card, D. (1995). Using geographic variation in college proximity to estimate the return to schooling. In L. Christofides, E. Grant, and R. Swidinsky (Eds.), *Aspects of Labor Market Behaviour: Essays in Honour of John Vanderkamp*, pp. 201–222. Toronto: University of Toronto Press.

Card, D., and A. B. Krueger (1994). Minimum wages and employment: A case study of the fast-food industry in new jersey and pennsylvania. *American Economic Review 84*, 772–793.

Card, D., D. S. Lee, Z. Pei, and A. Weber (2015). Inference on causal effects in a generalized regression kink design. *Econometrica 83*, 2453–2483.

Card, D., and T. Lemieux (2001). Going to college to avoid the draft: The unintended legacy of the Vietnam War. *American Economic Review 91*, 97–102.

Caria, S., M. Kasy, S. Quinn, S. Shami, and A. Teytelboym (2020). An adaptive targeted field experiment: Job search assistance for refugees in Jordan. CESifo Working Paper 8535, Munich.

Carneiro, P., J. J. Heckman, and E. J. Vytlacil (2011). Estimating marginal returns to education. *American Economic Review 101*, 2754–2781.

Cattaneo, M. D. (2010). Efficient semiparametric estimation of multi-valued treatment effects under ignorability. *Journal of Econometrics 155*, 138–154.

Cattaneo, M. D., B. R. Frandsen, and R. Titiunik (2015). Randomization inference in the regression discontinuity design: An application to party advantages in the U.S. Senate. *Journal of Causal Inference 3*, 1–24.

Cattaneo, M. D., M. Jansson, and W. K. Newey (2018). Alternative asymptotics and the partially linear model with many regressors. *Econometric Theory 34*, 277–301.

Cattaneo, M. D., L. Keele, R. Titiunik, and G. Vazquez-Bare (2016). Interpreting regression discontinuity designs with multiple cutoffs. *Journal of Politics 78*, 1229–1248.

Cattaneo, M. D., and R. Titiunik (2021). Regression discontinuity designs. *arXiv preprint 2108.09400.*

Chabé-Ferret, S. (2017). Should we combine difference in differences with conditioning on pre-treatment outcomes. Working paper, Toulouse School of Economics, Toulouse, France.

Chan, T. J. (2014). Jmisc: Julian miscellaneous function. *R package.*

Chang, N.-C. (2020, 02). Double/debiased machine learning for difference-in-differences models. *Econometrics Journal 23*, 177–191.

Chen, X., and C. A. Flores (2015). Bounds on treatment effects in the presence of sample selection and noncompliance: The wage effects of Job Corps. *Journal of Business & Economic Statistics 33*, 523–540.

Chen, X., C. A. Flores, and A. Flores-Lagunes (2018). Going beyond late: Bounding average treatment effects of Job Corps training. *Journal of Human Resources 53*, 1050–1099.

Chen, X., H. Hong, and A. Tarozzi (2008). Semiparametric efficiency in GMM models with auxiliary data. *Annals of Statistics 36*, 808–843.

Chernozhukov, V., D. Chetverikov, M. Demirer, E. Duflo, C. Hansen, W. Newey, and J. Robins (2018). Double/debiased machine learning for treatment and structural parameters. *Econometrics Journal 21*, C1–C68.

Chernozhukov, V., C. Cinelli, W. Newey, A. Sharma, and V. Syrgkanis (2021). Omitted variable bias in machine learned causal models. *arXiv preprint 2112.13398.*

Chernozhukov, V., I. Fernández-Val, and B. Melly (2013). Inference on counterfactual distributions. *Econometrica 81*, 2205–2268.

Chernozhukov, V., and C. Hansen (2005). An IV model of quantile treatment effects. *Econometrica 73*, 245–261.

Chernozhukov, V., C. Hansen, M. Spindler, and V. Syrgkanis (2022). *Applied Causal Inference Powered by ML and AI* [Manuscript submitted for publication].

Chernozhukov, V., H. Hong, and E. Tamer (2007). Estimation and confidence regions for parameter sets in econometric models. *Econometrica 75*, 1243–1284.

Chernozhukov, V., S. Lee, and A. M. Rosen (2013). Intersection bounds: Estimation and inference. *Econometrica 81*, 667–737.

Chernozhukov, V., K. Wüthrich, and Y. Zhu (2021). An exact and robust conformal inference method for counterfactual and synthetic controls. *Jounral of the American Statistical Association 116*, 1849–1864.

Chetty, R., J. N. Friedman, T. Olsen, and L. Pistaferri (2011). Adjustment costs, firm responses, and micro vs. macro labor supply elasticities: Evidence from Danish tax records. *Quarterly Journal of Economics 126*, 749–804.

Chipman, H. A., E. I. George, and R. E. McCulloch (2010). BART: Bayesian additive regression trees. *Annals of Applied Statistics 4*, 266–298.

Conley, T., and C. Taber (2011). Inference with "difference in differences" with a small number of policy changes. *Review of Economics and Statistics 93*, 113–125.

Coppock, A., A. Guess, and J. Ternovski (2016). When treatments are tweets: A network mobilization experiment over Twitter. *Political Behavior 38*, 105–128.

Cortes, C., and V. Vapnik (1995). Support-vector networks. *Machine Learning 20*, 273–297.

Cox, D. (1958). *Planning of Experiments*. Wiley.

Crépon, B., E. Duflo, M. Gurgand, R. Rathelot, and P. Zamora (2013). Do labor market policies have displacement effects? Evidence from a clustered randomized experiment. *Quarterly Journal of Economics 128*, 531–580.

Crump, R., J. Hotz, G. Imbens, and O. Mitnik (2009). Dealing with limited overlap in estimation of average treatment effects. *Biometrika 96*, 187–199.

Cunningham, S. (2021). *Causal Inference: The Mixtape*. New Haven, CT: Yale University Press.

Daw, J. R., and L. A. Hatfield (2018). Matching in difference-in-differences: Between a rock and a hard place. *Health Services Research 53*, 4111–4117.

Deaton, A. S. (2010). Instruments, randomization, and learning about development. *Journal of Economic Literature 48*, 424–455.

de Chaisemartin, C. (2017). Tolerating defiance? Local average treatment effects without monotonicity. *Quantitative Economics 8*, 367–396.

de Chaisemartin, C., and X. D'Haultfeuille (2018). Fuzzy differences-in-differences. *Review of Economic Studies 85*, 999–1028.

de Chaisemartin, C., and X. D'Haultfeuille (2020). Two-way fixed effects estimators with heterogeneous treatment effects. *American Economic Review 110*, 2964–2996.

de Chaisemartin, C., and X. D'Haultfoeuille (2022). Two-way fixed effects and differences-in-differences with heterogeneous treatment effects: A survey. *arXiv preprint 2112.04565*.

Dehejia, R. H., and S. Wahba (1999). Causal effects in non-experimental studies: Reevaluating the evaluation of training programmes. *Journal of American Statistical Association 94*, 1053–1062.

de Luna, X., and P. Johansson (2014). Testing for the unconfoundedness assumption using an instrumental assumption. *Journal of Causal Inference 2*, 187–199.

De Moivre, A. (1967). *The doctrine of chances*. Chelsea Publishing Company. (Original work published 1738)

Deuchert, E., M. Huber, and M. Schelker (2019). Direct and indirect effects based on difference-in-differences with an application to political preferences following the Vietnam draft lottery. *Journal of Business & Economic Statistics 37*, 710–720.

D'Haultfoeuille, X., S. Hoderlein, and Y. Sasaki (2021). Nonparametric difference-in-differences in repeated cross-sections with continuous treatments. *arXiv preprint 2104.14458*.

Diamond, A., and J. S. Sekhon (2013). Genetic matching for estimating causal effects: A general multivariate matching method for achieving balance in observational studies. *Review of Economics and Statistics 95*, 932–945.

Dimmery, D. (2016). rdd: Regression discontinuity estimation. *R package.*

DiNardo, J. E., N. M. Fortin, and T. Lemieux (1996). Labor market institutions and the distribution of wages, 1973–1992: A semiparametric approach. *Econometrica 64*, 1001–1044.

Donald, S., and K. Lang (2007). Inference with difference-in-differences and other panel data. *Review of Economics and Statistics 89*, 221–233.

Donald, S. G., and Y. C. Hsu (2014). Estimation and inference for distribution functions and quantile functions in treatment effect models. *Journal of Econometrics 178*, 383–397.

Donald, S. G., Y.-C. Hsu, and R. P. Lieli (2014). Testing the unconfoundedness assumption via inverse probability weighted estimators of (L)ATT. *Journal of Business & Economic Statistics 32*, 395–415.

Dong, Y. (2014). Jumpy or kinky? Regression discontinuity without the discontinuity. Working paper, University of California Irvine.

Dong, Y. (2015). Regression discontinuity applications with rounding errors in the running variable. *Journal of Applied Econometrics 30*, 422–446.

Dong, Y., and A. Lewbel (2015). Identifying the effect of changing the policy threshold in regression discontinuity models. *Review of Economics and Statistics 97*, 1081–1092.

Dorn, J., K. Guo, and N. Kallus (2021). Doubly-valid/doubly-sharp sensitivity analysis for causal inference with unmeasured confounding. *arXiv preprint 2112.11449.*

Doudchenko, N., and G. W. Imbens (2016). Balancing, regression, difference-in-differences and synthetic control methods: A synthesis. NBER Working Paper 22791.

Dudík, M., J. Langford, and L. Li (2011). Doubly robust policy evaluation and learning. *Procceedings of the 28th International Conference on Machine Learning*, 1097–1104.

Efron, B. (1979). Bootstrap methods: Another look at the jackknife. *Annals of Statistics 7*, 1–26.

Eicker, F. (1967). Limit theorems for regressions with unequal and dependent errors. In *Proceedings of the Fifth Berkeley Symposium on Mathematical Statistics and Probability*, pp. 59–82. University of California Press.

Fan, Q., Y.-C. Hsu, R. P. Lieli, and Y. Zhang (2020). Estimation of conditional average treatment effects with high-dimensional data. *Journal of Business & Economic Statistics 40*, 1–15.

Farbmacher, H., R. Guber, and S. Klaassen (2020). Instrument validity tests with causal forests. *Journal of Business & Economic Statistics 40*, 605–614.

Farbmacher, H., R. Guber, and J. Vikström (2018). Increasing the credibility of the twin birth instrument. *Journal of Applied Econometrics 33*, 457–472.

Farrell, M. H. (2015). Robust inference on average treatment effects with possibly more covariates than observations. *Journal of Econometrics 189*, 1–23.

Farrell, M. H., T. Liang, and S. Misra (2021). Deep neural networks for estimation and inference. *Econometrica 89*, 181–213.

Ferman, B., and C. Pinto (2019). Inference in differences-in-differences with few treated groups and heteroskedasticity. *Review of Economics and Statistics 101*, 452–467.

Ferman, B., and C. Pinto (2021). Synthetic controls with imperfect pre-treatment fit. *arXiv preprint 1911.08521.*

Fernandes, A., M. Huber, and G. Vaccaro (2021). Gender differences in wage expectations. *PLOS ONE 16*, 1–24.

Ferracci, M., G. Jolivet, and G. J. van den Berg (2014). Evidence of treatment spillovers within markets. *Review of Economics and Statistics 96*, 812–823.

Firpo, S. (2007). Efficient semiparametric estimation of quantile treatment effects. *Econometrica 75*, 259–276.

Fisher, R. (1935). *The Design of Experiments*. Oliver and Boyd.

Flores, C. A. (2007). Estimation of dose-response functions and optimal doses with a continuous treatment. Working paper, University of California, Berkeley.

Flores, C. A., and X. Chen. (2018). *Average Treatment Effect Bounds with an Instrumental Variable: Theory and Practice*. Springer.

Flores, C. A., and A. Flores-Lagunes (2013). Partial identification of local average treatment effects with an invalid instrument. *Journal of Business & Economic Statistics 31*, 534–545.

Flores, C. A., A. Flores-Lagunes, A. Gonzalez, and T. C. Neumann (2012). Estimating the effects of length of exposure to instruction in a training program: The case of Job Corps. *Review of Economics and Statistics 94*, 153–171.

Forastiere, L., E. M. Airoldi, and F. Mealli (2021). Identification and estimation of treatment and interference effects in observational studies on networks. *Journal of the American Statistical Association 116*, 901–918.

Forastiere, L., F. Mealli, and T. J. VanderWeele (2016). Identification and estimation of causal mechanisms in clustered encouragement designs: Disentangling bed nets using Bayesian principal stratification. *Journal of the American Statistical Association 111*, 510–525.

Frandsen, B. R., M. Frölich, and B. Melly (2012). Quantile treatment effects in the regression discontinuity design. *Journal of Econometrics 168*, 382–395.

Frangakis, C., and D. Rubin (1999). Addressing complications of intention-to-treat analysis in the combined presence of all-or-none treatment-noncompliance and subsequent missing outcomes. *Biometrika 86*, 365–379.

Freedman, D. A. (2008). On regression adjustments to experimental data. *Advances in Applied Mathematics 40*, 180–193.

Freund, Y., and R. E. Schapire (1997). A decision-theoretic generalization of on-line learning and an application to boosting. *Journal of Computer and System Sciences 55*, 119–139.

Fricke, H. (2017). Identification based on difference-in-differences approaches with multiple treatments. *Oxford Bulletin of Economics and Statistics 79*, 426–433.

Fricke, H., M. Frölich, M. Huber, and M. Lechner (2020). Endogeneity and non-response bias in treatment evaluation–nonparametric identification of causal effects by instruments. *Journal of Applied Econometrics 35*, 481–504.

Frölich, M. (2004). Finite sample properties of propensity-score matching and weighting estimators. *Review of Economics and Statistics 86*, 77–90.

Frölich, M. (2005). Matching estimators and optimal bandwidth choice. *Statistics and Computing 15*, 197–215.

Frölich, M. (2007). Nonparametric IV estimation of local average treatment effects with covariates. *Journal of Econometrics 139*, 35–75.

Frölich, M., and M. Huber (2017). Direct and indirect treatment effects—causal chains and mediation analysis with instrumental variables. *Journal of the Royal Statistical Society Series B 79*, 1645–1666.

Frölich, M., and M. Huber (2019). Including covariates in the regression discontinuity design. *Journal of Business and Economic Statistics 37*, 736–748.

Frölich, M., and B. Melly (2013). Unconditional quantile treatment effects under endogeneity. *Journal of Business & Economic Statistics 31*, 346–357.

Frölich, M., and S. Sperlich (2019). *Impact Evaluation: Treatment Effects and Causal Analysis*. Cambridge University Press.

Galvao, A. F., and L. Wang (2015). Uniformly semiparametric efficient estimation of treatment effects with a continuous treatment. *Journal of the American Statistical Association 110*, 1528–1542.

Ganong, P., and S. Jäger (2018). A permutation test for the regression kink design. *Journal of the American Statistical Association 113*, 494–504.

Gauss, C. F. (1809). *Theoria motus corporum coelestium*, Volume 1. FA Perthes.

Gelman, A., and G. Imbens (2018). Why high-order polynomials should not be used in regression discontinuity designs. *Journal of Business & Economic Statistics 37*, 447–456.

Glymour, C., K. Zhang, and P. Spirtes (2019). Review of causal discovery methods based on graphical models. *Frontiers in Genetics 10*, 1–15.

Goodman-Bacon, A. (2018). Difference-in-differences with variation in treatment timing. Working paper, Vanderbilt University, Nashville.

Graham, B., C. Pinto, and D. Egel (2012). Inverse probability tilting for moment condition models with missing data. *Review of Economic Studies 79*, 1053–1079.

Graham, N., M. Arai, and B. Hagströmer (2016). multiwayvcov: Multi-way standard error clustering. *R package*.

Haavelmo, T. (1943). The statistical implications of a system of simultaneous equations. *Econometrica 11*, 1–12.

Hadad, V., D. A. Hirshberg, R. Zhan, S. Wager, and S. Athey (2021). Confidence intervals for policy evaluation in adaptive experiments. *arXiv preprint 1911.02768*.

Hahn, J. (1998). On the role of the propensity score in efficient semiparametric estimation of average treatment effects. *Econometrica 66*, 315–331.

Hahn, J., P. Todd, and W. van der Klaauw (2001). Identification and estimation of treatment effects with a regression-discontinuity design. *Econometrica 69*, 201–209.

Hainmueller, J. (2012). Entropy balancing for causal effects: A multivariate reweighting method to produce balanced samples in observational studies. *Political Analysis 20*, 25–46.

Halloran, M. E., and C. J. Struchiner (1991). Study designs for dependent happenings. *Epidemiology 2*, 331–338.

Hastie, T., R. Tibshirani, and J. Friedman (2008). *The Elements of Statistical Learning*. Springer Series in Statistics. Springer New York.

Hayfield, T., and J. S. Racine (2008). Nonparametric econometrics: The np package. *Journal of Statistical Software 27*.

Heckman, J., H. Ichimura, J. Smith, and P. Todd (1998). Characterizing selection bias using experimental data. *Econometrica 66*, 1017–1098.

Heckman, J. J. (1976). The common structure of statistical models of truncation, sample selection and limited dependent variables and a simple estimator for such models. *Annals of Economic and Social Measurement 5*, 475–492.

Heckman, J. J. (1979). Sample selection bias as a specification error. *Econometrica 47*, 153–161.

Heckman, J. J., H. Ichimura, and P. Todd (1998). Matching as an econometric evaluation estimator. *Review of Economic Studies 65*, 261–294.

Heckman, J. J., L. Lochner, and C. Taber (1998). General-equilibrium treatment effects: A study of tuition policy. *American Economic Review 88*, 381–386.

Heckman, J. J., and R. Pinto (2018). Unordered monotonicity. *Econometrica 86*, 1–35.

Heckman, J. J., and E. Vytlacil (1999). Local instrumental variables and latent variable models for identifying and bounding treatment effects. *PNAS 96*, 4730–4734.

Heckman, J. J., and E. Vytlacil (2001). Local instrumental variables. In C. Hsiao, K. Morimune, and J. Powell (Eds.), *Nonlinear Statistical Inference: Essays in Honor of Takeshi Amemiya*. Cambridge: Cambridge University Press, 1–46.

Heckman, J. J., and E. Vytlacil (2005). Structural equations, treatment effects, and econometric policy evaluation 1. *Econometrica 73*, 669–738.

Hernandez-Diaz, S., E. F. Schisterman, and M. A. Hernan (2006). The birth weight "paradox" uncovered? *American Journal of Epidemiology 164*, 1115–1120.

Hilbe, J. M. (2016). Count: Functions, data and code for count data. *R package*.

Hirano, K., and G. W. Imbens (2005). *The Propensity Score with Continuous Treatments* In *Applied Bayesian Modeling and Causal Inference from Incomplete-Data Perspectives: An Essential Journey with Donald Rubin's Statistical Family*, pp. 73–84. Wiley-Blackwell.

Hirano, K., G. W. Imbens, and G. Ridder (2003). Efficient estimation of average treatment effects using the estimated propensity score. *Econometrica 71*, 1161–1189.

Hirano, K., and J. Porter (2009). Asymptotics for statistical treatment rules. *Econometrica 77*, 1683–1701.

Hirano, K., and J. R. Porter (2012). Impossibility results for nondifferentiable functionals. *Econometrica 80*, 1769–1790.

Ho, D. E., K. Imai, G. King, and E. A. Stuart (2011). MatchIt: Nonparametric preprocessing for parametric causal inference. *Journal of Statistical Software 42*, 1–28.

Ho, T. K. (1995). Random decision forests. In *Proceedings of Third International Conference on Document Analysis and Recognition*, Volume 1 (278–282). IEEE.

Hodges, J. L., and E. L. Lehmann (1963). Estimates of location based on rank tests. *Annals of Mathematical Statistics 34*, 598–611.

Hoerl, A. E., and R. W. Kennard (1970). Ridge regression: Biased estimation for nonorthogonal problems. *Technometrics 12*, 55–67.

Holland, P. (1986). Statistics and causal inference. *Journal of American Statistical Association 81*, 945–970.

Hong, G. (2010). Ratio of mediator probability weighting for estimating natural direct and indirect effects. In *Proceedings of the American Statistical Association* (2401–2415) (pp. 2401–2415. Alexandria, VA: American Statistical Association).

Hong, G., and S. W. Raudenbush (2006). Evaluating kindergarten retention policy. *Journal of the American Statistical Association 101*, 901–910.

Hong, G., X. Qin, and F. Yang (2018). Weighting-based sensitivity analysis in causal mediation studies. *Journal of Educational and Behavioral Statistics 43*, 32–56.

Hong, S.-H. (2013). Measuring the effect of Napster on recorded music sales: Difference-in-differences estimates under compositional changes. *Journal of Applied Econometrics 28*, 297–324.

Horowitz, J. L. (2019). Bootstrap methods in econometrics. *Annual Review of Economics 11*, 193–224.

Horvitz, D., and D. Thompson (1952). A generalization of sampling without replacement from a finite population. *Journal of American Statistical Association 47*, 663–685.

Hosmer, D. W., and S. Lemeshow (2000). *Applied Logistic Regression*. John Wiley and Sons.

Huber, M. (2014a). Identifying causal mechanisms (primarily) based on inverse probability weighting. *Journal of Applied Econometrics 29*, 920–943.

Huber, M. (2014b). Sensitivity checks for the local average treatment effect. *Economics Letters 123*, 220–223.

Huber, M., L. Laffers, and G. Mellace (2017). Sharp IV bounds on average treatment effects on the treated and other populations under endogeneity and noncompliance. *Journal of Applied Econometrics 32*, 56–79.

Huber, M., M. Lechner, and C. Wunsch (2013). The performance of estimators based on the propensity score. *Journal of Econometrics 175*, 1–21.

Huber, M., and G. Mellace (2015). Testing instrument validity for late identification based on inequality moment constraints. *Review of Economics and Statistics 97*, 398–411.

Huber, M., M. Schelker, and A. Strittmatter (2020). Direct and indirect effects based on changes-in-changes. *Journal of Business & Economic Statistics 40*, 432–443.

Huber, M., and A. Steinmayr (2021). A framework for separating individual-level treatment effects from spillover effects. *Journal of Business & Economic Statistics 39*, 422–436.

Huber, M., and K. Wüthrich (2019). Local average and quantile treatment effects under endogeneity: A review. *Journal of Econometric Methods 8*, 1–28.

Huber, P. J. (1967). The behavior of maximum likelihood estimates under nonstandard conditions. In *Proceedings of the Fifth Berkeley Symposium on Mathematical Statistics and Probability*, Volume 5, pp. 221–233. University of California Press.

Hudgens, M. G., and M. E. Halloran (2008). Toward causal inference with interference. *Journal of the American Statistical Association 103*, 832–842.

Huntington-Klein, N. (2022). *The Effect: An Introduction to Research Design and Causality*. Chapman and Hall/CRC.

Ichimura, H. (1993). Semiparametric least squares (SLS) and weighted SLS estimation of single-index models. *Journal of Econometrics 58*, 71–120.

Ichimura, H., and O. Linton (2005). Asymptotic expansions for some semiparametric program evaluation estimators. In D. Andrews and J. Stock (Eds.), *Identification and Inference for Econometric Models*, pp. 149–170. Cambridge University Press.

Ichino, A., F. Mealli, and T. Nannicini (2008). From temporary help jobs to permanent employment: What can we learn from matching estimators and their sensitivity? *Journal of Applied Econometrics 23*, 305–327.

Imai, K., Z. Jiang, and M. Li (2019). experiment: R package for designing and analyzing randomized experiments. *R package*.

Imai, K., Z. Jiang, and A. Malani (2021). Causal inference with interference and noncompliance in two-stage randomized experiments. *Journal of the American Statistical Association 116*, 632–644.

Imai, K., L. Keele, and T. Yamamoto (2010). Identification, inference and sensitivity analysis for causal mediation effects. *Statistical Science 25*, 51–71.

Imai, K., and M. Ratkovic (2014). Covariate balancing propensity score. *Journal of the Royal Statistical Society: Series B (Statistical Methodology) 76*, 243–263.

Imai, K., and D. A. van Dyk (2004). Causal inference with general treatment regimes. *Journal of the American Statistical Association 99*, 854–866.

Imai, K., and T. Yamamoto (2013). Identification and sensitivity analysis for multiple causal mechanisms: Revisiting evidence from framing experiments. *Political Analysis 21*, 141–171.

Imbens, G. W. (2000). The role of the propensity score in estimating dose-response functions. *Biometrika 87*, 706–710.

Imbens, G. W. (2003). Sensitivity to exogeneity assumptions in program evaluation. *The American Economic Review 93*, 126–132.

Imbens, G. W., and J. Angrist (1994). Identification and estimation of local average treatment effects. *Econometrica 62*, 467–475.

Imbens, G., and K. Kalyanaraman (2012). Optimal bandwidth choice for the regression discontinuity estimator. *Review of Economic Studies 79*, 933–959.

Imbens, G. W., and T. Lemieux (2008). Regression discontinuity designs: A guide to practice. *Journal of Econometrics 142*, 615–635.

Imbens, G. W., and C. F. Manski (2004). Confidence intervals for partially identified parameters. *Econometrica 72*(6), 1845–1857.

Imbens, G. W., and D. Rubin (1997). Estimating outcome distributions for compliers in instrumental variables models. *Review of Economic Studies 64*, 555–574.

Imbens, G. W., and D. B. Rubin (2015). *Causal inference in statistics, social, and biomedical sciences*. Cambridge University Press.

Imbens, G. W., and S. Wager (2019). Optimized regression discontinuity designs. *Review of Economics and Statistics 101*, 264–278.

Imbens, G. W., and J. M. Wooldridge (2009). Recent developments in the econometrics of program evaluation. *Journal of Economic Literature 47*, 5–86.

James, G., D. Witten, T. Hastie, and R. Tibshirani (2013). *An Introduction to Statistical Learning: with Applications in R*. Springer.

Jiang, K., R. Mukherjee, S. Sen, and P. Sur (2022). A new central limit theorem for the augmented IPW estimator: Variance inflation, cross-fit covariance and beyond. *arXiv preprint 2205.10198.*

Kalisch, M., and P. Bühlmann (2014). Causal structure learning and inference: A selective review. *Quality Technology & Quantitative Management 11*, 3–21.

Kallus, N. (2017). Balanced policy evaluation and learning. Working paper, Cornell University, Ithaca, NY.

Kallus, N., X. Mao, and A. Zhou (2019). Interval estimation of individual-level causal effects under unobserved confounding. In K. Chaudhuri and M. Sugiyama (Eds.), *Proceedings of the Twenty-Second International Conference on Artificial Intelligence and Statistics*, Volume 89 of *Proceedings of Machine Learning Research*, pp. 2281–2290. Proceedings of Machine Learning Research.

Kang, H., and G. Imbens (2016). Peer encouragement designs in causal inference with partial interference and identification of local average network effects. *arXiv preprint 1609.04464.*

Kassambara, A. (2019). datarium: Data bank for statistical analysis and visualization. *R package.*

Keane, M., and T. Neal (2021). A practical guide to weak instruments. UNSW Economics Working Paper No. 2021-05a, University of New South Wales, Australia.

Keele, L. J. (2014). rbounds: Perform Rosenbaum bounds sensitivity tests for matched and unmatched data. *R package.*

Keele, L. J., and R. Titiunik (2015). Geographic boundaries as regression discontinuities. *Political Analysis 23*, 127–155.

Kennedy, E. H. (2020). Optimal doubly robust estimation of heterogeneous causal effects. *arXiv preprint 2004.14497.*

Kennedy, E. H. (2021). npcausal: Nonparametric causal inference methods. *R package.*

Kennedy, E. H., Z. Ma, M. D. McHugh, and D. S. Small (2017). Non-parametric methods for doubly robust estimation of continuous treatment effects. *Journal of the Royal Statistical Society Series B 79*, 1229–1245.

Kettlewell, N., and P. Siminski (2020). Optimal model selection in RDD and related settings using placebo zones. IZA Discussion Paper No. 13639.

Khan, S., and E. Tamer (2010). Irregular identification, support conditions, and inverse weight estimation. *Econometrica 78*, 2021–2042.

Kiel, K., and K. McClain (1995). House prices during siting decision stages: The case of an incinerator from rumor through operation. *Journal of Environmental Economics and Management 28*, 241–255.

Kitagawa, T. (2015). A test for instrument validity. *Econometrica 83*, 2043–2063.

Kitagawa, T., and A. Tetenov (2018). Who should be treated? Empirical welfare maximization methods for treatment choice. *Econometrica 86*, 591–616.

Kleiber, C., and A. Zeileis (2008). *Applied Econometrics with* R. Springer-Verlag.

Klein, R. W., and R. H. Spady (1993). An efficient semiparametric estimator for binary response models. *Econometrica 61*, 387–421.

Kleven, H. J., and M. Waseem (2013). Using notches to uncover optimization frictions and structural elasticities: Theory and evidence from Pakistan. *Quarterly Journal of Economics 128*, 669–723.

Knaus, M. C. (2021). Double machine learning based program evaluation under unconfoundedness. *The Econometrics Journal 25*, 602–627

Koenker, R., and G. Bassett (1978). Regression quantiles. *Econometrica 46*, 33–50.

Kolesár, M. (2021). Rdhonest: Honest inference in regression discontinuity designs. *R package*.

Kolesár, M., and C. Rothe (2018). Inference in a regression discontinuity design with a discrete running variable. *American Economic Review 108*, 2277–2304.

Kreider, B., and J. V. Pepper (2007). Disability and employment. *Journal of the American Statistical Association 102*, 432–441.

Kreif, N., and K. DiazOrdaz (2019). Machine learning in policy evaluation: New tools for causal inference. *arXiv preprint 1903.00402*.

Kueck, J., Y. Luo, M. Spindler, and Z. Wang. Estimation and inference of treatment effects with L2-boosting in high-dimensional settings. *Journal of Econometrics* (forthcoming).

Lai, T. L., and H. Robbins (1985). Asymptotically efficient adaptive allocation rules. *Advances in Applied Mathematics 6*, 4–22.

Lalive, R. (2008). How do extended benefits affect unemployment duration? A regression discontinuity approach. *Journal of Econometrics 142*, 785–806.

LaLonde, R. (1986). Evaluating the econometric evaluations of training programs with experimental data. *American Economic Review 76*, 604–620.

Landais, C. (2015). Assessing the welfare effects of unemployment benefits using the regression kink design. *American Economic Journal: Economic Policy 7*, 243–278.

Lechner, M. (2001). Identification and estimation of causal effects of multiple treatments under the conditional independence assumption. In M. Lechner and F. Pfeiffer (Eds.), *Econometric Evaluations of Active Labor Market Policies in Europe*. Physica.

Lechner, M. (2009). Sequential causal models for the evaluation of labor market programs. *Journal of Business and Economic Statistics 27*, 71–83.

Lechner, M. (2011). The estimation of causal effects by difference-in-difference methods. *Foundations and Trends in Econometrics 4*, 165–224.

Lechner, M., and R. Miquel (2010). Identification of the effects of dynamic treatments by sequential conditional independence assumptions. *Empirical Economics 39*, 111–137.

Lechner, M., R. Miquel, and C. Wunsch (2011). Long-run effects of public sector sponsored training in West Germany. *Journal of the European Economic Association 9*, 742–784.

Lechner, M., and A. Strittmatter (2019). Practical procedures to deal with common support problems in matching estimation. *Econometric Reviews 38*, 193–207.

LeCun, Y., L. Bottou, Y. Bengio, and P. Haffner (1998). Gradient-based learning applied to document recognition. *Proceedings of the IEEE 86*, 2278–2324.

Lee, D. (2008). Randomized experiments from non-random selection in U.S. House elections. *Journal of Econometrics 142*, 675–697.

Lee, D. S. (2009). Training, wages, and sample selection: Estimating sharp bounds on treatment effects. *Review of Economic Studies 76*, 1071–1102.

Lee, D., and D. Card (2008). Regression discontinuity inference with specification error. *Journal of Econometrics 142*, 655–674.

Lee, D., and T. Lemieux (2010). Regression discontinuity designs in economics. *Journal of Economic Literature 48*, 281–355.

Lee, K., F. J. Bargagli-Stoffi, and F. Dominici (2020). Causal rule ensemble: Interpretable inference of heterogeneous treatment effects. *arXiv preprint 2009.09036*.

Lévy, P. (1937). *Théorie de l'addition de variables aléatoires*. Gauthier-Villars.

Li, K. T. (2020). Statistical inference for average treatment effects estimated by synthetic control methods. *Journal of the American Statistical Association 115*, 2068–2083.

Li, Q., J. Racine, and J. Wooldridge (2009). Efficient estimation of average treatment effects with mixed categorical and continuous data. *Journal of Business and Economics Statistics 27*, 206–223.

Liaw, A., and M. Wiener (2002). Classification and regression by randomForest. *R News 2*, 18–22.

Lieli, R. P., Y.-C. Hsu, and Á. Reguly (2022). The use of machine learning in treatment effect estimation. Working paper, Central European University, Budapest.

Lindeberg, J. W. (1922). Eine neue Herleitung des exponentialgesetzes in der Wahrscheinlichkeitsrechnung. *Mathematische Zeitschrift 15*, 211–225.

Liu, L., and M. G. Hudgens (2014). Large sample randomization inference of causal effects in the presence of interference. *Journal of the American Statistical Association 109*, 288–301.

Liu, L., M. G. Hudgens, B. Saul, J. D. Clemens, M. Ali, and M. E. Emch (2019). Doubly robust estimation in observational studies with partial interference. *Stat 8*, e214.

Ludwig, J., and D. L. Miller (2007). Does head start improve children's life chances? Evidence from a regression discontinuity design. *Quarterly Journal of Economics 122*, 159–208.

Lundqvist, H., M. Dahlberg, and E. Mörk (2014). Stimulating local public employment: Do general grants work? *American Economic Journal: Economic Policy 6*, 167–92.

Lyapunov, A. (1901). Nouvelle forme du théoreme sur la limite des probabilités. *Mémoires de l'Académie impériale des sciences de St Pétersbourg 12*, 1–24.

MacKinnon, J. G. (2006). Bootstrap methods in econometrics. *Economic Record 82*, S2–S18.

Manski, C. F. (1989). Anatomy of the selection problem. *Journal of Human Resources 24*, 343–360.

Manski, C. F. (1990). Nonparametric bounds on treatment effects. *American Economic Review, Papers and Proceedings 80*, 319–323.

Manski, C. F. (1997). Monotone treatment response. *Econometrica 65*, 1311–1334.

Manski, C. F. (2004). Statistical treatment rules for heterogeneous populations. *Econometrica 72*, 1221–1246.

Manski, C. F. (2013). Identification of treatment response with social interactions. *Econometrics Journal 16*, S1–S23.

Manski, C. F., and J. Pepper (2000). Monotone instrumental variables: With an application to the returns to schooling. *Econometrica 68*, 997–1010.

Masten, M. A., and A. Poirier (2018). Identification of treatment effects under conditional partial independence. *Econometrica 86*, 317–351.

Mavrokonstantis, P. (2019). bunching: Estimate bunching. *R package.*

McCrary, J. (2008). Manipulation of the running variable in the regression discontinuity design: A density test. *Journal of Econometrics 142*, 698–714.

McCulloch, W., and W. Pitts (1943). A logical calculus of ideas immanent in nervous activity. *Bulletin of Mathematical Biophysics 5*(4), 115–133.

Mealli, F., G. Imbens, S. Ferro, and A. Biggeri (2004). Analyzing a randomized trial on breast self-examination with noncompliance and missing outcomes. *Biostatistics 5*, 207–222.

Melly, B., and R. Lalive (2020). Estimation, inference, and interpretation in the regression discontinuity design. Discussion Paper 20-16, University of Bern, Switzerland.

Melly, B., and G. Santangelo (2015). The changes-in-changes model with covariates. Working paper, Department of Economics, University of Bern, Switzerland.

Miguel, E., and M. Kremer (2004). Worms: Identifying impacts on education and health in the presence of treatment externalities. *Econometrica 72*, 159–217.

Miquel, R. (2002). Identification of dynamic treatment effects by instrumental variables. *University of St. Gallen Economics Discussion Paper Series 2002-11.*

Moffitt, R. (2001). Policy interventions, low-level equilibria and social interactions. In S. Durlauf and H. Young (Eds.), *Social Dynamics*. MIT Press.

Mogstad, M., A. Torgovitsky, and C. R. Walters (2020). Policy evaluation with multiple instrumental variables. NBER Working Paper 27546.

Molinari, F. (2020). Microeconometrics with partial identification. In S. N. Durlauf, L. P. Hansen, J. J. Heckman, and R. L. Matzkin (Eds.), *Handbook of Econometrics, Volume 7A*, pp. 355–486. Elsevier.

Morgan, J. N., and J. A. Sonquist (1963). Problems in the analysis of survey data, and a proposal. *Journal of the American Statistical Association 58*, 415–434.

Moss, J., and M. Tveten (2020). kdensity: Kernel density estimation with parametric starts and asymmetric kernels. *R package.*

Mourifié, I., and Y. Wan (2017). Testing late assumptions. *Review of Economics and Statistics 99*, 305–313.

Nadaraya, E. A. (1964). On estimating regression. *Theory of Probability & Its Applications 9*, 141–142.

Neyman, J. (1923). On the application of probability theory to agricultural experiments. Essay on principles. *Statistical Science Reprint, 5*, 463–480.

Neyman, J. (1959). *Optimal asymptotic tests of composite statistical hypotheses*. In Ulf Grenander (Ed.), *Probability and Statistics: The Harald Cramer Volume*, 213–234. Almqvist and Wiksell.

Nie, X., and S. Wager (2020, 09). Quasi-oracle estimation of heterogeneous treatment effects. *Biometrika 108*, 299–319.

Noack, C. (2021). Sensitivity of late estimates to violations of the monotonicity assumption. *arXiv preprint 2106.06421.*

Noack, C., T. Olma, and C. Rothe (2021). Flexible covariate adjustments in regression discontinuity designs. *arXiv preprint 2107.07942.*

Papay, J. P., J. B. Willett, and R. J. Murnane (2011). Extending the regression-discontinuity approach to multiple assignment variables. *Journal of Econometrics 161*, 203–207.

Parzen, E. (1962). On estimation of a probability density and mode. *Annals of Mathematical Statistics 33*, 1065–1076.

Pearl, J. (2000). *Causality: Models, Reasoning, and Inference.* Cambridge University Press.

Pearl, J. (2001). Direct and indirect effects. In *Proceedings of the Seventeenth Conference on Uncertainty in Artificial Intelligence*, San Francisco, pp. 411–420. Morgan Kaufman.

Peters, J., D. Janzing, and B. Schölkopf (2017). *Elements of Causal Inference: Foundations and Learning Algorithms.* MIT Press.

Porter, J. (2003). Estimation in the regression discontinuity model. Working paper, University of Wisconsin Madison, Madison, WI.

Poterba, J. M., S. F. Venti, and D. A. Wise (1998). Personal retirement saving programs and asset accumulation: Reconciling the evidence. In *Frontiers in the Economics of Aging*, pp. 23–124. University of Chicago Press.

Qian, M., and S. A. Murphy (2011). Performance guarantees for individualized treatment rules. *Annals of Statistics 39*, 1180–1210.

Qu, Z., R. Xiong, J. Liu, and G. Imbens (2021). Efficient treatment effect estimation in observational studies under heterogeneous partial interference. *arXiv preprint 2107.12420.*

R Core Team (2015). *R: A Language and Environment for Statistical Computing.* R Foundation for Statistical Computing.

Rambachan, A., and J. Roth (2020). An honest approach to parallel trends. Working paper, Department of Economics, Harvard University, Cambridge, MA.

Raskutti, G., M. J. Wainwright, and B. Yu (2011). Minimax rates of estimation for high-dimensional linear regression over $\ell_q$-balls. *IEEE Transactions on Information Theory 57*, 6976–6994.

Reguly, A. (2021). Heterogeneous treatment effects in regression discontinuity design. *arXiv preprint arXiv:2106.11640.*

Richardson, T. S., and J. M. Robins (2010). Analysis of the binary instrumental variable model. In R. Dechter, H. Geffner, and J. Y. Halpern (Eds.), *Heuristics, Probability and Causality: A tribute to Judea Pearl*, pp. 415–440. College Publications.

Ripley, B. D. (1996). *Pattern Recognition and Neural Networks.* Cambridge University Press.

Robins, J. M. (1986). A new approach to causal inference in mortality studies with sustained exposure periods—application to control of the healthy worker survivor effect. *Mathematical Modelling 7*, 1393–1512.

Robins, J. M. (1989). The analysis of randomized and non-randomized AIDS treatment trials using a new approach to causal inference in longitudinal studies. In L. Sechrest, H. Freeman, and A. Mulley (Eds.), *Health Service Research Methodology: A Focus on AIDS*, pp. 113–159. Washington, DC: US Public Health Service.

Robins, J. M. (2000). Marginal structural models versus structural nested models as tools for causal inference. In M. E. Halloran and D. Berry (Eds.), *Statistical Models in Epidemiology, the Environment, and Clinical Trials*, pp. 95–133. Springer New York.

Robins, J. M. (2003). Semantics of causal DAG models and the identification of direct and indirect effects. In P. Green, N. Hjort, and S. Richardson (Eds.), *In Highly Structured Stochastic Systems*, pp. 70–81. Oxford University Press.

Robins, J. M., and S. Greenland (1992). Identifiability and exchangeability for direct and indirect effects. *Epidemiology 3*, 143–155.

Robins, J. M., M. A. Hernan, and B. Brumback (2000). Marginal structural models and causal inference in epidemiology. *Epidemiology 11*, 550–560.

Robins, J. M., S. D. Mark, and W. K. Newey (1992). Estimating exposure effects by modelling the expectation of exposure conditional on confounders. *Biometrics 48*, 479–495.

Robins, J. M., and T. Richardson (2010). Alternative graphical causal models and the identification of direct effects. In P. Shrout, K. Keyes, and K. Omstein (Eds.), *Causality and Psychopathology: Finding the Determinants of Disorders and Their Cures*, pp. 103–158. Oxford University Press.

Robins, J. M., and A. Rotnitzky (1995). Semiparametric efficiency in multivariate regression models with missing data. *Journal of the American Statistical Association 90*, 122–129.

Robins, J. M., A. Rotnitzky, and L. Zhao (1994). Estimation of regression coefficients when some regressors are not always observed. *Journal of the American Statistical Association 90*, 846–866.

Robins, J. M., A. Rotnitzky, and L. Zhao (1995). Analysis of semiparametric regression models for repeated outcomes in the presence of missing data. *Journal of the American Statistical Association 90*, 106–121.

Robinson, P. (1988). Root-N-consistent semiparametric regression. *Econometrica 56*, 931–954.

Romano, J. P., and A. M. Shaikh (2008). Inference for identifiable parameters in partially identified econometric models. *Journal of Statistical Planning and Inference 138*, 2786–2807.

Rosenbaum, P. (1984). The consequences of adjustment for a concomitant variable that has been affected by the treatment. *Journal of Royal Statistical Society, Series A 147*, 656–666.

Rosenbaum, P. (1995). *Observational Studies*. Springer Verlag.

Rosenbaum, P. R., and D. B. Rubin (1983a). Assessing sensitivity to an unobserved binary covariate in an observational study with binary outcome. *Journal of the Royal Statistical Society: Series B (Methodological) 45*, 212–218.

Rosenbaum, P. R., and D. B. Rubin (1983b). The central role of the propensity score in observational studies for causal effects. *Biometrika 70*, 41–55.

Rosenbaum, P. R., and D. B. Rubin (1985). Constructing a control group using multivariate matched sampling methods that incorporate the propensity score. *The American Statistician 39*, 33–38.

Rosenblatt, M. (1956). Remarks on some nonparametric estimates of a density function. *Annals of Mathematical Statistics 27*, 832–837.

Rosenzweig, M. R., and K. I. Wolpin (2000). Natural "natural experiments" in economics. *Journal of Economic Literature 38*, 827–874.

Roth, J., and P. H. C. Sant'Anna (2021). When is parallel trends sensitive to functional form? *arXiv preprint 2010.04814.*

Roth, J., P. H. C. Sant'Anna, A. Bilinski, and J. Poe (2022). What's trending in difference-in-differences? A synthesis of the recent econometrics literature. *arXiv preprint 2201.01194.*

Rothe, C., and S. Firpo (2013). Semiparametric estimation and inference using doubly robust moment conditions. IZA Discussion Paper No. 7564.

Roy, A. (1951). Some thoughts on the distribution of earnings. *Oxford Economic Papers 3*, 135–146.

Rubin, D. (1980). Comment on "Randomization analysis of experimental data: The Fisher randomization test" by D. Basu. *Journal of American Statistical Association 75*, 591–593.

Rubin, D. B. (1974). Estimating causal effects of treatments in randomized and nonrandomized studies. *Journal of Educational Psychology 66*, 688–701.

Rubin, D. B. (1976). Inference and missing data. *Biometrika 63*, 581–592.

Rubin, D. B. (1979). Using multivariate matched sampling and regression adjustment to control bias in observational studies. *Journal of the American Statistical Association 74*, 318–328.

Russo, D., B. V. Roy, A. Kazerouni, I. Osband, and Z. Wen (2020). A tutorial on Thompson sampling. *arXiv preprint 1707.02038.*

Saez, E. (2010). Do taxpayers bunch at kink points? *American Economic Journal: Economic Policy 2*, 180–212.

Sant'Anna, P. H. C., and J. B. Zhao (2018). Doubly robust difference-in-differences estimators. Working paper, Vanderbilt University, Nashville.

Sawada, M. (2019). Non-compliance in randomized control trials without exclusion restrictions. *arXiv preprint 1910.03204.*

Schochet, P. Z., J. Burghardt, and S. Glazerman (2001). *National Job Corps Study: The Impacts of Job Corps on Participants' Employment and Related Outcomes.* Washington, DC: Mathematica Policy Research, Inc.

Schochet, P. Z., J. Burghardt, and S. McConnell (2008). Does Job Corps work? Impact findings from the National Job Corps Study. *American Economic Review 98*, 1864–1886.

Sekhon, J. S. (2011). Multivariate and propensity score matching software with automated balance optimization: The Matching package for R. *Journal of Statistical Software 42*, 1–52.

Semenova, V. (2020). Better lee bounds. *arXiv preprint 2008.12720.*

Semenova, V. (2021). leebounds: Nonparametric bounds on average treatment effect in the presence of endogenous selection. *R package.*

Semenova, V., and V. Chernozhukov (2021). Debiased machine learning of conditional average treatment effects and other causal functions. *Econometrics Journal 24*, 264–289.

Shah, V., N. Kreif, and A. M. Jones (2021). Machine learning for causal inference: Estimating heterogeneous treatment effects. In *Handbook of Research Methods and Applications in Empirical Microeconomics.* Edward Elgar Publishing.

Shea, J. M. (2021). wooldridge: 115 data sets from "Introductory econometrics: A modern approach" by Jeffrey M. Wooldridge. *R package*.

Sianesi, B. (2004). An evaluation of the Swedish system of active labor market programs in the 1990s. *Review of Economics and Statistics 86*, 133–155.

Simonsen, M., L. Skipper, and N. Skipper (2016). Price sensitivity of demand for prescription drugs: Exploiting a regression kink design. *Journal of Applied Econometrics 31*, 320–337.

Sjölander, A. (2009). Bounds on natural direct effects in the presence of confounded intermediate variables. *Statistics in Medicine 28*, 558–571.

Slutsky, E. (1925). Über stochastische Asymptoten und Grenzwerte. *Metron 5*, 3–89.

Small, D. S., Z. Tan, R. R. Ramsahai, S. A. Lorch, and M. A. Brookhart (2017). Instrumental variable estimation with a stochastic monotonicity assumption. *Statistical Science 32*, 561–579.

Smith, J., and P. Todd (2005). Rejoinder. *Journal of Econometrics 125*, 365–375.

Snow, J. (1855). On the mode of communication of cholera. *Edinburgh Medical Journal 1*, 668.

Sobel, M. E. (2006). What do randomized studies of housing mobility demonstrate? *Journal of the American Statistical Association 101*, 1398–1407.

Staiger, D., and J. H. Stock (1997). Instrumental variables regression with weak instruments. *Econometrica 65*, 557–586.

Stock, J. H., J. H. Wright, and M. Yogo (2002). A survey of weak instruments and weak identification in generalized method of moments. *Journal of Business & Economic Statistics 20*, 518–529.

Stone, C. (1974). Cross-validatory choice and assessment of statistical predictions. *Journal of Royal Statistical Society, Series B 36*, 111–147 (with discussion).

Stoye, J. (2009). Minimax regret treatment choice with finite samples. *Journal of Econometrics 151*, 70–81.

Sutton, R., and A. Barto (1998). *Reinforcement Learning: An Introduction*. MIT Press.

Tamer, E. (2010). Partial identification in econometrics. *Annual Review of Economics 2*, 167–195.

Tan, Z. (2006). Regression and weighting methods for causal inference using instrumental variables. *Journal of the American Statistical Association 101*, 1607–1618.

Tchetgen Tchetgen, E. J., and I. Shpitser (2012). Semiparametric theory for causal mediation analysis: Efficiency bounds, multiple robustness, and sensitivity analysis. *Annals of Statistics 40*, 1816–1845.

Tchetgen Tchetgen, E. J. T., and T. J. VanderWeele (2012). On causal inference in the presence of interference. *Statistical Methods in Medical Research 21*, 55–75.

Tchetgen Tchetgen, E. J., and T. J. VanderWeele (2014). On identification of natural direct effects when a confounder of the mediator is directly affected by exposure. *Epidemiology 25*, 282–291.

Thistlethwaite, D., and D. Campbell (1960). Regression-discontinuity analysis: An alternative to the ex post facto experiment. *Journal of Educational Psychology 51*, 309–317.

Thompson, W. R. (1933). On the likelihood that one unknown probability exceeds another in view of the evidence of two samples. *Biometrika 25*, 285–294.

Tibshirani, J., S. Athey, and S. Wager (2020). grf: Generalized random forests. *R package*.

Tibshirani, R. (1996). Regresson shrinkage and selection via the lasso. *Journal of the Royal Statistical Society 58*, 267–288.

Tihonov, A. N. (1963). Solution of incorrectly formulated problems and the regularization method. *Soviet Mathematics 4*, 1035–1038.

Tran, L., C. Yiannoutsos, K. Wools-Kaloustian, A. Siika, M. van der Laan, and M. Petersen (2019). Double robust efficient estimators of longitudinal treatment effects: Comparative performance in simulations and a case study. *International Journal of Biostatistics 15*(2), 1–27.

Ugander, J., B. Karrer, L. Backstrom, and J. Kleinberg (2013). Graph cluster randomization: Network exposure to multiple universes. In *Proceedings of the 19th ACM SIGKDD International Conference on Knowledge Discovery and Data Mining*, New York, pp. 329–337. Association for Computing Machinery.

van der Laan, M. J. (2014). Causal inference for a population of causally connected units. *Journal of Causal Inference* (1), 13–74.

van der Laan, M. J., E. C. Polley, and A. E. Hubbard (2007). Super learner. *Statistical Applications in Genetics and Molecular Biology 6*.

van der Laan, M., and D. Rubin (2006). Targeted maximum likelihood learning. *International Journal of Biostatistics 2*, 1–38.

VanderWeele, T. J. (2010). Direct and indirect effects for neighborhood-based clustered and longitudinal data. *Sociological Methods & Research 38*(4), 515–544.

VanderWeele, T. J., and Y. Chiba (2014). Sensitivity analysis for direct and indirect effects in the presence of exposure-induced mediator-outcome confounders. *Epidemiology, Biostatistics, and Public Health 11*.

Vansteelandt, S., and T. J. VanderWeele (2012). Natural direct and indirect effects on the exposed: Effect decomposition under weaker assumptions. *Biometrics 68*, 1019–1027.

Vytlacil, E. (2002). Independence, monotonicity, and latent index models: An equivalence result. *Econometrica 70*, 331–341.

Waernbaum, I. (2012). Model misspecification and robustness in causal inference: Comparing matching with doubly robust estimation. *Statistics in Medicine 31*, 1572–1581.

Wager, S., and S. Athey (2018). Estimation and inference of heterogeneous treatment effects using random forests. *Journal of the American Statistical Association 113*, 1228–1242.

Wald, A. (1940). The fitting of straight lines if both variables are subject to error. *Annals of Mathematical Statistics 11*, 284–300.

Wand, M. P., and M. C. Jones (1994). *Kernel smoothing*. CRC Press.

Watson, G. S. (1964). Smooth regression analysis. *Sankhya: The Indian Journal of Statistics, Series A (1961–2002) 26*, 359–372.

Welch, B. L. (1947). The generalization of "student's" problem when several different population variances are involved. *Biometrika 34*, 28–35.

White, H. (1980). A heteroskedasticity-consistent covariance matrix estimator and a direct test for heteroskedasticity. *Econometrica 48*, 817–838.

Wickham, H., J. Hester, and W. Chang (2021). devtools: Tools to make developing R packages easier. *R package.*

Wickham, H., and E. Miller (2021). haven: Import and export SPSS, Stata and SAS files. *R package.*

Wilcox, A. J. (2001). On the importance—and the unimportance—of birthweight. *International Journal of Epidemiology 30*, 1233–1241.

Wittwer, J. (2020). Der Erfolg von Videospielen—eine empirische Untersuchung möglicher Erfolgsfaktoren. Bachelor thesis, Faculty of Management, Economics and Social Sciences, University of Fribourg.

Wright, P. G. (1928). *The Tariff on Animal and Vegetable Oils.* Macmillan Company.

Wu, C. F. J. (1986). Jackknife, bootstrap and other resampling methods in regression analysis. *Annals of Statistics 14*, 1261–1295.

Xia, F., and K. C. G. Chan (2021). Identification, semiparametric efficiency, and quadruply robust estimation in mediation analysis with treatment-induced confounding. *Journal of the American Statistical Association.*

Yamamoto, T. (2013). Identification and estimation of causal mediation effects with treatment noncompliance. Unpublished manuscript, Department of Political Science, Massachusetts Institute of Technology, Cambridge, MA.

Zeileis, A., and T. Hothorn (2002). Diagnostic checking in regression relationships. *R News 2*, 7–10.

Zeileis, A., S. Köll, and N. Graham (2020). Various versatile variances: An object-oriented implementation of clustered covariances in r. *Journal of Statistical Software 95*.

Zetterqvist, J., and A. Sjölander (2015). Doubly robust estimation with the R package drgee. *Epidemiologic Methods 4*, 69–86.

Zhang, B., A. A. Tsiatis, M. Davidian, M. Zhang, and E. Laber (2012). Estimating optimal treatment regimes from a classification perspective. *Stat 1*, 103–114.

Zhang, J., and D. B. Rubin (2003). Estimation of causal effects via principal stratification when some outcomes are truncated by death. *Journal of Educational and Behavioral Statistics 28*, 353–368.

Zhao, Z. (2004). Using matching to estimate treatment effects: Data requirements, matching metrics, and Monte Carlo evidence. *Review of Economics and Statistics 86*, 91–107.

Zheng, W., and M. J. van der Laan (2011). Cross-validated targeted minimum-loss-based estimation. *In Targeted Learning*, pp. 459–474. New York: Springer.

Zhou, Z.-H. (2012). *Ensemble Methods: Foundations and Algorithms.* Boca Raton, FL: Chapman & Hall/CRC.

Zhou, X. (2020). localiv: Estimation of marginal treatment effects using local instrumental variables. *R package.*

Zhou, X., N. Mayer-Hamblett, U. Khan, and M. R. Kosorok (2017). Residual weighted learning for estimating individualized treatment rules. *Journal of the American Statistical Association 112*, 169–187.

Zhou, Z., S. Athey, and S. Wager (2018). Offline multi-action policy learning: Generalization and optimization. Working paper, Stanford University, Stanford, CA.

Zimmert, M. (2020). Efficient difference-in-differences estimation with high-dimensional common trend confounding. *arXiv preprint 1809.01643.*

Zimmert, M., and M. Lechner (2019). Nonparametric estimation of causal heterogeneity under high-dimensional confounding. *arXiv preprint 1908.08779.*

Zonszein, S., C. Samii, and P. Aronow (2021). interference: Design-based estimation of spill-over effects. *R package.*

Zubizarreta, J. R. (2015). Stable weights that balance covariates for estimation with incomplete outcome data. *Journal of the American Statistical Association 110*, 910–922.

# Index